East Asian Cartographic Print Culture

East Asian Cartographic Print Culture

The Late Ming Publishing Boom and its Trans-Regional Connections

Alexander Akin

Routledge
Taylor & Francis Group
LONDON AND NEW YORK

First published in 2021 by Amsterdam University Press Ltd.

Published 2025 by Routledge
4 Park Square, Milton Park, Abingdon, Oxon OX14 4RN
605 Third Avenue, New York, NY 10158

Routledge is an imprint of the Taylor & Francis Group, an informa business

ISBN: 9789463726122 (hbk)
ISBN: 9781041178538 (pbk)
ISBN: 9781003694472 (ebk)
NUR 684 | 692

Cover illustration: Yunnan province as shown in the *Guang yutu* (Enlarged Atlas), 1566 edition (*juan* 2, pp. 87b-88a)

Courtesy of the Harvard-Yenching Library

Cover design: Coördesign, Leiden

DOI 10.5117/9789463726122

For Product Safety Concerns and Information please contact our EU representative:
GPSR@taylorandfrancis.com
Taylor & Francis Verlag GmbH, Kaufingerstraße 24, 80331 München, Germany

Table of Contents

List of Illustrations

Acknowledgments

This has been an effort of many years, and I gratefully recognize the support and constructive criticism of the following individuals in particular.

My first thanks go to Peter K. Bol, Mark Elliott, and Sarah Schneewind, who challenged me with their close reading and extraordinary useful criticisms, and especially to Kären Wigen, who played a singular role in supporting this project.

I am indebted to those who read and gave advice on particular chapters, including Joshua Hill, Eugene Anderson, Mark Byington, Wayne De Fremery, Hilde De Weerdt, Sun Joo Kim, Matthew Mosca, Mike Pincus, Jaeyoon Song, and Sixiang Wang. Librarians who were especially helpful include Ma Xiaohe, Nanni Deng, Miao-lin Hsu, Raymond Lum, and Ellen McGill at Harvard, and Bruce Williams, Deborah Rudolph, and Hisayuki Ishimatsu at UC Berkeley. My indispensable friend and colleague Matthew Fraleigh has been equal parts sounding board and 'shrink.'

I am extraordinarily grateful to those at Amsterdam University Press who so adroitly handled the process of editing and production even in the midst of a global pandemic; in particular I wish to thank Shannon Cunningham, Victoria Blud, and Jaap Wagenaar.

To Tracy Ying Cui Akin, who at our first encounter in Wangfujing saw me perusing a street map and recognized that I must need help, and to my inspirations in everything, Ella and Eli, I owe the world.

Introduction

The year 1975 marked five hundred years since a hefty tome published in Lübeck, the *Rudimentum Novitorum*, had incorporated a simple map of the world printed from two woodblocks. This quincentennial of the first printing of a map in a European book was widely celebrated among bibliophiles. The cartographic historian Arthur H. Robinson termed it a milestone in the annals of communication:

> To appreciate properly this momentous event, we must remember that in all preceding time maps had existed only in manuscript form. That basic fact allows two important assertions: first, there could only be a few maps and, second, one could never be sure whether the content of a map was the work of the original maker or merely reflected the independence, or carelessness, of a copyist. Obviously, both inhibited scholarship.
> The capability of printing maps immediately opened the way for countless numbers of exact duplicates that, for the first time, allowed scholars easily to compare many geographical portrayals, consider their characteristics, and plan ways to produce even better images of the emerging world. No doubt it also had a very considerable psychological impact on mapmakers, since the realization that their work could be widely subject to critical review probably served as an incentive to some and an inhibition to others. To the age-old art and science of mapmaking a tremendously significant new element had been added – the printer.[1]

Scholars of Asian cartography might point out that this anniversary celebration was off by hundreds of years, given that East Asian printed maps survive from as early as the twelfth century. Despite this fact, European maps remain the sole focus of much work on cartographic history. What is perhaps more surprising than Robinson's omission, however, is the degree to which the effects of publishing on cartography in East Asia differ from what we find on the European scene.

1 Robinson, 'Mapmaking and Map Printing: The Evolution of a Working Relationship,' p. 1.

Akin, Alexander, *East Asian Cartographic Print Culture: The Late Ming Publishing Boom and its Trans-Regional Connections*. Taylor & Francis Group, 2021
DOI: 10.5117/9789463726122_INTRO

We know a fair amount about the competition that arose between printers in some parts of Europe, and the ways such rivalry propelled the evolution of cartographic representation. For example, James Elliot has discussed the 'peer pressure' that Sebastian Münster felt after he illustrated a number of cities with woodblock prints in the 1544 edition of his popular *Cosmographia*. Only four years later, Johannes Stumpf's *Schweizer Chronik* raised the stakes by presenting 'the first realistic views of Swiss cities, the accuracy of which represented a great advance in geographical illustration.' Münster responded by revising the *Cosmographia*, including 'substantial improvements in both the quantity and quality of the town views [...] thus providing Renaissance scholars with their first comprehensive view of urban Europe.'[2] Furthermore, not only increasingly sophisticated maps, but also books about *making* maps came onto the market. The 1559 *Cosmographical Glasse* discussed the principles one would need to master for cartographic compilation.[3]

In East Asia, no such dramatic technical transformations in map publication appeared around this time, nor did more than very cursory materials on surveying or map composition see public circulation. As we shall see, the form and function of printed maps changed little from a wide range of precedents that had already been established by the twelfth century. A very large proportion of maps in late Ming books were simply copied from earlier works. And yet, by the beginning of the seventeenth century, when the printed word (and image) was unprecedentedly accessible to a general audience, we find that something has indeed changed. This transformation was not so much in the nature of the maps themselves as in the ways they were deployed, contrasted, and combined.

This book examines how the 'publishing boom' of the late Ming dynasty, beginning in the mid-sixteenth century, affected the nature and circulation of cartographic materials in East Asia. Rather than focusing on the beautiful, colorful, and rare large-format maps that (understandably) form the core of most cartographic histories, it looks instead at the smaller, often smudgily reproduced black and white maps that were carved into wooden blocks to be printed as illustrations in books, and occasionally, as stand-alone works – in short, maps for the masses. By examining contemporaneous developments in neighboring Chosŏn Korea and Japan, highlighting local responses to Ming publications as well as differences from late Ming publishing culture, the study demonstrates that it is imperative to consider the broader East Asian sphere in the early modern period as a network of communication and

2 Elliott, *The City in Maps: Urban Mapping to 1900*, p. 19.

3 Cunningham, *The Cosmographical Glasse*.

publication, rather than focusing on countries as discrete units with separate cartographic histories. It also reexamines the place of Jesuit cartographic materials across these same political and cultural boundaries, arguing that these Catholic missionaries printing maps on Ming soil should be seen as participants in the local cartographic publishing boom and its trans-regional repercussions.

In the course of examining a series of pathbreaking woodblock-printed works from the late sixteenth and early seventeenth centuries, in genres including general geographical education, military affairs, local administration, and history, we get a sense of the ways in which maps achieved unprecedented penetration among published materials, even in the absence of transformative theoretical or technological changes. We see no panoply of technical or stylistic transformations of the sort that dramatically changed the appearance of European maps over the same period; nor was there any late Ming equivalent of the horizon-widening explorations sponsored by various European courts, or the subsequent waves of colonization, that generated continuously updated representations of far-flung lands. Instead, most new maps were drawn on the basis of older charts and textual records. By the late Ming, certain oft-reprinted maps had become part of the patrimony of civilization, handed down to scholars of the time to be shared and borrowed.

Most of the techniques used to create maps in the Ming, and the models these maps followed, can be traced to Southern Song examples from centuries earlier. Maps dating back to the Song period that were used to elucidate the Classics, or to clarify dynastic history, attest to the importance of the civil service examination system: Students needed such resources to aid their studies. However, many of these maps continued to be reprinted right through the Ming and Qing, often with little modification. Classic early works, supplemented by a relatively small number of pathbreaking texts first published during the Ming, formed the core reservoir from which the majority of later printed maps were copied.

Despite this repetition of certain cartographic themes, there was a late Ming profusion of publishing genres that touched upon geographical issues, accompanied by what many scholars believe to have been a broadening audience of readers. New types of texts began to include cartographic illustrations. Moreover, works in the same genre often retained earlier maps even as they added new ones. For example, treatises on military affairs reflected both historical events and emerging threats. In the 1550s the focus of such works shifted from the northern Mongol frontier to the eastern seaboard, where piracy and tumult in coastal towns inspired distress.

In the late sixteenth and early seventeenth century, attention moved up to the northeastern frontier where the Manchu polity was taking shape, foreshadowing disaster for the Ming dynasty. Each period of focus on a particular region left its own cartographic tradition that became part of a cumulative store of maps that were reprinted in later works on similar topics. By the same token, local prefectural or county gazetteers, which sometimes included maps as early as the Song period, became dramatically more likely to include cartographic elements by the late Ming. The types of maps that might be included in a single gazetteer, their styles and reasons for compilation, became more diverse.

In the printed record we find, again and again, that existing works were cannibalized, with or without citation, and their maps reused.[4] In Chapter One, which addresses the issue of intertextual cartographic repetition in more detail, we see numerous examples of both text and maps taken from earlier publications. Rather than a struggle to supersede and replace earlier images, we often see a cumulative recapitulation of earlier maps that were considered authoritative, with the presentation of diverse perspectives in a single work.

Why were people drawing maps in the first place? The answers to this question cannot simply be transplanted from the research that has been done on European or colonial contexts. Despite the recent academic emphasis on cartography as a function of early modern state formation, many cartographic traditions derive from a framework of scholasticism and textual analysis rather than military or political needs. Beginning in the Song dynasty, a Classics-based curriculum for examination candidates had a significant influence on map production. In order to help students interpret the geography of the texts of the Confucian tradition and the dynastic histories, cartographic illustrations were printed, copied by hand by students, or engraved in stone on monuments at schools. Such historical maps, atlases, and treatises coexisted with other types of mapping that fit more comfortably into the narrative of state building and governance, as well as with other genres that had different purposes.

4 Craig Clunas makes a point in passing that is directly relevant to the study of cartography in the late Ming. In discussing the authorship of the *Chang wu zhi* 長物志 (*Treatise on Superfluous Things*) by Weng Zhenheng 文震亨, Clunas notes that much of its text is copied from an earlier work by Tu Long 屠隆. Clunas remarks that 'Post-Renaissance European concepts of "plagiarism" and "originality" are here of very little help as tools of analysis.' He describes Ming writing on connoisseurship as 'constituting a single "text", repeated and reaffirmed by a number of separate individual writers' (Clunas, *Superfluous Things*, p. 28).

What changes in the late Ming publishing boom is therefore less the technology or typological repertoire of maps than the ways in which they are presented, combined, contrasted, and analyzed. In Zhang Huang's late sixteenth-century *Tushu bian* 圖書編 (Compendium of illustrations and texts), we find in one collection a distillation of the entire cartographic record, from historical maps depicting the age of the Sage Kings to contemporary maps of the Grand Canal, a Buddhist depiction of the continent of Jambudvīpa, and even copies of newly introduced Jesuit world maps. Zhang is not content, however, with the novelty of simply presenting Jesuit and Buddhist worldviews alongside the Sinocentric and state-centric tradition of official cartography. He uses these maps as illustrations for his discussion of larger themes, debating the contrasting epistemological foundations of different worldviews and offering one of the earliest extant serious considerations by a non-convert of the Jesuit claims about the form of the earth. Zhang's example highlights the often-complex reaction of individual authors and editors to both longstanding indigenous traditions and newly imported Jesuit theories.

As the publishing boom made this wide variety of cartographic materials more accessible, a broad range of conceptual frames – contending cartographies – came into direct contact. This book will illustrate how this diversity of coexisting schools of cartographic thought and practice rode the crest of the late Ming publishing wave. The proliferation of new works as well as reprints facilitated exposure to competing theories. This book will not attempt to form a master scheme of 'traditional Chinese cartography' because it was no less than the sum of many diverse and often contradictory threads, represented by individual human beings and their relationships with one another in person or on the page, shaped by the sources to which they had access and the ideologies that informed their worldviews. It has been argued that the long-standing dominance of 'Confucian geography' was first seriously challenged only at the end of the Qing.[5] As the works discussed in this book reveal, however, diversity in cartographic thought was in fact much older, ranging from historical reconstructions based on revered texts, to statements of Buddhist religious dogma, to compilations based on actual ground surveys.

Chapter Two singles out one genre, historical cartography, for deeper examination, tracing its development from the earliest extant examples to the end of the Ming dynasty to show how the conditions of the publishing boom fostered their recombination and adaptation to new contexts. This

5 Tang Xiaofeng, *From Dynastic Geography to Historical Geography*, Chapter 2.

chapter examines historical maps in a broad range of formats, including works compiled for administration, for education, and for religious purposes. Because of its utility in contextualizing classical works, this genre was intimately intertwined with the publishing industry that emerged around the national examination system, but historical maps could also challenge and undermine administrative and Sinocentric perspectives.

The third chapter takes a Ming-centric approach to the Jesuit cartographic project under Matteo Ricci and his immediate successors. I discuss reactions to the missionaries' maps and their citation in works published by Ming scholars. Of all the empires in which the Jesuits set foot, the Ming was the first in which they encountered an already highly developed cartographic culture, leading to an unusually prominent role for cartography in their proselytization. Examining the issues they addressed through maps, and their methods of production, distribution, and influence, I argue for an understanding of their publications as part of the late Ming publishing boom.

The two final chapters follow exported Ming publications, and the cartographic trends they incorporated, into Korea and Japan. The preeminent gazetteer of Chosŏn, the *Tongguk yŏji sŭngnam* (Newly augmented geographical survey of the Land of the East), was directly inspired by the *Da Ming yitong zhi* (Gazetteer of the Great Ming's unification), a widely circulated state-sanctioned gazetteer of the Ming. Even long before the emergence of a substantial commercial publishing industry, the maps of the *Sŭngnam* penetrated the consciousness of Chosŏn's *yangban* gentry through the widespread hand-copying of atlases passed from scholar to scholar. The chaotic variety of later Ming works tended to be rejected in Chosŏn, however, in favor of adherence to orthodox precedent. For reasons I touch upon in Chapter Five, Japan was not as bound by established cartographic models, but even in the context of Japan's comparatively direct contact with Dutch, Portuguese and other European cartographic traditions, influential Ming texts such as the *Sancai tuhui* (Illustrated compendium of the three fields of knowledge) and the *Huang Ming zhifang ditu* (Administrative atlas of the Imperial Ming) appeared in both printed and manuscript Japanese editions, and their maps are copied in Japanese works. Ming works about Japan were assiduously collected and republished there, even works that reflected rather crude biases.

Although these last two chapters focus mostly on works that responded to Ming and European precedents, they also discuss the existence of an inverse phenomenon, the importation of materials from Japan and Korea into Ming China. Chapter Four examines one such work in some detail, the *Chaoxian tushuo* (Annotated maps of Chosŏn), printed in 1600 by a Ming military officer returning from the Korean peninsula. Thus we see that, although Ming works set precedents for many Chosŏn and Japanese publications, the flow of cartographic texts was to some extent multidirectional, linking nodes of cartographic and geographic synthesis within a broader web of influence.

Methodology

This project began with an attempt to address a few simple questions about local cartography in gazetteers from Jinhua prefecture in Zhejiang province. In the course of this research I discovered that scholars working with early modern Chinese maps had often drawn conclusions based on the map images themselves without taking into consideration the relevant textual explanations or prefaces contained within the works concerned, which often explain the compiler's intent (which can vary substantially from our expectations of modern maps). The first of my methodological considerations has therefore been to consider maps as integrally related to textual material, the *latter* usually being the author's or editor's primary tool in conveying detailed information.

The inclusion of textual information together with maps reflects a manner of reading in which details are sought in words rather than in the illustrations; maps in books usually serve as illustrations to the text rather than as independent documents. This presumed incompleteness of maps without textual context is demonstrated in Ming works by the use of notes like 'Where there is an image, there must be an explanation to express its meaning' (有圖必有說以發其義也).[6] This nesting within a larger text is one of the major features distinguishing most printed cartography from large, colorfully painted maps. While a map painted on a roll of silk can be considered as a work complete unto itself, a map contained within a printed text should not – even if it is a copy of that painted scroll – because its new context almost always provokes new readings.

Despite such caveats, much cartographic history has focused on the maps alone, without close examination of the text. This can lead to pitfalls such as evaluating the degree to which Western maps were understood in East Asia by the apparent skill (or lack thereof) in copying them in books. While these maps are indeed poorly copied in crudely woodblock-printed books, so are the copies of their indigenous counterparts. What does the accompanying *text* tell us about the ways in which they were understood?

6 Zhang Huang, *Tushu bian, fanli* p. 1b.

When long textual passages copied from Matteo Ricci's preface to his world map get incorporated into Ming works, even in the Japanese *Wa-Kan sansai zue* at a time when citation of Christian works was theoretically a capital offense in Japan, it becomes clear that we cannot make such judgments based on imagery alone. Examining both maps and text becomes particularly important when addressing the encyclopedic works of the late Ming, with their extensive sections on geography.

My study of Jinhua gazetteers also revealed how frequently maps were copied from earlier publications, and how poorly the assumed motivation of state building explained the actual diversity of maps included in gazetteers. When new maps were provided, an accompanying textual explanation sometimes noted how and why this happened. It became clear that cartographic materials moved both up and down the administrative chain, with county maps being gathered together and reproduced in prefectural gazetteers, while text about the locales cited in the county gazetteers often came from authoritative higher-level works. When private publishing in turn took these maps and texts out of their official contexts, source materials were drawn from every type of publication, both 'highbrow' and 'low,' not to mention countless locally available manuscript resources and military maps that were available for consultation at the time but have now been lost.

In search of a way to track down the sources of uncited text, I started with the *Siku quanshu* full-text database. Despite the fragmentary nature of the record preserved by the *Siku quanshu*, and its biases against many late Ming works on border affairs (which often contained maps, but were censored if they offended Manchu sensibilities), the database remains useful in tracing possible sources for later works, as well as identifying later texts that cited works I was studying. In the *Siku quanshu* the maps themselves are poor copies that are of little use in studying the original representations, but the searchability of the textual parts of these works is very handy. Another method for examining the changes wrought by the publishing boom has been to examine different editions of important late Ming works to which I had access, searching for changes made by those who republished them. In some cases entirely new maps were added to later editions due to the influence of a new work that had appeared in the interval between printings.

The least systematic methodology, but one that was serendipitously fruitful time and time again, was to search the relevant stacks of the Harvard-Yenching library for cartographic works by the armful, paging through to find superficial trends or anomalies that could then be examined more systematically. The sheer volume of material available, even if limited to texts from the Ming and earlier, could easily supply the raw material for dozens of books before even the foundations of the topic are sketched out.

It is ironic that international scholars working with cartographic materials – maps held hostage to concerns about 'national security,' even if centuries old – may most fruitfully pursue research in this field at libraries outside of the People's Republic of China. One outcome of this unfortunate situation, which ended up becoming a silver lining for my larger project, was that it forced me to shift my focus away from the magnificent court-sponsored works, military charts, and other manuscripts of limited circulation that researchers ache to see in Beijing and at various provincial archives. I turned instead to the crudely produced monochromatic woodblock prints that were used as illustrations in mass-produced books, the types of maps that far more people saw at the time. This fundamentally changed the direction of my work: Rather than taking a genre of maps, such as national maps, and examining all of the different forms in which they appear, my goal has been to examine all of the major genres in which maps were *printed*, from the highest registers to the humblest, and trace their interconnections.

Why maps matter

Why were maps significant to people of the Ming dynasty? Why are Ming maps important to scholarship today? To address such questions we must first decide what we mean by 'map.' This is a sufficiently contentious question in English, but it becomes more so when working across linguistic boundaries. J.B. Harley and David Woodward have proposed a transcultural answer to this question in their preface to the monumental series *The History of Cartography*, in which they define maps as 'graphic representations that facilitate a spatial understanding of things, concepts, conditions, processes, or events in the human world.'[7] While these very broad criteria may be vague, they are explicitly designed to include artifacts typically excluded from the category of 'maps' due to their incongruence with modern Western expectations of that category. Thus, geomantic diagrams or depictions of inner space are treated as worthy of discussion, alongside the atlases of Abraham Ortelius.

Mary Elizabeth Berry, a scholar of Japanese history whose work on cartography is seminal to the field, defines a map as 'a form of graphic representation that takes as its frame of reference the physical environment,

7 Harley and Woodward, 'Preface,' in *The History of Cartography, Volume 1*, p. xvi.

which it normally treats from an aerial perspective, with some attention to verifiable spatial distribution. Furthermore, a map relies on a combination of codes – particularly an iconic code – to construct that environment.'[8] Clearly closer to the expectations most modern readers have of their maps, this definition incorporates a bird's-eye view and adheres to some sort of scale. The use of 'codes' can be explicit or intuitive. Among Ming works the symbols used on maps for mountains, town walls, bridges and other features are generally treated as self-explanatory; it was unusual (though not unheard of) to provide an explicit key to standardized symbols.

What were the classical Chinese terms for 'map,' and how closely did they correspond with the definitions above? Examining the lists of map titles from the *Tushu bian* (Compendium of Illustrations and Texts) and the *Sancai tuhui* (Illustrated compendium of the three fields of knowledge), two late Ming encyclopedias, we see that one of the few consistencies is that *yutu* 輿圖 is only used for the most 'map-like' maps, while *tu*, on a lower register of specificity, can be applied to any type of illustration, sometimes with an identifying marker such as *biantu* 邊圖 (map of the frontier) or *shantu* 山圖 (a more painterly depiction of a mountain). I suggest that the ancient term *yuditu* 輿地圖 and its abbreviated variants *yutu* 輿圖 and *ditu* 地圖 can be translated as 'map' without falling far afield of Berry's definition, even if the stylistic characteristics or intended uses of such works do not exactly correspond with the expectations of a modern viewer.

Just as there were various types of 'maps' in early modern China, there were a variety of types of 'publication.' Stone stelae with maps engraved upon their faces were sometimes erected on the grounds of schools, enabling students or visitors to make their own rubbings on paper. A map carved on an immobile stone could thus circulate around the empire through its paper impressions. When Mao Huang cites the *Yujitu*'s depiction of the Heishui River in the twelfth century, or the later Yuan writer Chen Shikai cites the same stele, they were almost certainly writing with their eye on a paper rubbing, rather than examining the stone itself.[9] These stelae, which might be considered as heftier versions of woodblocks, often stood at the same prefectural schools that might have stored the printing blocks for local gazetteers. I suggest that they should be considered part of cartographic

8 Berry, *Japan in Print*, p. 61.

9 Mao Huang 毛晃, *Yugong zhinan* 禹貢指南, *juan* 2, p. 30b; Chen Shikai 陳師凱, *Shucai shi zhuanpang tong* 書蔡氏傳旁通, *juan* 1中, p. 21a. The list of works used at the beginning of the latter work includes it alongside numerous books and other maps (*yinyong shumu* 引用書目, p. 1b). There were other maps with this generic title, but the use of the prefix Chang'an appears to identify the reference specifically.

print culture. The earliest extant examples of the 'published' world map, if thus considered, would be the highly detailed *Yujitu* 禹跡圖 (Tracks of Yu) of 1136 and the slightly later *Huayitu* 華夷圖 (Map of Chinese and Barbarian [Lands]) engraved on the reverse of the same stone. The practice continued for centuries, well into the period I address in this book; even Matteo Ricci's world map in Chinese was committed to a stone block in the 1590s (sadly long lost).

Given the endless array of features, natural and artificial, that might be included on a map, the first problem for a cartographer is to narrow down the categories of data to be included. Such choices are informed by cultural and ideological factors, whether overt or subliminal. A cartographer's construction of a map involves choices between available strategies for creating a two-dimensional rendering that will be not only recognizable, but also useful to the reader. Richard Smith has enumerated many of the purposes for which maps were designed, including education, either for the student, or for a magistrate familiarizing himself with a new post; intelligence for the emperor or the battlefield commander; the planning of engineering projects; assertion of territorial claims, and even for undertaking a 'spiritual journey.'[10] Most maps have multiple functions, and if we wish to judge them critically we should do so not based on how closely they resemble modern cartographic styles, but by how well they suit their intended purpose, to the extent that this purpose is discernable today.

The question of 'accuracy'

Theoretical discussions of East Asian cartography, including Chinese cartography, have occasionally expanded their purview to consider spatial depictions that differ from what David Woodward has called the 'Western model of scaled orthogonal representations of the physical world.'[11] Woodward himself suggests that 'It is our preoccupation with our view of reality that inhibits Westerners' understanding of "Eastern" cartographies on their own terms.'[12] I have a certain degree of sympathy with this position, but would suggest that there is too much discussion of cartographic accuracy in early Chinese sources for us to dismiss it, even if it raises questions of privileging 'Western' expectations. 'Accuracy' in its modern cartographic

10 Smith, 'Mapping China's World: Cultural Cartography in Late Imperial Times,' p. 58.

11 Woodward, 'Preface' in *The History of Cartography: Volume Two, Book Two*, xxiv.

12 Ibid.

sense was not an alien concept. The 1136 *Yujitu* reflects far greater exactitude over a vast area of the earth's surface than any other map of its time from anywhere in the world.[13] Numerous Chinese cartographers and mathematicians from Pei Xiu (223-271) onward sought better ways of measuring and visually representing the types of data that are fundamental to modern cartography, and contemporary critics discussed the perceived accuracy of maps along with other features such as their aesthetic quality. Witness, for example, the frustration expressed by the famous Southern Song writer Hong Mai, who searched an historical atlas for its depiction of his native region only to be dismayed at its errors. He notes that the distances given are wrong, and claims that anyone who has traveled to the places it depicts will notice its mistakes.[14]

Luo Hongxian tells us in his mid-sixteenth century preface to the *Guang yutu* 廣輿圖 (Enlarged atlas) that he was so dissatisfied with the inexactitude of contemporary maps of the realm, which were replete with errors and inconsistent scale, that when he discovered a technically sophisticated grid-scale map (now lost) compiled in the Yuan by Zhu Siben, he took it as the model for his own work.[15] Both Joseph Needham and Ullrich Libbrecht have helped to provide important understanding of specific technical practices related to the construction of such maps. Libbrecht's annotated translation of a textbook by Qin Jiushao, originally published in 1247, includes sample problems of precisely the sort encountered in mapping a town or determining the distance and relative heights of natural features.[16] Rather than dismissing 'accuracy' as a term laden with Eurocentric connotations, it makes more sense to deduce what accuracy meant in different contexts.

13 For a discussion of this map, including mathematical analysis of its distortion, see Akin and Mumford, '"Yu Laid out the Lands": Georeferencing the Chinese *Yujitu* [Map of the Tracks of Yu] of 1136.'

14 Hong Mai, *Rongzhai suibi, juan* 10, 14b-15a.

15 Lu Liangzhi 卢良志, *Zhongguo ditu xue shi* 中国地图学史, p. 102; Fuchs, *The 'Mongol Atlas' of China by Chu Ssu-Pen and the Kuang-Yü-Tu*, pp. 7-8. When emperors appointed geographers, surveyors, and cartographers to undertake major expeditions such as the attempts to locate the source of the Yellow River during the Yuan and Ming, accuracy was not disregarded as an ideal trait of the eventual report, though this was more likely to be reflected in the textual record of such expeditions rather than in their graphic component. The map portion of the 1280 *Huang He yuan tu*, for example, is very simplistic in its layout, although the brothers who compiled it surveyed the route in person and recorded their journey in great textual detail in their report to the emperor.

16 Ullrich Libbrecht notes mistakes in the calculations used in sample problems for finding height and distance of a mountain (Libbrecht, *Chinese Mathematics in the Thirteenth Century*, p. 123).

In modern maps we often judge accuracy by the consistency of a map's depiction to scale, but this is not the only measure of precision. On a subway map, for instance, scale might be inconsistent, but riders certainly do demand that the stations be shown in the right sequence on the right lines. In a period when geodetic surveys were not only extraordinarily costly but also beyond the technical skills of most cartographers – and therefore not expected by the audience – 'accuracy' was more likely to mean the correct relative positions of landmarks. The most frequently performed measurement done in preparation for a new edition of a local gazetteer or other reference – usually the *only* measurement – was a tally of the distances along roads. This was done, not so much with maps in mind, as for the gazetteer's textual list of directions and distances to other towns and landmarks. A reader would not turn to the gazetteer's map for such information but would search for it in written form.

Though it falls outside the temporal scope of this study, the 1692 *Yiwu xianzhi* offers some insights into perceptions of accuracy in gazetteer maps. The 1692 work praises the high quality of the illustrations incorporated in Xiong Renlin's 1640 edition, noting a division of labor in their preparation (those who drafted the general scheme and calculated the distances gave their work to a professional engraver for commitment to woodblock plates). The method used for calculating distances is described as having relied on topography rather than the distance via roads.[17] A number of years later, the Kangxi emperor was willing to appoint European missionaries to manage cartographic surveys, demonstrating that he sought precise results, not merely a new mandala of spatial power upon which to meditate.[18] In cases where cartographers state that their goal is to create rigorously objective maps, it is fully appropriate for modern scholars to evaluate their accuracy, as it helps us to understand the cartographer's data-gathering resources and theoretical tools. Most maps, however, evince little effort to spread cognitive space over a precisely mathematically proportioned framework. In such cases it is clear that mapmakers had other purposes foremost in mind.

Spatial illustrations that are based on cultural beliefs or that are otherwise not evidentially anchored belong to a different category of illustration in most modern typologies, but in some of the works addressed in this study

17 '計里之法以形不以塗.' Wang Tingzeng 王廷曾, ed., *Yiwu xianzhi* 義烏縣志, *juan* 1, *tu shuo*: p. 1a-b.

18 Yee's 1992 essay 'A Cartography of Introspection: Chinese Maps as Other Than European' takes the latter position to an extreme; his articles in *The History of Cartography* are more evenhanded.

it is difficult to discern where one ends and the other begins. Furthermore, because maps of even imaginary space possessed real meaning in the minds of many as tools for understanding their 'place' in the cosmos, such maps have much to tell us about their makers' understandings of the world.

Given this study's focus on woodblock printed cartography, it is also important to note that the process of publication, transferring the draft of a map to a carved woodblock, and the subsequent printing of prints from this engraving, introduced many opportunities for variance from the cartographer's original vision. When a later publication copied these maps, even more changes could be introduced, sometimes as dramatic as the omission of spatial grids, or the inversion of images.

Problems of cartographic preservation

While this study attempts to survey the diversity of the printed cartographic record in the late Ming context, it is important to recognize that any effort to create a representative sample of the extant record will encounter a monumental problem: the differential survival of various types of cartographic materials. Wall-mounted maps deteriorate far faster than maps kept in less-exposed conditions; certain types of maps may be intentionally destroyed when they become out of date or when political changes forbid their preservation. Unbound paper ephemera like sheet maps tend to deteriorate much faster than pages bound in books, resulting in a disproportionately large sample of the often-crude little maps that were used to illustrate texts. Useless materials may lie undisturbed in storage and ultimately survive in greater numbers than dog-eared references that were thumbed to death. In a field like cartography, much information, such as trade or military secrets, was kept confined within a certain class of users rather than reproduced for a broader audience in ways that might boost its chances of surviving into the present. The destructive effects of war and the selective patterns of collectors and libraries have all affected the formation of the extant cartographic record. It is important to keep in mind that some works that were highly respected in their time are inaccessible to us today.

As R.A. Skelton has noted, 'Generalizations founded by historians of cartography on surviving examples must be taken with a pinch of salt. Many links in the chain of transmission are lost. The discovery of a hitherto unknown map may, like the turning of a kaleidoscope, recast the accepted pattern of thought and hypothesis or provide a "missing link" whose existence had been conjectured.'[19] James Hargett's figures on the survival rates of early geographical treatises provide some fodder for consideration. Of sixteen Tang-period *tujing* known from records, only fragments of two are extant, and those survive only in later copies.[20] Of sixty *tujing* known to have existed in the Northern Song, only one survives essentially complete.[21] Even the much later large-format sheet maps of the Jesuits, printed in large numbers as part of their missionary effort around the turn of the seventeenth century, survive in just a handful of copies representing several editions, some of them actually locally printed imitations.

When considering Ming maps, the imbalance of the record can be interrogated to some degree by consideration of the factors affecting the survival of different types of materials. Not only the physical properties of cartographic media come into play here; political and religious factors play a role in the shaping of the record. Examples include the suppression of Jesuit works in 1616-1617, and the grand censorial project that accompanied the creation of the *Siku quanshu*. The latter dramatically influenced the survival of texts related to military affairs on the Jurchen-Manchu frontier, many of which included maps, because works considered offensive to the Manchus were (with varying degrees of success) purged from the historical record. Aside from the basic question of whether an item endures or is destroyed, and whether even discarded items are in some way accessible to later researchers, we face the problem of where to find those that have survived. The usefulness of taking formation processes into consideration can be concretely illustrated by Li Xiaocong's efforts to track down maps in European institutions by determining which European groups would have been likely to obtain what maps at which times and places.[22]

The thinness of the earlier cartographic record plays a significant role in Mary Elizabeth Berry's argument in her chapter on maps in *Japan in Print*, her highly regarded discussion of the cultural transformations enacted and reflected by the publishing industry in Tokugawa Japan. She posits that a profound shift took place around the seventeenth century, a move from medieval to early modern cartography, based on 'the ability of ideologues to think generically about the space of the nation.'[23] The power of the shogunate

19 Skelton, *Maps: A Historical Survey of Their Study and Collecting*, p. 26.

20 Hargett, 'Song Dynasty Local Gazetteers and their Place in the History of *Difangzhi* Writing,' p. 411.

21 Ibid., pp. 412, 414.

22 Li Xiaocong, *Ouzhou shoucang bufen Zhongwen gu ditu xu lu = A Descriptive Catalogue of Pre-1900 Chinese Maps Seen in Europe*, p. 16.

23 Berry, *Japan in Print*, p. 60.

to limit and shape the ways in which space was understood through maps, Berry argues, hinged on this transformation in public understanding.

Berry sums up in a few words the essence of much recent scholarship when she writes that, 'Because mapmaking is code-making, and because communicable codes rely on social conventions, historical cartography is very deeply the study of spatial ideology.'[24] She notes that 'Maps, and all taxonomic schemes, require order and focus, dominant and subordinate motifs, the elision of unruly material. They require, in effect, organizing ideas that discipline evidence into normative structures of meaning.'[25]

When Berry asks why 'the regime was actually able to impose a cartographic logic on Japan,'[26] she privileges the state as a node of cartographic generation, echoing one of the dominant themes of current cartographic theory – the map as imposer of a mensurate framework on an otherwise tumultuous and colorful reality, speaking with a logic that inherently favors the regime in power. In short, early modern Japanese cartography imposed a regime of universality on what had been the particularity and idiosyncrasy of classical and medieval maps, but it did so only because the transformation of society and politics, first through the chaos of war and then through the reconstitution of unity under a newly unified Tokugawa state, had made such an imposition possible.

An anecdote about a colleague searching for a medieval map of Kyoto launches Berry's discussion of the lack of early city maps. Berry argues that these lacunae are not due so much to loss of the cartographic record, as to an absence of such a record in the first place. She states point-blank, 'I don't think maps of Kyoto were made at all in the medieval period.'[27] Among Berry's arguments for why maps would *not* have been made during the centuries when Kyoto was the capital are the difficulty of penetrating the complexity of the city's 'spatial politics,' and the danger that 'cartographic belligerence' could endanger the fragile balance of power among its elites.[28] Only after this local competition had been overwhelmed by the general unification of the nation as a whole did anyone dare to map Kyoto.

While I don't disagree with Berry's general argument that mapmaking was 'selective and irregular, far from comprehensive and routine, until rather recent times,'[29] and that we should not be surprised to find that maps

24 Ibid., p. 69.
25 Ibid., p. 48.
26 Ibid., p. 60.
27 Ibid., p. 54.
28 Ibid., p. 77.
29 Ibid., p. 56.

were not made of some subjects that might appear natural and obvious to us today, her thesis triggers concerns about the nature of negative evidence and the processes that form the cartographic record. Given that a similar perspective has inspired the cartographic theorist Denis Wood and others, it merits critical attention.[30]

In the parts of East Asia that felt the influence of the Tang, the construction of new capitals on the model of Chang'an was heavily laden with both cosmological and practical significance, from the selection of a site to the layout of streets. The city was a functioning representation of both the domestic realm and its universal context. This is evident in the layout of the Heian capital, now Kyoto. Certainly the layout of streets, the siting of markets and military institutions and their positioning relative to the main palace, would have been done with written plans or maps. Once captured on paper or silk, is it plausible that no one, either state actors or private individuals, dared to reproduce or update these maps in the Kamakura or Ashikaga periods? Is it more plausible to accept that maps may have been lost, than to view Kyoto's denizens as consciously deciding, repeatedly, for generation after generation, not to depict their city for fear of social consequences?

The fact that a handful of isolated examples survive – some in later copies – of a scattering of paddy maps, outlines of Japan, and even a Buddhist worldview depicting the route of a Tang pilgrim to India, shows that cartography as such was not alien to the medieval Japanese experience. The cosmography that placed Japan at the fringe of the Jambudvīpa continent suggests something of a 'totalizing vision,' a systematic organization of the world in which Japan and its subsidiary parts form a concrete portion. The cautious observer must admit, however, that the record is so clearly incomplete, its surviving fragments shaped by the contingencies of history, that it can never be definitively reconstructed. The discovery of a handful of ancient manuscript maps from Nara in 1833, maps that had been stored away and largely forgotten for centuries, has turned Japanese cartographic historiography in a direction it could not possibly have followed in the absence of such evidence. If a single box of Kamakura maps had survived in some unremembered alcove, what would our story of the medieval period look like? One danger to Berry's argument, unlikely though it may be at this late date, is that such evidence might yet appear some day.

To remind us of how unstable a form of evidence is the *lack* of surviving maps, we need only to consider the mapping of the Tang capital, Chang'an. No even remotely contemporary street maps of Chang'an exist on paper. In

30 See Wood, *Rethinking the Power of Maps.*

1080 Lü Dafang 吕大防 had a map of Tang-era Chang'an engraved in stone based on an old, presumably Tang-period, map supplemented with data from observations of the surviving layout of the original city. Despite having been engraved on stone – the very epitome of permanence – this 1080 map is only known today because its broken fragments were unearthed during the Qing period. Thus the only reason we can say with confidence that the Tang mapped the streets of its capital is because of a few fragments of a stone stele that was smashed and buried, barely escaping the oblivion of total disappearance. If not for this chance discovery of the fragments of a stone tablet that was itself a Northern Song scholar's salvaging of an older map of Chang'an, Berry's argument could easily be transferred to Chang'an with no solid evidence to disprove it.[31]

In addition to the normal destructive factors that pare down the cartographic record, there may be more locally relevant explanations for the paucity of earlier Japanese maps. For example, Japan's well-known tradition of tradesmen's secret transmission of knowledge tended to make specialized skills something to be guarded and passed down rather than broadly circulated.[32] The notion that maps should be printed for general education (perhaps inspired in part by the maps increasingly used to illustrate late Ming texts) appeared comparatively late, much as did the printing of other types of texts that had existed in only small circles before, written in manuscript form if at all. Maps preserved only in manuscript are, in simple statistical terms, far more likely to disappear from the record than are printed maps.[33]

31 Images of the 1080 map fragments can be found in Cao Wanru's *Atlas of Ancient Maps in China* 中國古代地圖集*: From the Warring States Period to the Yuan Dynasty*, Plates 45-48. For a broader discussion of Chang'an's layout, and city planning during the period, see Steinhardt, *Chinese Imperial City Planning*, Chapter 5.

32 On the role of secret transmission in the pre- and early Tokugawa periods, and the role of physician/printers in unraveling it, see Bartholomew, *The Formation of Science in Japan: Building a Research Tradition*, pp. 43-45. Unno notes more specifically that the system of master-apprentice secret transmission accounts for the paucity of texts explaining survey methods before the eighteenth century (Unno, 'Cartography in Japan,' p. 359).

33 Berry argues for a high point of cartography during the classical period, followed by a flat or declining curve in production during the medieval period; this analysis is based on her statistics for extant manuscript maps: thirty-one maps from the Nara period (710-784), thirteen from Heian (794-1185), fifty-five from Kamakura (1185-1333), fifteen Northern and Southern courts (1336-1392), and 111 Muromachi and Warring states periods (1392-1600) (Berry, *Japan in Print*, p. 58). However, almost all of the Nara maps are from one cache discovered at the Shōsōin, the storehouse of the Tōdaiji temple. The Tōdaiji in Nara was, as Unno notes, 'one of the major land reclaimers in the eighth century,' and this secular role led to the collection of paddy maps in the temple's archives (Unno, 'Cartography in Japan,' pp. 351-352). Without the role of Buddhist temples in landholding during this period, and the resulting storage of paddy field maps at a

The cultural marketplace vs. the State

In 'The "Spatial vernacular" in Tokugawa Maps,' Marcia Yonemoto describes a process whereby standardization developed in commercial maps as a way of making space 'comprehensible and usable to the reader by visually decoding cities, roads, and the Japanese archipelago itself.'[34] This perspective challenges the conception that cartography predominantly served the state's desire to shoehorn the diversity of reality into a standardized form. We are forced to recognize that 'standardization' is a double-edged sword; it may indeed serve government interests, but without it maps simply cannot fulfill the needs of their popular audiences. The usefulness of a map derives not only from its *presentation* of information, but from its *elimination* of extraneous information.

Yonemoto's focus on commercial printing, and on maps as artifacts sharing a cultural stage with the *ukiyo-e* and other prints for the developing market, leads her to emphasize the demand-driven side of this question, as well as some of the more playful and unexpected aspects of Tokugawa cartography. She observes that 'appreciating and cultivating knowledge was not only utilitarian, for one might argue that knowledge-seeking was, in a word, pleasurable.'[35]

In contrast to approaching maps as ideological tools imparting claims of durability to the state or other institutions, an approach which has its merits but is easily oversimplified and overextended, Yonemoto discusses maps that are self-consciously transitory, particularly those that depicted Edo and other major urban centers. Of course the nature of the map has much to do with the degree to which novelty is valued: City maps were marketed in seventeenth century Japan on the basis of up-to-date revisions, whereas antiquarian Gyōki-style maps intentionally retained their outdated forms.[36] Yonemoto observes that administrative cartography was shared with publishers with the consequence that 'The free flow of geographic information had the unintended effect of leaving the discursive

longstanding Buddhist institution, these maps would most likely have been dispersed and lost, as have any government maps or other commercial or legal charts of the same period. The number of surviving maps is too small, too localized, and too subject to the vagaries of preservation to draw any statistically meaningful conclusions about vectors of change in map production.

34 Yonemoto, 'The "Spatial Vernacular" in Tokugawa Maps,' p. 648.

35 Ibid., p. 661.

36 In his discussion of print commercialization, Peter Kornicki briefly addresses the marketing of city maps and their updating through alterations to woodblocks (*The Book in Japan: A Cultural History from the Beginnings to the Nineteenth Century*, pp. 60-62).

field of mapping open to the innovations and interpretations of non-elites. Ultimately, then, it was not the shogunal government or local officials but artists, writers, mapmakers and their commercial publishers who were most effective at spreading the texts and images of mapping to the public.'[37]

Though Berry and Yonemoto never challenge one another in so many words, their visions of the driving forces behind cartography diverge more dramatically than they might at first appear. In her conclusion Yonemoto challenges Benedict Anderson: 'Mapping was not highly politicized, it was not dominated by governing authorities, nor was it the vehicle of hegemonic power [...] The power dynamic of the modern map as described by Anderson is thus inverted in the early modern Japanese map.'[38] Though the manner and degree to which the state was imbricated in cartographic production varied not only between northeast Asian states but over time within each state, the reader will find in the rest of this book that my own reading of cartographic history generally endorses the spirit of Yonemoto's conclusions. The desire of the state to impose its spatial vision on the reading public accounts for a significant subset of printed cartography, but there were many other forces at play.

Perspectives on the late Ming publishing boom

As Kai-wing Chow has written, 'The cultural history of printing and the history of reading are now on the agenda of China historians.'[39] A coterie of scholars has focused specifically on the late Ming publishing boom and what it might signify about social and economic changes of the time.[40] Some of these studies address late Ming phenomena that are directly relevant to

37 Yonemoto, *Mapping Early Modern Japan: Space, Place, and Culture in the Tokugawa Period, 1603-1868*, p. 4.

38 Ibid., p. 177.

39 Kai-Wing Chow, 'Writing for Success: Printing, Examinations, and Intellectual Change in Late Ming China,' p. 120.

40 K.T. Wu, Cynthia Brokaw, Lucille Chia, Craig Clunas, and many others have linked the proliferation of printed texts to social and economic changes. An entire volume of the journal *Late Imperial China* was devoted to discussions of the place of publishing in late imperial society (Volume 17, no. 1, June 1996). None of these articles address cartography, however; nor does the subsequent revised collection that emerged from this issue (Cynthia J. Brokaw and Kai-wing Chow, eds, *Printing and Book Culture in Late Imperial China*). A collection edited by Lucille Chia and Hilde De Weerdt, *Knowledge and Text Production in an Age of Print: China, 900-1400*, includes an essay by De Weerdt on cartographic printing of an earlier period (De Weerdt, 'The Cultural Logics of Map Reading: Text, Time and Space in Printed Maps of the Song Empire').

cartography, but without drawing maps under their umbrella of discussion. For example, Julia K. Murray describes the role of printed illustrations in developing a common and widely distributed 'Chinese' culture during the sixteenth and seventeenth centuries, but without focusing on maps.[41] Robert E. Hegel examines illustrated fiction of the Ming and Qing, perforce addressing a different category of illustration. He traces book illustrations to two different sources, namely, popular religious art and traditions of elite painting.[42] Though the influence of landscape painting can be found in gazetteer maps in some late Ming works, and such depictions blossom in the high Qing, elite painting is generally quite far removed from the sorts of maps we will see in this book.[43] Nor do maps come from a tradition of religious painting, though the Buddhist works addressed in Chapter Two and in their Japanese iterations in Chapter Five clearly do demonstrate linkages to such a heritage. For the influential printed maps of the Song, the depictions of the *Yugong* or of shifting dynastic administrations, it is clear that there was a third source: the explication of historiography for an audience largely, if not primarily, consisting of students of the Classics and those preparing for examinations. A tradition of administrative cartography as applied to both civil and military purposes became wedded to this historiographical genre with some interesting results, a merger we see especially clearly in the Ming.

Lucille Chia, one of the foremost scholars of the nuts and bolts of the Ming publishing industry, has raised doubts as to the accuracy of modern impressions of an ***unprecedented*** publishing boom in the late Ming. She notes that destruction of the record makes it difficult to determine whether the larger number of extant late Ming works, as opposed to Song or Yuan publications, is the result of increased publication or of an accident of preservation.[44] What is beyond a doubt is a change ***within*** the Ming, with an increase of commercial publishing towards the end of the dynasty, but again Chia challenges a common assumption by questioning whether this increase correlates with a higher literacy rate.[45] Although cartographic works do not help to resolve the latter question, and much of the cartography in

41 Murray, 'Didactic Illustrations in Printed Books,' p. 417.

42 Discussed in Hegel, *Reading Illustrated Fiction in Late Imperial China*, Chapter Four.

43 An example of the painterly depiction of a country seat, immediately following a very different schematic map of its administrative buildings, can be found in the 1637 edition of the Jiaxing gazetteer (Luo Kai 羅炌, *Chongzhen Jiaxing xianzhi* 崇禎嘉興縣志), maps following *juan* 1, p. 37.

44 Chia, 'Mashaben: Commercial Publishing from the Song to the Ming,' p. 302.

45 Ibid., p. 284.

Ming publications can ultimately be traced to earlier sources, there is a clear expansion of the cartographic repertoire visible in late Ming publications.

This study has benefited from Kai-wing Chow's exploration of transformations wrought by commercial publishing on the production of literary culture, as discussed in *Publishing, Culture, and Power in Early Modern China*. Though he only mentions maps in passing, a number of Chow's arguments are relevant to cartographic questions. Chow argues that the 'literary public sphere' was profoundly impacted by printing, especially after printed works superseded manuscript in the sixteenth century; he unearths evidence of changes in the social status of the *shishang* (merchant literati) involved in publishing, changes that have been hidden by a pervasive rhetoric obscuring their commercial interests. Chow places the turning point for 'print culture' in the Wanli reign of the Ming dynasty. By the time Matteo Ricci arrived as an outside observer in the late sixteenth century, he expressed surprise at the widespread ownership of books in contrast to the places he had visited before. Liu Chengfan recorded him as having said in 1592, 'I have visited more than one hundred lands in ten years and yet I did not see a single person carrying a book.'[46]

Though he rejects the concept of a 'public sphere' in Habermas' sense, Chow convincingly argues for the interpretation of the term *gong* 公 as 'a public space shared by the literate population,' emphasizing the 'centrality of print in constituting the public – a community of readers scattered over immense geographical space of China and yet connected by means of their ability to read and access printed texts.'[47] Though this process can certainly be seen before the Ming, most of the changes in literary culture discussed by Chow take place from the 1550s to the 1630s. It is no coincidence that these dates are close to those I have chosen based on the publication dates of influential works in the field of cartography.

To borrow a page from Mary Elizabeth Berry and her discussion in *Japan in Print* of the materials available to a man-about-town in Tokugawa Japan, let us consider what someone interested in cartography might have found in circulation at the beginning of the Ming's last, and most culturally tumultuous, century. In the 1540s, a student would already have had a broad range of cartographies with which to contend. The standard histories, those monumental evaluations of each dynasty compiled by their successors, did not contain maps; however, the *Da Ming yitong zhi*, a national gazetteer of the current dynasty first published in 1461, had already provided a model for

46 Cited in Hsia, *Matteo Ricci and the Catholic Mission to China*, p. 75.

47 Chow, *Publishing, Culture and Power in Early Modern China*, p. 12. I would add that the range of circulation included more than China alone.

combining province-by-province summaries of geographical information in a systematic set of categories, illustrated with basic sketch maps to outline the broad contours of the empire and its subdivisions. This gazetteer was widely disseminated to schools and other public institutions around the empire, reprinted numerous times, and its influence was strongly felt beyond Ming borders.

Our hypothetical student could supplement this state-oriented, mostly textual work containing a mere handful of maps by finding a recent reprint of the *Lidai dili zhizhang tu* 歷代地理指掌圖, a densely illustrated historical atlas first published more than three centuries earlier (discussed below in Chapter Two). This atlas vividly depicts the changing boundaries and shifting urban centers of past regimes. Other more recent additions to the repertoire available to our student would have included the handful of illustrated defense-related texts that had already been published, books such as Xu Lun's *Jiubian tulun* 九邊圖論, which discussed the problems of the northern borderlands.

Our student could focus on smaller areas by examining maps in local gazetteers depicting individual prefectures, counties, or towns. An increasing number of gazetteers were by this time issued with simple maps, a trend which became more pronounced later in the dynasty. On an even more local level, if the student's clan had a genealogy, he could crack it open in hopes of finding a geomantic map of the family's ancestral graves.[48] Turning from the local to the cosmic scale, he could seek out a *fenye* 分野 map to see how the principles of cosmic resonance used for siting graves were applied to determining subtle linkages between the twenty-eight constellations of the zodiac and the regions of the earth.

With sufficient dedication our student might have finagled a way to see a copy of the extraordinarily large world map compiled in the early years of the Ming dynasty from Yuan sources, the *Da Ming hunyi tu* 大明混一圖 (Amalgamated map of the Great Ming), depicting lands as far away as western Europe with names transliterated from Persian into Chinese characters.[49] If our student passed through an urban center, he might take a trip to the prefectural school to see if there were any educational maps engraved on stone monuments from which a rubbing could be made to order

48 For a discussion of geomancy with notes on its cartographic representation, see Bennett, 'Patterns of the Sky and Earth: A Chinese Science of Applied Cosmology.'

49 Today this map is known only from an example in the Number One Historical Archives in Beijing (probably a later copy) and Chosŏn Korean versions, but some version of it must have circulated more widely because, as discussed below, a portion of it showing Africa is reproduced in several late Ming atlases.

and taken home – a depiction of the empire's river network, for example, or even an image of the town as it once appeared, preserved for posterity in lines incised on stone. Perhaps our student is not a would-be literatus but instead a Buddhist who dismisses the Sinocentric bias present in almost all of these maps, finding solace instead in the worldviews depicted in a copy of the *Fozu tongji*, a Song Buddhist encyclopedia which placed the Buddha's homeland at the center.

But our student is dissatisfied. These books, sheet maps and paintings are rather difficult to get hold of. Furthermore, in no single book can one find everything of interest. One solution might be to compile a new work, copying sources out as they become available and organizing the contents in a manner that follows reasonable precedent but tweaks it a bit to fit one's own purposes. Of this broad range of precedents, which to follow? How to combine them? These questions became central to cartography after the 1550s, as the passionate students of the age became the authors of new texts, and the market for books developed new channels for their circulation. The publishing of maps in the late Ming marked not so much a rupture from cartographic traditions as a continuation in new forms. Much of the new material would trace its lineage from preexisting models, but by 1610, we can find examples of almost every type of map cited above *combined in a single work*.

The proliferation of maps in gazetteers is a striking trend over the course of the Ming. As early as the Southern Song (1127-1279), it was not unusual to include a map in the local gazetteer, a textual record of a county or prefecture's history, administration, and culture. Out of twenty-nine extant Song gazetteers, nine contain maps; Hu Bangbo's examination of these gazetteers leads him to argue that this phenomenon emerged in the Song because the increasing centralization of power produced a need for 'the kind of books which could provide comprehensive reference about each local division for both local and central governments.'[50] This perspective grants precedence to the state, and the needs of its actors, above other factors.

Admittedly, the stereotype of the newly assigned district magistrate picking up a copy of the most recent gazetteer to get a handle on the local conditions of his posting did not become a cliché by accident. Joseph Dennis has explored in nuanced detail the role of gazetteers in linking the local to the imperial, and vice versa.[51] James M. Hargett has, on the other hand,

50 Hu Bangbo, *Cartography in Chinese Administrative Gazetteers of the Song Dynasty (A.D. 960-1279)*, p. 41.

51 Dennis, *Writing, Publishing, and Reading Local Gazetteers in Imperial China, 1100-1700*, especially Part I.

drawn attention to the role of the gazetteer as a 'scholarly monograph' rather than an administrator's tool, the audience for which 'now extended to virtually everyone who was literate and interested in local affairs.'[52] Peter Bol has enumerated a diverse range of reasons for gazetteer compilation, ranging from reasons of civic pride and utility, to the literary enthusiast's desire to impart his writings with local flavor from never-visited locales, as in the case of the *Fangyu shenglan* 方輿勝覽 (Exhaustive survey of the realm).[53] Rather than assuming that the maps in gazetteers are always, or even generally, tools or tokens of state power, a more fruitful analysis is possible if we consider a broader range of uses and audiences.[54]

In the late Ming we see an increasing number of maps in print, not only because more books that included maps were printed in the first place, but also because more types of works had come to include maps, and the average number of maps in such texts had increased. The typical local gazetteer might have included only one or two general maps in the mid-1500s, but by the end of the Ming it was not unusual to have a dozen or more maps, some of them adopting fundamentally different modes of representation. It is possible in a single late Ming gazetteer to find a map of the county in the present, paired with one showing it a century earlier; a map showing the location of the state apparatus in the center of town, followed by a map in which the government buildings are nowhere to be found but the turrets of scattered pagodas, shrines and monasteries pepper the suburbs around a simple representation of the town. At the prefectural level or above, a gazetteer also becomes more likely to include separate maps of the counties or other administrative units subsumed within its administration.

While such striking cartographic diversity was new to the local gazetteer in the late Ming, only in rare cases did it exhibit anything new to printed Chinese cartography in general. Everything from the blueprint-like

52 Hargett, 'Song Dynasty Local Gazetteers and their Place in the History of *Difangzhi* Writing,' p. 427.

53 See Bol, 'The Rise of Local History: History, Geography, and Culture in Southern Song and Yuan Wuzhou.'

54 The assumption that gazetteers were made for the aid of the state holds more truth in the Song, where this is a specific purpose cited in prefaces (Hu Bangbo, *Cartography in Chinese administrative Gazetteers*, p. 55) than in the late Ming, when it becomes common to see prefaces cite the need to commemorate the worthies of a locale or to present models for future generations of a local region. Local gazetteers, particularly from the county level, are much more idiosyncratic than those from the higher levels of administration, for which government agency more broadly applies. Some local gazetteers were compiled by scholars independently and submitted to the local magistrate as *faits accompli*, while this is not the case with provincial gazetteers.

depictions of urban centers to the concept of the historical map series can be found in extant maps from the Song, many of which may in turn be based on now-lost earlier precedents. Fairly accurate and labor-intensive means of surveying were not unheard of, and have left their traces in some extraordinary monuments such as the *Yujitu* maps, but these were very rarely used, and there do not appear to have been any formal institutions for transmitting more than basic skills to students for the use of property owners or the state. Thus, it bears repeating: What changes in late Ming cartography is context and quantity more than technology.

There are exceptions, of course. Multicolor printing did more than make the printed page prettier when it was introduced in the latter decades of the Ming dynasty; it had practical applications, as with the depiction of current and former place names in different colors on historical maps. The use of red and black to denote past and present place names had been used before in *manuscript* maps since at least the time of Jia Dan in the Tang, but the technique for publishing large numbers of printed books using this method appears to have arrived only in the Ming.[55] Conventions of labeling also became better established, with the *Guang yutu* offering for the first time an extensive and systematic key to symbols used on the map. Though most later cartographers did not trouble themselves with such details as a key, the use of consistent symbols was considered one of the features distinguishing a better map, and the replacement of older maps with ones that adopted such features was seen as something to boast of in the prefaces or *fanli* of later-edition gazetteers.

In these prefatory pages I have laid out some basic points that will be addressed in greater detail in the following pages. Even without significant technological changes to the process of creating or printing maps, cartography proliferated on the printed page to an unprecedented extent during the late Ming. The purposes of maps reflected the diversity of their users' social interests, spreading far beyond the administration of the state or the training of future functionaries. In the following chapter, I will turn to a sequence of works that might be considered landmarks in Ming printed cartography, exploring the ways they overlapped and intersected to make possible the weaving of a new tapestry by the final decades of the dynasty.

55 The *Jiu Tang shu*'s biography of Jia Dan 賈耽 states that he used this method on an important painted map. Liu Xu 劉昫, *Jiu Tang shu* 舊唐書, *juan* 138, p. 3786.

Works cited

Primary sources

Chen Shikai 陳師凱. *Shu Cai shi zhuanpang tong* 書蔡氏傳旁通. Wenyange facsimile edition. [Taibei]: Taiwan shangwu yinshuguan, [1983].

Hong Mai 洪邁. *Rong zhai sui bi* 容齋隨筆. Shanghai: Shanghai guji chubanshe, 1978.

Liu Xu 劉昫, ed. *Jiu Tang shu* 舊唐書. Beijing: Zhonghua shuju 中華書局, 1975.

Luo Kai 羅炌. *Chongzhen Jiaxing xianzhi* 崇禎嘉興縣志. Facsimile of 1637 edition. Beijing: Shumu wenxian chubanshe 書目文獻出版社, 1991.

Mao Huang 毛晃. *Yugong zhinan* 禹貢指南. Wenyange facsimile edition. [Taibei]: Taiwan shangwu yinshuguan, [1983].

Wang Tingzeng 王廷曾 et al. *Yiwu xianzhi* 義烏縣志. Yiwu: 1692.

Zhang Huang 章潢. *Tushu bian* 圖書編. Facsimile of Wanli 41 [1613] edition. Taibei: Chengwen chubanshe 成文出版社, 1971.

Secondary sources

Akin, Alexander, and David Mumford. 'Yu Laid out the Lands: Georeferencing the Chinese *Yujitu* [Map of the Tracks of Yu] of 1136.' *Cartography and Geographic Information Science* 39, no. 3 (March 2013): 154-169.

Bartholomew, James R. *The Formation of Science in Japan: Building a Research Tradition*. New Haven, CT: Yale University Press, 1989.

Bennett, Steven J. 'Patterns of the Sky and Earth: A Chinese Science of Applied Cosmology.' *Chinese Science* 3 (1978): 1-26.

Berry, Mary Elizabeth. *Japan in Print: Information and Nation in the Early Modern Period*. Berkeley: University of California Press, 2006.

Bol, Peter K. 'The Rise of Local History: History, Geography, and Culture in Southern Song and Yuan Wuzhou.' *Harvard Journal of Asiatic Studies* 61, no. 1 (June 2001): 37-76.

Brokaw, Cynthia J., and Kai-wing Chow, editors. *Printing and Book Culture in Late Imperial China*. Berkeley: University of California Press, 2005.

Cao Wanru, ed. *Atlas of Ancient Maps in China* 中國古代地圖集*: From the Warring States Period to the Yuan Dynasty*. Beijing: Wenwu chubanshe 文物出版社, 1990.

Chia, Lucille. 'Mashaben: Commercial Publishing from the Song to the Ming.' In *The Song-Yuan-Ming Transition in Chinese History*, edited by Paul Jakov Smith and Richard von Glahn, 284-328. Cambridge, MA: Harvard University Asia Center, 2003.

Chow, Kai-Wing. *Publishing, Culture, and Power in Early Modern China*. Stanford: Stanford University Press, 2004.

—. 'Writing for Success: Printing, Examinations, and Intellectual Change in Late Ming China.' *Late Imperial China* 17, no. 1 (June 1996): 120-157.

Clunas, Craig. *Superfluous Things: Material Culture and Social Status in Early Modern China*. Urbana: University of Illinois Press, 1991.

Cunningham, William. *The Cosmographical Glasse*. Amsterdam, Theatrum Orbis Terrarum. New York: Da Capo Press, 1968.

De Weerdt, Hilde. 'The Cultural Logics of Map Reading: Text, Time and Space in Printed Maps of the Song Empire.' In *Knowledge and Text Production in an Age of Print: China, 900-1400*, edited by Lucille Chia and Hilde De Weerdt, 239-270. Leiden: Brill, 2011.

Dennis, Joseph. *Writing, Publishing, and Reading Local Gazetteers in Imperial China, 1100-1700*. Cambridge, MA: Harvard University Asia Center, 2015.

Elliott, James. *The City in Maps: Urban Mapping to 1900*. London: British Library, 1987.

Fuchs, Walter. *The 'Mongol Atlas' of China by Chu Ssu-Pen and the Kuang-Yü-Tu*. Peiping: The Catholic University Press, 1946.

Hargett, James M. 'Song Dynasty Local Gazetteers and their Place in the History of *Difangzhi* Writing.' *Harvard Journal of Asiatic Studies* 56, no. 2 (1996): 405-442.

Harley, J. B., and David Woodward. 'Preface.' In *The History of Cartography, Volume 1: Cartography in Prehistoric, Ancient, and Medieval Europe and the Mediterranean*, edited by J. B. Harley and David Woodward, xv-xxi. Chicago: University of Chicago Press, 1987.

Hegel, Robert E. *Reading Illustrated Fiction in Late Imperial China*. Stanford, CA: Stanford University Press, 1998.

Hsia, R. Po-chia. *Matteo Ricci and the Catholic Mission to China: A Short History With Documents*. Indianapolis: Hackett Publishing Company, 2016.

Hu Bangbo. *Cartography in Chinese administrative gazetteers of the Song dynasty (A.D. 960-1279)*. PhD dissertation in Geography, University of Wisconsin at Milwaukee, 1994.

Kornicki, Peter. *The Book in Japan: A Cultural History from the Beginnings to the Nineteenth Century*. Leiden and Boston: Brill, 1998.

Li Xiaocong. *Ouzhou shoucang bufen Zhongwen gu ditu xu lu = A Descriptive Catalogue of Pre-1900 Chinese Maps Seen in Europe*. Beijing: Guoji wenhua chuban gongsi, Xinhua shudian jingxiao, 1996.

Libbrecht, Ullrich. *Chinese Mathematics in the Thirteenth Century*. Cambridge, MA: MIT Press, 1973.

Lu Liangzhi 卢良志. *Zhongguo ditu xue shi* 中国地图学史. Beijing: Cehui chubanshe, 1984.

Murray, Julia K. 'Didactic Illustrations in Printed Books.' In *Printing and Book Culture in Late Imperial China*, edited by Cynthia J. Brokaw and Kai-wing Chow, 417-450. Berkeley: University of California Press, 2005.

Robinson, Arthur H. 'Mapmaking and Map Printing: The Evolution of a Working Relationship.' In *Five Centuries of Map Printing*, edited by David Woodward, 1-9. Chicago: University of Chicago Press, 1975.

Skelton, Raleigh Ashlin. *A Historical Survey of Their Study and Collecting*. Chicago: Published for the Hermon Dunlap Smith Center for the History of Cartography, the Newberry Library, by the University of Chicago Press, 1975.

Smith, Richard J. 'Mapping China's World: Cultural Cartography in Late Imperial Times.' In *Landscape, Culture, and Power in Chinese Society*, edited by Wen-hsin Yeh, 52-109. Berkeley, CA: Institute of East Asian Studies, 1998.

Steinhardt, Nancy Shatzman. *Chinese Imperial City Planning*. Honolulu: University of Hawaii Press, 1990.

Tang Xiaofeng. *From Dynastic Geography to Historical Geography: A Change in Perspective Towards the Geographical Past of China*. Beijing: Commercial Press International, 2000.

Unno Kazutaka. 'Cartography in Japan.' In *The History of Cartography, Volume 2, Book 2: Cartography in the Traditional East and Southeast Asian Societies*, edited by J.B. Harley and David Woodward, 346-477. Chicago and London: University of Chicago Press, 1994.

Wood, Denis, John Fels, and John Krygier. *Rethinking the Power of Maps*. New York: Guilford Press, 2010.

Woodward, David. 'Preface.' In *The History of Cartography, Volume 2, Book 2: Cartography in the Traditional East and Southeast Asian Societies*, edited by J.B. Harley and David Woodward, xxiii-xxvii. Chicago and London: University of Chicago Press, 1994.

Yee, Cordell. 'A Cartography of Introspection: Chinese Maps as Other Than European.' *Asian Art* 5, no. 4 (1992): 29-47.

—. 'Chinese Cartography among the Arts: Objectivity, Subjectivity, Representation.' In *The History of Cartography, Volume 2, Book 2: Cartography in the Traditional East and Southeast Asian Societies*, edited by J.B. Harley and David Woodward, 128-169. Chicago and London: University of Chicago Press, 1994.

Yonemoto, Marcia. *Mapping Early Modern Japan: Space, Place, and Culture in the Tokugawa Period, 1603-1868*. Berkeley: University of California Press, 2003.

—. 'The "Spatial Vernacular" in Tokugawa Maps.' *The Journal of Asian Studies* 53, no. 3 (August 2000): 647-666.

1 Printed Cartography in the Late Ming

Old Typologies, New Audiences

Abstract
This chapter examines a sequence of major Ming publications in the categories of national gazetteer, atlas, maritime defense treatise, and encyclopedia, chosen for their influence and for their illustration of important trends or turning points in printed cartography. Outlining the ways in which these works organize information, it examines their relationship to earlier precedent-setting works, demonstrating that one feature of the Ming publishing boom was the reproduction and greatly expanded circulation of earlier maps, some from as early as the Song and Yuan. By the time the illustrated encyclopedias of the early seventeenth century were being compiled, a large proportion of their maps derived directly from the works discussed in this chapter.

Keywords: Chinese cartography, *Da Ming yitong zhi*, *Guang yutu*, *Chouhai tubian*, *Tushu bian*, *Sancai tuhui*

I have chosen the mid-sixteenth century as the starting date for this discussion of the late Ming cartographic publishing boom. What begins to change significantly around this point is the range of ways in which maps circulate, the intellectual contexts into which they are placed, and the audiences they reach. From this period forward, maps and their accompanying texts become a significant part of the reading matter of broader sectors of society.

In this chapter we turn to a sequence of major Ming publications, each of which has been chosen for its influence and for its illustration of important trends or turning points in printed cartography. In each case I will outline the ways in which these works organize information, and their relationship to earlier precedent-setting works. We shall see that part of the publishing boom was the reproduction and greatly expanded circulation of earlier materials,

Akin, Alexander, *East Asian Cartographic Print Culture: The Late Ming Publishing Boom and its Trans-Regional Connections*. Taylor & Francis Group, 2021
DOI: 10.5117/9789463726122_CH01

some from as early as the Song and Yuan.[1] By the time the illustrated encyclopedias of the early seventeenth century were being compiled, a large proportion of their maps derived directly from the works discussed in this chapter.

As more books featuring maps made their way into print, an increasing number of works featured maps as a self-sufficient component. By the mid-sixteenth century there also appeared an increasing number of large maps made by pasting together multiple printed sheets, which though still heavily annotated with text, were expected to stand on their own for both edification and decoration. The increasing availability of maps though mass printing also brought conflicting views into contact with one another; by the late Ming we see authors trying to balance fundamentally different worldviews. A prime example discussed in this chapter is the presentation of both Jesuit and Buddhist worldviews alongside the Sinocentric and administration-centered tradition of official cartography, as seen in the *Tushu bian* 圖書編 (Compendium of illustrations and texts).

By the mid-Ming, concerns about defense against the Mongols on the northern frontier had led to the publication of a number of books on the problem, many of which drew upon published memorials and other records of anti-Mongol campaigns and borderland defenses. It became common to include maps as illustrations in such texts. While some like the *Jiubian tushuo* 九邊圖說 (Illustrated treatise on the Nine Frontier Zones) of 1569 were published by the Ministry of War or other official bodies, others were composed by individuals who were driven by passion about the issue, and a sense that their personal experiences or access to documents should be shared with others. In the 1550s, the problem of piracy brought more attention to the seaboard and Japan, with later decades seeing Chosŏn Korea and the territory of the Jurchens (later Manchus) joining the regions of concern and the subjects of cartography. This array of published maps became a cumulative thematic repository from which later compilers drew, so that the sections on border security in later encyclopedic works gathered

1 While he does not discuss maps in particular, Kai-Wing Chow notes this trend with books in general, briefly discussing the career of Mao Jin 毛晉, whose Jiguge 汲古閣 house was one of the most prolific republishers of antique books and also put some older manuscripts into print for the first time (Chow, *Publishing, Culture, and Power in Early Modern China*, pp. 72-73). The sheer number of books in print expanded dramatically, as it became common to circulate writings in print rather than in manuscript form. On the switch from manuscript to print in the Ming, see McDermott, 'The Ascendance of the Imprint in China.' For a discussion of the phenomenon in Japan, see Kornicki, *The Book in Japan*.

together maps that were copied directly from specialized books that were in some cases more than half a century old.

One striking aspect of cartography in this period is that a new map was rarely drawn up from scratch if an old one could serve the purpose. Regional gazetteers borrowed text from other works that were believed to be authoritative, and freely used maps from earlier editions. Imitation was at least partly an appeal to authority, as expressed by Chen Zushou, who anticipates in the preface to his *Huang Ming zhifang ditu* that some critics might say his maps are too similar to those published in Luo Hongxian's *Guang yutu*: 'Maps depict the land, so how dare I make them differ?' (圖所以圖乎地豈敢有異).[2]

In European printing, the problem of cartographic conservatism or ossification came about in part because of the durability of copperplate engravings. Maps remained in print even after they were obsolete if more money was to be made using the old plates. Captain Cook famously complained that profitability came before considerations of the inconvenience that an obsolete map might pose.[3] The situation in China was quite different, in that maps were printed using carved wooden blocks, with a significantly shorter work life. Not only did they wear down, but they were vulnerable to fire and insect damage. Reprints were thus made in many cases from newly carved blocks. Yet, well-established images continue to resurface in print long after alternate representations have become available.

In the discussion that follows I will introduce the cartographic and textual characteristics of major works from four important genres of publishing. The state gazetteer, a mostly textual document, used a scattering of maps to orient the reader. The atlas, which had maps as its central theme, hosted a vastly larger ratio of images to text. Treatises on maritime affairs, a genre that exploded during the era of *Wokou* raids, remained important up through Hideyoshi's attempt to take the mainland via the Korean peninsula, followed by the rise of the Manchus in the northeast. The encyclopedia, a thematically organized survey of knowledge that included a section on geography, drew together material from all of the genres that came before.

The first work discussed here comes from before the late Ming publishing boom; it is the *Da Ming yitong zhi* 大明一統志 (Gazetteer of the Great Ming's unification), a national gazetteer first published in 1461. This gazetteer, often referred to in Ming sources simply as the *Yitong zhi*, served as a model of administrative geography even into the Qing dynasty. It systematically surveys

2 Chen Zushou 陳組綬, *Huang Ming zhifang ditu* 皇明職方地圖, *xu* p. 1b.

3 Skelton, *Maps: A Historical Survey of their Study and Collecting*, p. 16.

the details of individual prefectures, moving prefecture-by-prefecture within each province, to gradually build up a comprehensive picture of the Ming empire (as well as a handful of places beyond its frontiers) over the course of ninety volumes. Maps play only a small supporting role in this compendium, the vague sketch of each province merely serving as an outline upon which the details of the text unfold. Although it is included here as an example of pre-publishing boom cartography, numerous reprints continued to appear during the period we will examine, and its influence remained strong.

The role of maps as mere sketches to orient the reader changes fundamentally in the *Guang yutu* 廣輿圖 (Enlarged atlas) of Luo Hongxian 羅洪先, perhaps the most influential atlas of the Ming period. It presents the Ming provinces and frontier areas using a spatial grid, adopts a standardized scale on its various maps, and employs a defined system of cartographic symbols. It reflects the military concerns of the period, with early editions focusing on the northern frontier and later revisions adding maps of the troubled coast. Though the representations of the Middle Kingdom in this atlas derive directly from a huge manuscript map compiled by the Yuan cartographer Zhu Siben 朱思本, the *Guang yutu* was the first publication to bring these representations to a broad audience through printing, and it is the only form in which Zhu's long-lost work survives. Its multiple editions, with their many changes, exemplify the transitions of the late Ming publishing boom.

The *Chouhai tubian* 籌海圖編 (Illustrated compendium on coastal strategy) edited by Zheng Ruozeng 鄭若曾 focuses the cartographic lens on the seaboard, which was plagued in the mid-sixteenth century by *Wokou* piracy. This compendium, largely an amalgamation of material from several works that Zheng had published separately, is the most important of the new wave of maritime security books; I illustrate below how one map of Japan from this work was reused for decades in Ming books, reprinted even in Japan itself. There appears to have been a similar surge of cartographic interest in the northeastern frontier in the waning years of the dynasty, but the suppression of anti-Jurchen and anti-Manchu Ming military texts during the Qing dynasty destroyed many of these works.

The last category to be considered in this chapter is the encyclopedia, or *leishu* 類書. The practice of adding maps to encyclopedias was not new to the Ming dynasty.[4] The encyclopedic works of the late Ming differ from

4 A Yuan edition of the *Shilin guangji* (Copious record of the forest of affairs) includes both historical and contemporary maps, the contemporary illustrations portraying the various provinces of the Yuan. (Chen Yuanjing 陳元靚, *Shilin guangji* 事林廣記, *qianji san* 前集三,

their predecessors, however, in the sheer scale of their illustrated component, and in their stated audience. Many works targeted a readership that sought knowledge and 'culture' for both social status and intellectual pleasure, regardless of relationship to officialdom or academe. There are various registers, high and low, of this rather broadly defined genre; *leishu* such as the *Tushu bian* 圖書編 (Compendium of illustrations and texts) and the *Sancai tuhui* 三才圖會 (Illustrated compendium of the three fields of knowledge) retain a more literary gloss than the popular household almanacs.

In the *Tushu bian* in particular, evaluates cartographic representations in comparative perspective, accompanied by textual discussion of the relative merits of different theories. While some such as Zhang felt compelled to give different perspectives a fair shake, by the late Ming even authors who were stridently advocating one particular worldview often felt compelled to defend it against other concepts they realized their audience might already know about. Thus we see Chen Zushou articulate a spirited defense of the Sinocentric administrative map tradition against the heretical imports of the Jesuits and their depiction of multiple continents.

The popular miscellany or household almanac may be considered a subset of the encyclopedia category. These works condensed the most basic geographical, historical, and cultural knowledge into a much-abbreviated form. The presence of a series of maps as a standard feature of these almanacs, crudely produced as they may be, reflects the depth of cartographic penetration into late Ming popular publishing.

Gazetteer

Da Ming yitong zhi 大明一統志 (Gazetteer of the Great Ming's unification)[5]

The *Da Ming yitong zhi* differs from the other books discussed in this chapter in a number of ways. Imperial directives prompted its compilation, and its organization expressly mirrored the hierarchical system of the imperial administration. There are similarities between the text of the *Da Ming yitong*

diyu 地輿 section.) For a discussion of the concerns surrounding encyclopedia composition and the 'orthodox epistemological agenda,' changing *leishu* typology, and an examination of Hu Wenhuan's 胡文煥 collectanea, see Elman, 'Comparing and Classifying: Ming Dynasty Compendia and Encyclopedias (*Leishu*).'

5 Li Xian 李賢, *Da Ming yitong zhi* 大明一統志.

zhi and the *Huanyu tongzhi* 寰宇通志 (Universal gazetteer), which was completed five years earlier, in 1456, but the earlier work was not illustrated.[6]

Although maps comprise only a handful of the pages of the *Da Ming yitong zhi* (hereafter *Yitong zhi*), functioning primarily to orient the reader, its cartographic importance derives both from its wide distribution and its longstanding influence as a standard model of the geographic reference.[7] Its academic credentials, backed by state support of important scholars in its composition, were such that it continued to be cited as a source of information about localities – even in local gazetteers composed within those localities – well after the end of the dynasty. Though published for the first time in 1461, long before the beginning of the publishing boom as defined here, this gazetteer reached a broad audience because it was distributed to public institutions with government support and became available even in far-flung corners of the realm. Copies were also exported, either as diplomatic gifts or as personal purchases by visitors, and this work became particularly influential in Chosŏn Korea.[8]

Despite the perceived reliability of its text, this gazetteer's maps are very crude (see Figure 1:1).[9] One of its prefaces expresses the purpose of the work and the function of the maps within it.[10] We are told that the officials of the Zhou had written annals for surveying affairs, and maps for scrutinizing

6 Wolfgang Franke, *An Introduction to the Sources of Ming History*, p. 237.

7 Brook notes that this work was one of the 'core texts one would expect to find in a prefectural or county school in the mid-Ming,' and it is the only geographical text he lists (Brook, 'Edifying Knowledge: The Building of School Libraries in Ming China,' table 2, p. 108). Brook writes that in addition to the Classics and moral exhortation texts, which comprised the main contents of school libraries, during the Jiajing reign the imperial government began to have the *Da Ming yitong zhi* produced and distributed to schools (p. 104).

8 Franke notes five different Ming editions, from the earliest 1461 palace edition to later versions issued by such publishers as the Guiren Zhai and the Wanshou Tang. In 1564 a Chosŏn edition appeared, while an edition was published in Japan in 1713 (Franke, *An Introduction to the Sources of Ming History*, p. 237). A Qing edition held at Harvard changes the title to *Tianxia yitong zhi* 天下一統志, thus replacing 'Great Ming' with the less dynasty-sensitive 'All under Heaven,' while still citing Li Xian as compiler.

9 Timothy Brook goes so far as to attribute the roughness of gazetteer maps in general to the example set by this book. See Brook, 'Communications and Commerce,' p. 659. Gari Ledyard derides the crudeness of the *Da Ming yitong zhi* maps in comparison to the 'superiority' of the maps in a Chosŏn gazetteer, but the maps in the former were not intended to represent the 'state of the art' in cartography, and the maps in the latter work were actually designed to imitate the style of those in the Ming gazetteer; their alleged supersedence is Ledyard's subjective stylistic judgment. I discuss this issue and the influence of the gazetteer in Chosŏn in Chapter Four.

10 Li Xian, *Da Ming yitong zhi*, Tianxia yitong zhi tuxu 天下一統志圖叙, pp. 1a-2b. The version of this text appearing in the *Sancai tuhui*, which has updated figures on the number of counties and other changes, is translated in full in Chapter Three.

Figure 1.1 General overview of Ming territory from the 1461 edition of the *Da Ming yitong zhi*

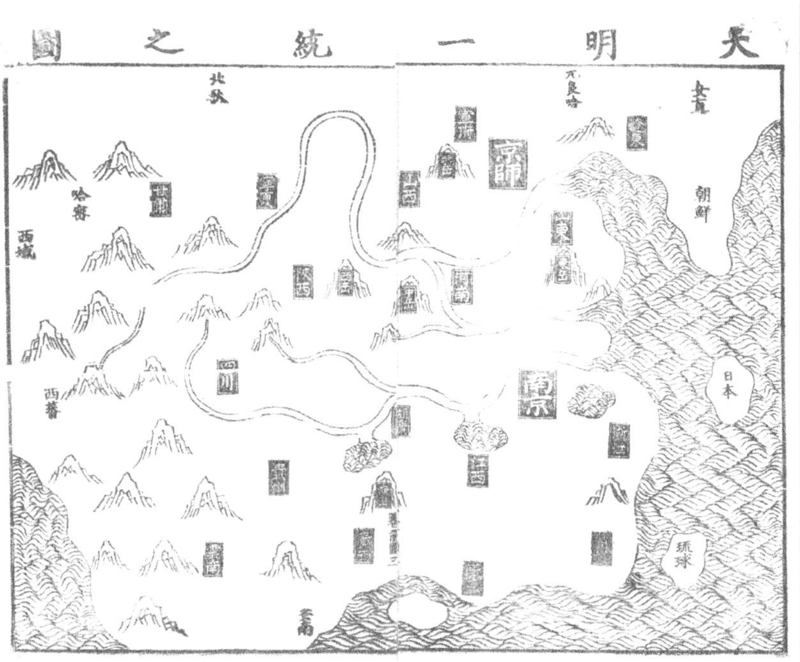

Gazetteer of the Great Ming's Unification. *DMYTZ*, Tianxia yitong zhi tuxu 天下一統志圖叔, pp. 1a-2b
Courtesy of the Harvard-Yenching Library

the land; the work at hand is a continuation of that tradition. Gazing at the map of the empire, the territory's vastness will be fully evident, sufficient to prove for a myriad generations the supremacy over past and present of the Imperial Ming's unity.[11] It is, in fact, a very basic depiction, but one reprinted many times over the course of the Ming dynasty.

In the following text, each province is broken down by prefecture, listing the establishment of the prefecture and historical changes in its territory, then moving on to topography, local customs, mountains and rivers, products, important institutions, local biographies, and other matters. Though prefectures are not illustrated with individual maps, the maps of provinces as a whole depict prefectural seats in black rectangles.

11 Ibid., p. 2b. The 1505 edition I have examined at UC Berkeley has its maps bound out of order, so this map is not first but second.

In the *Yitong zhi*, the broader world beyond the largest map's frontiers is acknowledged in perfunctory form as a list of 'tributary' states appended at the end, following the pattern of the standard dynastic history. Chapters 89 and 90 discuss Chosŏn, the Jurchen lands, Japan, the Ryukyus, and various Southeast Asian, Central Asian, and Indian sites, as well as Western Regions that had been visited by Zheng He's fleets as far away as Malindi, in East Africa. The descriptions of these places follow the format used for prefectures within the Ming dominion, listing each locale's mountains and rivers, local customs, products and so on, as if to deny the existence of any land truly exterior to the Ming's unification. Unlike some later Ming atlases, the *Yitong zhi* contains no maps to illustrate these far-flung places.

In later decades, this originally government-published work became an important text for the private publishing industry. Its respectability and longstanding authority gave it an automatic market. Lucille Chia notes two late Ming private editions published in Jianyang alone.[12] For examples of books in which maps were a more central component, however, we must move on to other texts, which exemplify the changes to cartographic publishing that were wrought by the late Ming publishing boom.

Atlas

Guang yutu 廣輿圖 (Enlarged Atlas)

The *Guang yutu*, compiled by Luo Hongxian 羅洪先 (1504-1564), is, as its title suggests, an expansion of the long-lost *Yutu* 輿圖 (Atlas) of the Yuan dynasty cartographer Zhu Siben. The *Guang yutu* marked a watershed in geographic publishing, spawning imitations and lending its maps to numerous later works, even in other East Asian countries and as far away as Europe.

Luo was an accomplished scholar but his cartographic interest might be viewed as more of a personal passion than a responsibility of office. As a student he had lamented that the maps he saw, even those that contained extensive detail, failed to accurately convey distances between the points depicted. After asking around for three years he found a copy of Zhu Siben's map, which used a graticule system – a grid of squares signifying distance – that Luo believed to offer the solution to this problem.[13]

12 Chia, *Printing for Profit*, p. 228.

13 Luo Hongxian, Preface to the *Guang yutu*, p. 2a.

Zhu Siben's original map no longer exists, but Luo copied its preface in his own book, giving us some insight into its creation. Zhu wrote that the map was the product of extensive travel combined with critical examination of earlier texts. His citation of the *Yujitu* 禹跡圖, a milestone of twelfth-century cartography engraved on stone with a regular grid marking its scale, clarifies Zhu's model for compiling such a map.[14] His fact-finding travels may have been under state sponsorship, but they satisfied a personal quest; as the cartographic historian Wang Yong puts it, he used an opportunity supplied by state business to realize a personal project.[15]

Zhu's original map had been projected over a grid for marking distances with a consistent scale. Its awkward dimensions (about seven feet along the longest edge) made it difficult to use, but by breaking it up into smaller sections that could be assembled into an easily portable book, Luo saw that he could make it more widely available.[16]

Aside from the use of a grid to ensure equivalent scale, the other feature most frequently discussed by modern scholars wishing to emphasize the scientific accomplishment of Luo's atlas is its formal key of twenty-four symbols.[17] While the use of implicitly standardized symbols dates back much further (for example, small crenulated walls representing fortified cities), this is the first known key explicitly enumerating them.[18] Even after this, however, the use of map keys did not become widespread. Given that most pictorial symbols were already intuitive, a formal key might have been seen as unnecessary for most purposes.

14 Zhu Siben, Preface to *Yutu*, copied in *Guang yutu*, *xu* p. 1a. Cordell Yee, who is reluctant to accept Needham's thesis of a continuous grid tradition, notes the citation of the *Yujitu* but says '*If* this map had a grid, it seems possible that Luo and Zhu were consciously following a tradition of grid mapping' (Yee, 'Reinterpreting Traditional Chinese Geographical Maps,' p. 51, emphasis mine). That Zhu's map had a grid seems beyond question, though Yee's larger point about the bias of Needham's assessment, and the vast number of maps that do not fit into his schema, seems justified. The use of grids for cartography in the Yuan has sometimes been linked to Islamic influence, such as Jamal ed-Din's presentation of a globe to Khubilai Khan (see Chapter Three). However, Zhu's citation of the famous *Yujitu* shows reliance on an even earlier, pre-Yuan precedent. The ultimate origin of the cartographic grid will probably never be resolved because the relevant evidence no longer survives, but for discussion of some very early appearances and their cultural context, see Lewis, *The Construction of Space in Early China*, especially the subsection 'Grids and Magic Squares' (pp. 247-273).

15 Wang Yong 王庸, *Zhongguo dilixue shi* 中國地理學史, p. 89.

16 Luo Hongxian, *Guang yutu*, *xu*.

17 See for example Needham, 'Geography and Cartography,' p. 555, and Yee, 'Reinterpreting Traditional Chinese Geographical Maps,' p. 51.

18 On early use of symbols see Hsu, 'The Han Maps and Early Chinese Cartography.'

The textual content of the *Guang yutu* continues the tradition of earlier national gazetteers, though with a greater emphasis on military matters. The book is divided into two halves, the first functioning as a national atlas and the second comprising a fairly extensive illustrated discussion of affairs along the northern frontier. In subsequent editions the focus broadened to incorporate contemporary concerns about maritime security along the coast, and later revisions inserted additional maps reflecting this change.

While a manuscript version was completed in the early 1540s, the first printing of Luo's atlas did not take place until after December 1553. Walter Fuchs has determined that the earliest extant edition, an undated version held at the time by a Chinese collector in Port Arthur, must have been printed between 1554 and 1558.[19] The illustrations of the maps in Fuchs' report are taken from this edition. If one compares these maps with those found in later reprints, there is often a dramatic contrast in quality due to the clumsy carving of later woodblocks. Aside from the distortion and elongation of the grid in the later maps, we see dramatic errors such as the transformation of clouds off the coast of Zhejiang into solid landforms in later versions. This is a reminder of the necessity to check various editions before making statements about Ming maps, particularly statements about quality or accuracy, because reprints could appear even during the author's lifetime with little control.

The first known reprint of the atlas to bear a date is the 1558 edition. Its printing blocks were carved at the Censorate at Nanjing,[20] reflecting the government's adoption and promotion of Luo's work. Unlike the *Yitong zhi*, commissioned from the top down, Luo's was a product of intellectual passion that moved up the chain of state recognition. After this 1558 edition a manuscript copy was made with some additions, a revision probably prepared by Luo Hongxian himself for presentation to the emperor.[21] Luo's work had now reached the ultimate audience.

A 1561 edition was the first to incorporate additional maps of Japan and the Ryukyu Islands.[22] This edition bears a preface by its reviser, Hu Song 胡松 (1503-1566), an adviser to Hu Zongxian, whose republication of the *Guang yutu* took place around the same time as Zheng Ruozeng's publication of the *Chouhai tubian* (Illustrated compendium on coastal strategy, discussed

19 Fuchs, *The 'Mongol Atlas' of China*, p. 15. This monograph remains the essential overview of the *Guang yutu*.

20 Ibid., p. 17.

21 Ibid., p. 18.

22 Ibid., p. 18.

later in this chapter).[23] The gathering of talented thinkers interested in geography that took place as a direct result of military concerns on the seaboard played a significant role in the fluorescence of defense-related cartography in the mid-sixteenth century.

A 1566 edition retains these additions and the preface by Hu Song, but the new editor, Han Jun'en 韓君恩, makes a dramatic change by adding brief essays after the heretofore rather dry administrative data listed for each province. These essays on the provinces are copied from a 1529 work by Gui E 桂萼,[24] whereas an essay on frontier defense is incorporated from Xu Lun's *Jiubian tulun* 九邊圖論.[25] In other words, the *Guang yutu*, now in at least its fourth printing, was expanding significantly; its basic core was supplemented in stages by significant additions of maps and text from the works of other scholars. A commercial reprint by the Yang 楊 family in Jian'an 建安, Fujian is attested for 1572, marking the fifth known edition, though this example appears to be lost.[26] The 1579 edition shows some updating of administrative statistics but a notable decline in cartographic quality. As this was the edition that had the broadest impact, however, it is the edition I discuss in detail below. A 1799 reprint also exists, which is almost identical to the 1579 version.[27]

In its organization the *Guang yutu* begins with the highest level of administration, presenting a general map (*Yudi zongtu*輿地總圖), followed by a textual enumeration of statistics for the empire as a whole. Its subdivision then echoes the organization of the *Da Ming yitong zhi* and other earlier works, though the text is extremely concise in comparison. The province, rather than the prefecture, is the lowest administrative level for which significant detail appears. Figure 1:2 is a sample provincial map, with a grid pattern representing squares of 100 *li* to a side, and a list of cartographic symbols in the text at right.

23 Huang, 'Cheng Jo-tseng,' in *Dictionary of Ming Biography*, p. 205.

24 Gui E (*jinshi* 1511, died 1531) had firsthand experience as a magistrate in Zhejiang before rising to higher office. He is known for his political opposition to Wang Yangming (Chou Tai-chi, 'Kuei O,' in *Dictionary of Ming Biography*, pp. 756-759).

25 Fuchs, *The 'Mongol Atlas' of China*, p. 20.

26 Ibid., p. 21.

27 The 1799 edition is available in a facsimile reprint (Taibei: Xuehai chubanshe 學海出版社, 1969). The 1579 edition is the one to which Matteo Ricci referred, as evidenced by his citation of tax rolls. See D'Elia, *Fonti ricciane*, vol. I, pp. 14-15; cited also in David E. Mungello, *Curious Land: Jesuit Accommodation and the Origins of Sinology*, p. 51. Fuchs notes (p. 24) that the dated preface to the Qing edition – the most commonly found – has often been removed from copies whose owners or sellers wished to make it appear to be the 1579 edition.

Figure 1.2 Huguang province as shown in the *Guang yutu*

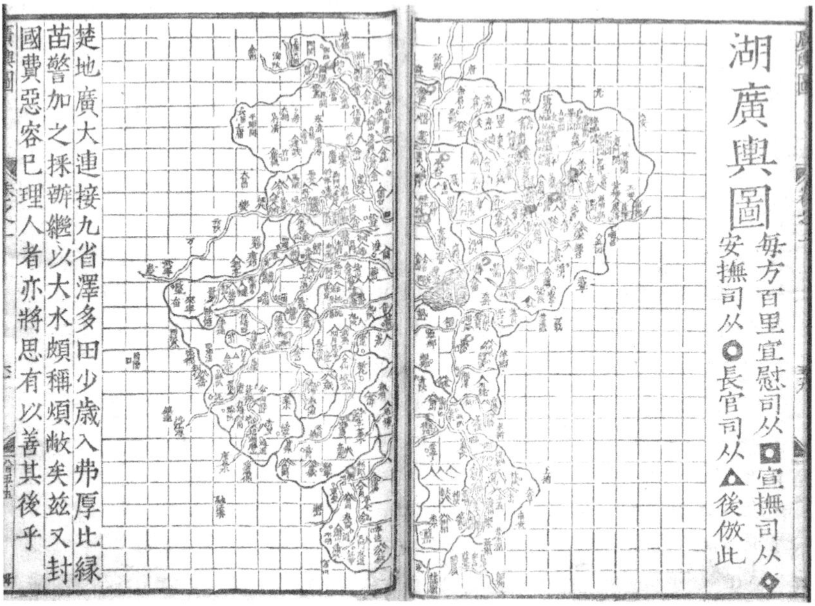

Enlarged Atlas, 1566 edition (*juan* 1, pp. 59b-60a)
Courtesy of the Harvard-Yenching Library

Each provincial listing begins with information on the administrative, military, and surveillance commissions governing the region, their subsidiary units, and the resources at their disposal. These data are followed by an administrative chart, which lists the province's circuits and their subdivision into prefectures and sub-prefectures. These prefectures are in turn subdivided into counties, some of which are listed as the sites of the province's Military Guard Units and Battalions. Under each county in the chart is a list of its former names, a brief description of its terrain ('mountainous with few fields' is a common description for Zhejiang, for example), and its administrative rating. In editions from 1566 on, this chart is followed by one of Gui E's essays describing the province's history and its recurrent problems, as well as its natural advantages.

In the second half of the *Guang yutu*, the focus shifts from administration to concerns of security and the frontier. After having used a copy of Zhu Siben's manuscript map for the first half, Luo applies a similar metric graticule to maps of the northern frontier.[28] By pairing a concise general

28 I have not been able to trace the source of these grid-based frontier maps; I suspect that a grid was simply placed over existing maps.

gazetteer in the first half of the book with timely discussion of more recent security concerns in the second half, Luo and his later revisers captured the spirit of the times: It seems that there was a significant audience for a basic reference to the empire that also incorporated current events.

This second half of the book begins with a 'General Map of the Nine Frontiers' (*Jiubian zongtu* 九邊總圖) with a stated scale of 500 *li* per square. This section draws to a close with maps of the Yellow River, a maritime route map, a map of the course of the Grand Canal, and depictions of Chosŏn, the lands of the Southeastern and the Southwestern Seas, Annam, the Western Regions (in these pre-Jesuit contact days a reference to western Asia), the Northern Deserts, the Ryukyu Islands, Japan, and lastly, a General Map of the Chinese and Barbarians (*Huayi zongtu* 華夷總圖). The depiction of the Barbarians of the Southwestern Seas is startling for its inclusion of a portion of the famous *Da Ming hunyi tu*, an early Ming manuscript map compiled in circa 1389 based on Yuan-era sources, which depicts Africa in the far west.[29] The place names in Africa vary somewhat between the early edition of the *Guang yutu* illustrated by Fuchs and the 1579 edition, but it is clear that Luo had access to one of perhaps many formerly extant copies of this map. Later the *Huangyu kao* and the *Tushu bian* would reprint this map from Luo's version, but because none of these works including the *Guang yutu* presents any textual annotation to this map, it is unclear that any reader would have realized that this was the same African continent visited by Zheng He's fleets.

The map of Japan, which bears no scale, has a note indicating that it is borrowed from Zheng Ruozeng (崑山鄭子若著), though the name of Zheng's original work is not indicated.[30] The addition of this and other new maps in later editions of the *Guang yutu* reminds us that in the late Ming the contents of the 'same' book were not static from edition to edition, often reflecting the influence of outside sources and changing contemporary interests. For the scholar of today, any such work should best be evaluated comparatively across editions, with attention paid when possible to the interests of the individuals involved in revising and republishing it.

In later editions of the *Guang yutu* the squares have been stretched into rectangles on many maps, possibly to fill the space on the rectangular leaf more completely, but indicating that the carvers of the printing blocks

29 A color reproduction of this manuscript map can be found as the first plate in Cao Wanru's *Atlas of Ancient Maps in China* 中國古代地圖集*: The Ming Dynasty 1368-1644*.

30 Maps of Japan became somewhat *de rigueur* in books on military security after the 1560s. Further examples are presented in the next section on Zheng's *Chouhai tubian*.

for later editions were careless about the grid's intended purpose as an indicator of scale.[31] As we shall see, other works that copied maps from this atlas sometimes did away with the grid altogether. The process of moving a map to woodblock print could thus divorce it from even its most notable characteristics. Some printed maps that lack grids were probably drafted using them, but had the background lines erased from the final version to enhance their aesthetic appeal.[32] The relative distances that can be shown with a grid can be recorded instead in textual form appended to the map.[33]

After the introduction of Jesuit maps, Chinese cartographers sometimes copied latitudinal and longitudinal lines on their translations and reprints, but in haphazard ways demonstrating little awareness of the function of these lines, which were likely mistaken for a spatial grid rather than geodetic coordinates.[34] For those who did attempt to use grids, flaws in their final product could point either to imperfect understanding of methodology, or as in the case of the *Guang yutu*'s later editions, poor copying by an engraver with no connection to the original cartographer.

Luo Hongxian was the rare example of a cartographer who had well-developed ideas and methodologies for applying survey technology to the creation of maps on a grid. Timothy Brook has noted the influence of Luo's mensurate system on the local gazetteer maps of Ye Chunji, who had met Luo in the 1550s.[35] Ye's maps are the only known surviving examples of Luo's grid system put to work at the local scale – a major undertaking that involved the training of junior staff after established bureaucrats repeatedly failed to fathom Ye's directions (at first they had simply copied earlier maps onto a

31 Fuchs, *The 'Mongol Atlas' of China*, p. 21; Yee 'Reinterpreting Traditional Chinese Geographical Maps,' p. 60.

32 Hu Bangbo discusses one map of Jiankang prefecture, for example, that originally included a network of grid lines but was later republished repeatedly without it. Hu, *Cartography in Chinese Administrative Gazetteers of the Song Dynasty*, pp. 123-127.

33 This is exemplified in a hydrological map illustrated by Yee, 'Reinterpreting Traditional Chinese Geographical Maps,' pp. 60-61. Though grid-scale maps look more technically precocious today, cartographers were trying to balance aesthetics with utility; evidence of their erasure in these cases means that the existence or lack of a grid on other maps is not necessarily a useful bellwether of accuracy or the care taken in composition.

34 Smith, 'Mapping China's World: Cultural Cartography in Late Imperial Times,' p. 75. For an example of a Ming map fitting this description, see Cao Wanru's *Atlas of Ancient Maps in China: The Ming Dynasty*, Plate 146. Smith also suggests that scholars of the *kaozheng* movement, whom one might expect to be eager advocates of empirical cartographic methods, were actually suspicious of what seemed to be excessive symmetry and regularity and thus tended to be dismissive of Jesuit cartography. Smith, ibid., p. 75.

35 See Brook, 'Mapping Knowledge in the Sixteenth Century: The Gazetteer Cartography of Ye Chunji.'

grid, rather than drafting new ones based on actual measurements).[36] Few gazetteers adopted systematic scale indicators until the Qing.[37]

Among later Ming works, the 1636 *Huang Ming zhifang ditu* 皇明職方地圖 (Administrative atlas of the Imperial Ming) by Chen Zushou 陳組綬 expressly states its descent from the *Guang yutu*.[38] Chen's atlas appeared in a radically transformed environment, though, one in which a new worldview introduced by the Jesuits posed a dramatic challenge. The *Huang Ming zhifang ditu* was a rebuttal to this outside influence, a hearkening back to traditions that Chen argued could not be lightly abandoned.[39]

The influence of the *Guang yutu* was not limited to Ming territory; indeed, it reached clear across the globe, leading to a radical revision of European maps of Asia through the mediation of the Jesuits. It was the most complete atlas of the Ming and East Asia to reach the hands of the early Jesuits in the 1580s, and Matteo Ricci provided the first translation of its title and part of its contents into a Western language.[40] The *Guang yutu* served as the foundation of Martino Martini's 1655 *Novus Atlas Sinensis* (New atlas of China), through which untold numbers of European readers were introduced to Chinese geography.[41]

Later in this chapter, as we examine encyclopedias like the *Tushu bian* 圖書編 (Compendium of illustrations and texts) and the *Sancai tuhui* 三才圖會 (Illustrated compendium of the three fields of knowledge), it will become apparent just how pervasive the maps from the *Guang yutu* had become. Luo's adaptation of the graticule system to maps of the frontier

36 Ibid., pp. 11-12. Aside from Luo Hongxian, Brook notes that Ye took inspiration from Xu Lun, the compiler of the *Jiubian tulun*, which the reader will recall provided the template for Luo's second half on frontier defense (p. 11).

37 The 1823 edition of the Jinhua county gazetteer, for example, portrays the county's landmass floating above a grid in which each square represents 10 *li* to a side. Significantly, after quoting Pei Xiu's six rules for accurate representational mapping, it cites as its model the 1718 atlas composed by French Jesuits under the Kangxi emperor's patronage (Huang Jinsheng 黄金聲, editor, *Daoguang Jinhua xianzhi* 道光金華縣志, *juan shou*, p. 6b). Such examples pose a challenge to Cordell Yee's argument that Jesuit cartography was not influential among native mapmakers, but they admittedly appear to be rather few. A much broader survey covering gazetteers from other regions would be necessary to draw a firmer conclusion.

38 This relationship has been discussed by Wang Yong in *Zhongguo dilixue shi*, pp. 93-94.

39 Due to its role in the anti-Jesuit movement I discuss Chen's work in Chapter Three rather than here. In addition to Chen's book, Wang Yong identifies a number of other works derived directly or indirectly from the contents of the *Guang yutu*: Wang Yong, *Zhongguo dilixue shi*, p. 94.

40 Szcześniak, 'Matteo Ricci's Maps of China,' pp. 127-129.

41 On this and other European maps derived from Chinese prototypes see Szcześniak's 'The Seventeenth Century Maps of China: An Inquiry into the Compilations of European Cartographers.'

conveyed such a clear logic of representation (in impression, even if the effectiveness of execution is disputable) that later compilers copied maps from the *Guang yutu* instead of from official sources such as the 1569 *Jiubian tushuo*, an atlas of the northern frontier that had been published by the Board of War.[42] The many reprints of Luo's book and its changing contents under the auspices of different editors serve as a microcosm of the changes precipitated in the cartographic arena by shifting public concerns in the late Ming, and exemplify the sharing of cartographic material between works that was a hallmark of this period.

At the same time as Luo's work was becoming widely available there was also a profusion of large-format sheet maps, usually printed on several pieces of paper which were then pasted together to form a large depiction that could be placed on a wall or spread out on a flat surface. These often took the form of a large central map with extensive text or commentary around the edges. One such work that survives is the *Gujin xingsheng zhi tu* 古今形勝之圖 (Map of topographic advantages, past and present), published in 1555 in Longqi County, Fujian.[43] Ricci's world maps would later take this form as well. Sheet maps appear to have become a more significant part of the vocabulary of publishing in the mid-Ming; they offered the display appeal of a large painted map, but could be produced in bulk.

Sheet maps were of interest not only to literati, but also to merchants and other late Ming seekers of intellectual stimulation. Some of our information about this comes from an unlikely source. The map captured by Captain John Saris and incorporated by Samuel Purchas into his 1625 book *Hakluytus Posthumus or Purchas His Pilgrimes* is an example of such a Ming sheet map. According to Paris, Saris was confiscating the property of a traveling Chinese merchant in Java as payment for debts, when the merchant was observed surreptitiously removing a box from among his other belongings, attempting to prevent its confiscation. This naturally sparked Saris's interest in its contents, which turned out to be a map. The merchant begged for the

42 Bing bu (Board of War) 兵部, *Jiubian tushuo* 九邊圖說. The maps in this work, done in a scroll-like linear format, far outweigh the text (which mainly lists the officials designated for each locale along the defensive corridor, the locales in each region being listed in three degrees of defensive importance). For example, the text preceding the section on Liaodong consists of only five leaves of text, three of them introducing the book as a whole, while the map of Liaodong and its various commanderies stretches over twenty-five leaves. There is no attempt at unified scale or orientation, and each regional map is followed by close-ups that are barely coherent with the larger-scale representation.

43 A color image of this map is reproduced as Plate 1 in Harley and Woodward, *The History of Cartography, Volume 2, Book 2*. It has survived thanks to its isolation in a Spanish archive.

box's return, apparently fearing punishment for allowing the map to fall into the hands of a foreign ship captain. Alas, his pleas only made it 'on the other side as earnestly desired and kept' – Saris wanted the map all the more.[44] The original sheet was four by five feet, of which one square yard was the map itself and the rest text. Purchas was unable to read the text but he copied the title characters in his miniature reproduction, allowing us to ascertain that the map is the *Huang Ming yitong fangyu beilan* 皇明一統方輿備覽, apparently unknown from any other source. Purchas notes its similarity to old European woodblock-printed maps but compares its style and execution unfavorably, at the same time asserting its great superiority to any map of China theretofore published in Europe.[45]

It is difficult to make even a rough estimate of the proportion of such sheet maps to the total number of maps in all formats, because these maps, if used as intended, were far more exposed to the elements and to the potential of damage by handling than those maps that were protected between the pages of bound volumes. What can be stated with certainty is that these printed sheet maps were firmly entrenched in the standard repertoire of cartographic sources by the late Ming. The regularized spatial grid of the *Guang yutu* offered a link to such maps; the promise of equal scale on each provincial map suggested that, just as Luo had taken a large map and broken it into sections to fit into a book, the pages of Luo's atlas could theoretically be removed from the binding and spread out together on a flat surface to create a larger map of the entire empire.

Maritime defense

Chouhai tubian 籌海圖編 (Illustrated compendium on coastal strategy)[46]

The development of the *Chouhai tubian*, a heavily illustrated work on maritime defense, shows some interdependence with the *Guang yutu*, but

44 Purchas, *Hakluytus Posthumus, or, Purchas his Pilgrimes: Contayning a History of the World in Sea Voyages and Lande Travells by Englishmen and Others*, p. 470.

45 Ibid., p. 478.

46 I have examined two Tianqi jiazi 甲子 [1624] copies held at UC Berkeley's East Asian Library. One is formerly from the Fryer collection and has been rebound in a Western-style binding, in two large volumes. The two copies are identical, indeed printed from the same woodblocks. The English title on the binding of Fryer's copy is given as 'The Coast of China and Japan.' A more literal translation, used by Benjamin Elman, is 'A Maritime Survey: A Compilation of

it marks a turning point in several important respects. Its compiler, Zheng Ruozeng 鄭若曾 (1503-1570), referred to in his collected works by his *hao*, Kaiyang 開陽, found inspiration in the *Guang yutu* and cites it as a source. Later editions of the *Guang yutu*, in turn, add maps taken from the works of Zheng Ruozeng, as we have noted in the case of the map of Japan. In contrast to the *Guang yutu*, the maps in this work are mostly quite stylized and bear no indication of scale. Its appeal to contemporaries, however, no doubt lay at least partly in its presentation of newly relevant and previously difficult-to-find information.

Before the 1550s, the Mongol frontier had been the primary focus of concern in the few frontier books that contained maps. Other zones of ongoing or recurrent military activity, such as the frontier with what is now Vietnam, also received attention.[47] The *Chouhai tubian* was an important milestone in the transformation of this situation.

The breakdown in domestic order that engulfed Japan in the mid-sixteenth century, as well as economic problems on the Ming coast exacerbated by the prohibition of regular trade, enabled the proliferation of piracy and coastal raids by the *wokou*, so-called Japanese pirates.[48] The Ming military response gave several prominent geographers and military planners the opportunity to meet and work together, and one of the fruits of this gathering was the circulation of maps among the officials involved. The *Chouhai tubian*, a printed compilation of several previously independent works, turned a new leaf with its shift of emphasis to the coast and its pairing of essays and personal observations on maritime security with extensive maps.[49] The *Chouhai tubian* is also distinctive because Zheng used maps that were actually employed by the military and the administrative bureaucracy.

Plans' (Elman, *On Their Own Terms*, p. 125). See Franke, *An Introduction to the Sources of Ming History*, pp. 223-224 for a descriptive list of five different Ming editions. Note that the *Siku quanshu* Wenyange edition has had significant forematter removed, including the list of sources consulted, and is thus inferior as a reference unless one is studying the evolution of this text in later periods.

47 The earliest extant military work on Vietnam to have been printed with cartographic illustrations may be the 1551 *Jiaoli jiaoping shilue*, based on a 1548-1549 campaign in Vietnam and Hainan (Zhang Ao 張鏊, *Jiaoli jiaoping shilue* 交黎剿平事略).

48 I say 'so-called' Japanese pirates because, as Kwan-wai So has shown through the citation of numerous contemporary sources (including the *Chouhai tubian*), there was significant involvement and even leadership by residents of the Ming coastal areas among the allegedly foreign *wokou*. See So, *Japanese Piracy in Ming China During the 16th Century*, especially Chapter 1.

49 Stanley Huang notes that the *Chouhai tubian* was instrumental in shifting the discussion of defense definitively from its previous obsession with the Mongol frontier (Huang, 'Cheng Jo-tseng,' in *Dictionary of Ming Biography*, pp. 206-207).

Zheng cites these maps, which had quite limited circulation, as sources alongside more commonly available published works.

Zheng was an important figure in the history of geography and military studies but never held high degree or office. Hu Zongxian 胡宗憲, a prominent appointee in the anti-*wokou* efforts, hired Zheng as an advisor in 1560, having seen one of his early publications, and Zheng stayed with him for about one year.[50] It was also in 1560 that Zheng completed the *Chouhai tubian*, which was first printed in 1562.[51] Because of this association between Hu and Zheng, some scholars have mistakenly attributed the text to Hu, particularly because his grandsons republished the book during the Tianqi period citing their grandfather rather than Zheng as compiler.[52] Because this Tianqi edition of 1624 is the earliest generally accessible Ming edition, this erroneous credit has been widely repeated, including in the *Siku quanshu* edition.[53]

Zheng's works on coastal defense survive in various permutations and differing stages of completion and revision, both as independent texts and as parts of collections, with some smaller works serving as chapters in larger compilations. For example, the *Chouhai tubian*'s materials on Japan borrow extensively from Zheng's own *Riben tuzuan* 日本圖纂, a shorter work focusing on Japan alone.[54] Aside from their synthesis in the *Chouhai tubian*, several of these shorter pieces that circulated as individual works were later recombined in a Qing collection, the *Zheng Kaiyang zazhu* 鄭開陽雜著, assembled by Zheng's descendants.[55]

50 Hu's anti-piracy efforts, his political maneuvering and eventual downfall are discussed in Kwan-wai So, *Japanese Piracy in Ming China During the 16th Century*, especially Chapter 4.

51 Zheng Ruozeng was a friend of Tang Shunzhi (two of whose illustrated works are listed as sources for the *Chouhai tubian*, see below). Tang provided funding to help Zheng collect materials about coastal defense beginning in the late 1540s. The work was completed in 1560, the year of Tang's death. After 1560 Zheng's support continued from Wang Daoxing 王道行, the prefect of Suzhou (Huang, 'Cheng Jo-tseng,' in *Dictionary of Ming Biography*, pp. 204-208).

52 Franke, *An Introduction to the Sources of Ming History*, p. 224.

53 Tanaka notes his surprise that Ishihara Michihiro 石原道博 had erroneously attributed authorship to Hu Zongxian, and takes this as a starting point to discuss the process of the work's compilation (Tanaka Takeo 田中健夫, 籌海圖編の成立 [The Formation of the *Chouhai tubian*], p. 18). An informal survey of library catalogs suggests that Hu today receives credit for the text more than half the time. To be fair, because Zheng was working for Hu at the time the work was published, it would not have been outside the norms of hierarchical claims of credit for the supervisor to claim the work of an underling. The fact that Zheng was involved in the work is not hidden in the preface to the 1624 edition published by Hu Weiji 胡惟極, Preface, p. 2a.

54 The *Riben tuzuan* is printed together with Xue Jun's 薛俊 *Riben kaolue* 日本考略 in *Nihon kigo no kenkyū* 日本寄語の研究.

55 Zheng Ruozeng 鄭若曽. *Zheng Kaiyang za zhu* 鄭開陽雜著.

The sources used in compiling the *Chouhai tubian* are far more varied and numerous than those relied upon by Luo Hongxian. The list of works consulted (*canguo tuji* 參過圖籍) runs on for several pages. It includes works on strategy, discussions of military engagements on the coast, and standard references like the Ming legal code.[56] Of these references, the works that are mainly cartographic in nature are set off in their own section at the beginning of the list, marked off at the end with the words *you tu* 右圖 ('the aforementioned are maps').[57] Quite helpfully for the many cases where the works are now lost, each listing includes the name of its author and, where applicable, his official title.

This list of primarily cartographic sources contains twenty works, together with the names and titles given for their authors, editors, or in some cases perhaps simply the individuals who were holding copies of the work. In Appendix 1 I cite them at length to give some sense of the breadth of resources that were available to Zheng, the majority of which are lost today.[58] It is likely that a number of these sources were manuscript maps with very limited circulation. The men listed in connection with most of them are officials with relevant responsibilities, suggesting that Zheng had access to materials kept on hand for reference by such officers themselves.

The first chapter of the *Chouhai tubian* presents a general overview of the empire followed by a series of coastal maps beginning in Guangdong and continuing up the coast through Fujian, Zhejiang, Zhili, Shandong and Liaoyang. Each of these provincial coastal maps is presented in the form of a multipage series tracing the shoreline from south to north. The book presents the entire course of the coastline in this format, one span after another, allowing the reader to survey along the shore in a counterclockwise sweep from the south to the northeast. Some are quite lengthy, the longest being Zhejiang (spanning twenty-one double-sided leaves). The maps are drawn from a bird's-eye perspective, facing out toward the ocean from inland, thereby mirroring the sea-gazing perspective of a land-based

56 Zheng Ruozeng, *Chouhai tubian*, *fanli* pp. 5a-8b.

57 Ibid., *fanli* p. 6a.

58 When future political changes make it easier to conduct cartographic research in Chinese archives, it will be enlightening to delve into the relationship between the manuscript maps belonging to military officers and the maritime charts used by traders, such as the Selden Map. On the latter, see Brook, *Mr Selden's Map of China: Decoding the Secrets of a Vanished Cartographer.* Maps used in the field by military men on the seaboard, and those made for and used by sailors, have left little trace in the printed record, but may have had heretofore unexplored connections with one another.

Figure 1.3 Bird's-eye view looking out from part of the Zhejiang coast toward the sea in the *Chouhai tubian*

Illustrated Compendium on Coastal Strategy, 1624 edition (map 19 in the Zhejiang series, *juan* 1, p. 40)
Courtesy of the Harvard-Yenching Library

official guarding against encroachment. There is special attention to coastal fortifications, as well as offshore islands that could afford shelter for pirates.

The second chapter leaps to the perceived source of the troubles on the coast, Japan. Zheng provides miscellaneous information not directly related to defense, such as sailing directions to Japan.[59] This information may have been included simply because Zheng had it; in any case it highlights the character of the *Chouhai tubian* as an assemblage of formerly discrete materials. Above the navigational narrative is a series of small depictions of the landmarks described in the text, rocky outcroppings drawn as they might appear from a passing ship.

In the interest of better understanding the enemy, this chapter concludes with a simple Japanese dictionary, using Chinese characters to represent the sounds of Japanese words. For example, the listing for 'hand' reads '手: 鐵,' the Chinese *tie* standing for the Japanese pronunciation *te*. One might imagine

59 These directions reference two works titled *Duhai fangcheng* 渡海方程 (Handbook for crossing the seas) and *Haidao zhenjing* 海道針經 (Compass-needle classic of the ocean lanes). Chap. 2, starting p. 6b.

Figure 1.4 Illustration of outcroppings visible while sailing to Japan

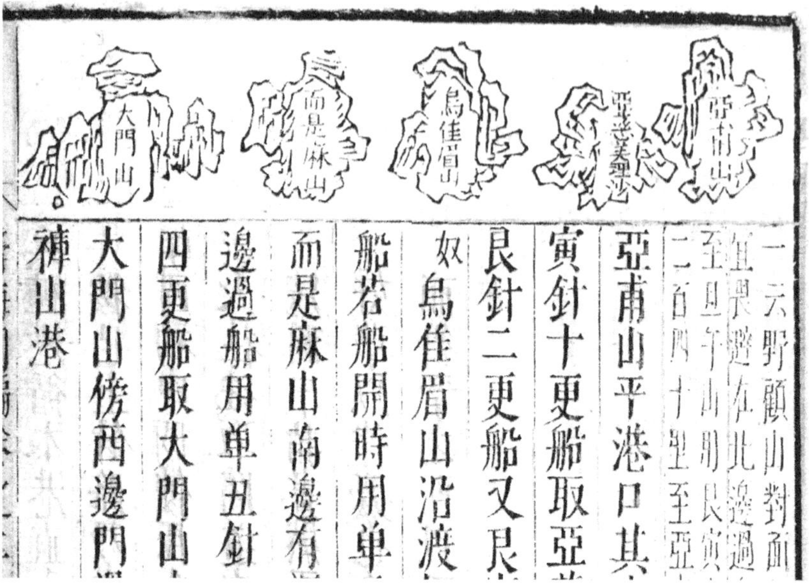

From the *Chouhai tubian* (*juan* 2, p. 9a)
Courtesy of the Harvard-Yenching Library

a glossary like this being of use for coastal officers in communicating with captured 'pirates,' though its original source is obscure.

The third chapter begins a series of overviews of the Ming's coastal provinces, each of which presents a map of the province as a whole followed by depictions of its prefectures and surrounding terrain. This series follows the same order as the coastal maps, covering Guangdong, Fujian, Zhejiang, Zhili, Shandong, and Liaoyang from south to north, but this time depicting inland prefectures as well as the coast. The text accompanying these maps records *wokou* raids, describes local defense measures, and includes various other military information depending on the province and its situation. The eighth chapter provides a chronology of *wokou* raids, and includes a map of raid routes reprinted from Zheng's *Riben tuzuan*.

Now that the geographical basics have been laid out, the book goes on to discuss specific past incidents and discourses on tactics. This is not to say the remainder is not illustrated – for example, the thirteenth chapter depicts weapons and equipment ranging from ships to arrows and flintlock guns – but the cartographic component is limited to the first half of the work.

Unlike the *Da Ming yitong zhi*, that national gazetteer that was far more important for its text than its crude maps, and differing from the *Guang yutu* in that it used a broad range of specialized sources, the *Chouhai tubian* formed a repository of complex maps and text that was recycled numerous times in later works. It had tackled an essential problem of the day, the issue of coastal defense, and provided readers with a combination of cartographic and textual resources to understand it. Its publication, first in 1562 and thence at least four more times before the end of the Ming, put these materials within reach of later thinkers on similar themes.[60] Tanaka Takeo observes that the *Chouhai tubian*'s notes on Japan were highly influential in such later books as Hou Jigao's 侯繼高 *Riben fengtu ji* 日本風土記 (Record of Japan's culture and climate) and the *Haifang leikao* 海防類考 (Investigation into maritime defense), while the 'Weiguo shilue' 倭國事略 (Outline of Japanese affairs) in the later *Tushu bian* is adapted from this book's 'Ribenguo xu' 日本國序 (Preface on Japan).[61] Mao Ruizheng 茅瑞徵 (fl. 1597-1636) wrote extensively on the military affairs of the Northeast at the beginning of the seventeenth century, yet his *Dongyi kaolue* 東夷考略 (Outline investigation of the Eastern Barbarians) reuses maps of Japan and the *wokou* raids that had been popularized much earlier by Zheng Ruozeng.[62] Figure 1:5 shows an example of the reuse of a map of Japan from Zheng's work, one instance of the broad reproduction of this iconic cartographic image. Other instances of the reuse of this map are in Mao Ruizheng's 茅瑞徵 *Dongyi kaolue* 東夷考略, and in the *Riben kao* 日本考 of Li Yangong 李言恭, where it is flipped so that East is at left.

By the end of the Ming dynasty, although the number of publications on security had increased dramatically, a disproportionate number of the maps used in these works can be found in two printed publications, late editions of the *Guang yutu* and the *Chouhai tubian*. The increase in publishing had made maps a more visible part of print culture, but without the results one might venture to predict based on the situation described by Arthur Robinson for Europe in this period: We do not find a milieu of increasing technical sophistication in which mapmakers compete to make technical improvements or factual corrections. Without access to the original materials it is unclear to what extent the situation differed at the higher echelons of military planning, where there were practical reasons for seeking to upgrade existing intelligence. Zheng gives us a tantalizing glimpse, though the copies

60 Franke enumerates the editions in his *Introduction to the Sources of Ming History*, p. 224.

61 Tanaka Takeo 田中健夫, 籌海圖編の成立 (The formation of the *Chouhai tubian*), pp. 18-23.

62 Mao Ruizheng 茅瑞徵, *Dongyi kaolue* 東夷考略. It was an early termination of his career in 1621, based on his alleged responsibility for Nurhaci's victory in battles of 1618-1619, that gave Mao time to write so many works (Yi and Goodrich, 'Mao Jui-cheng,' in *Dictionary of Ming Biography*, p. 1041).

Figure 1.5 Illustration of the reuse of a standardized map of Japan

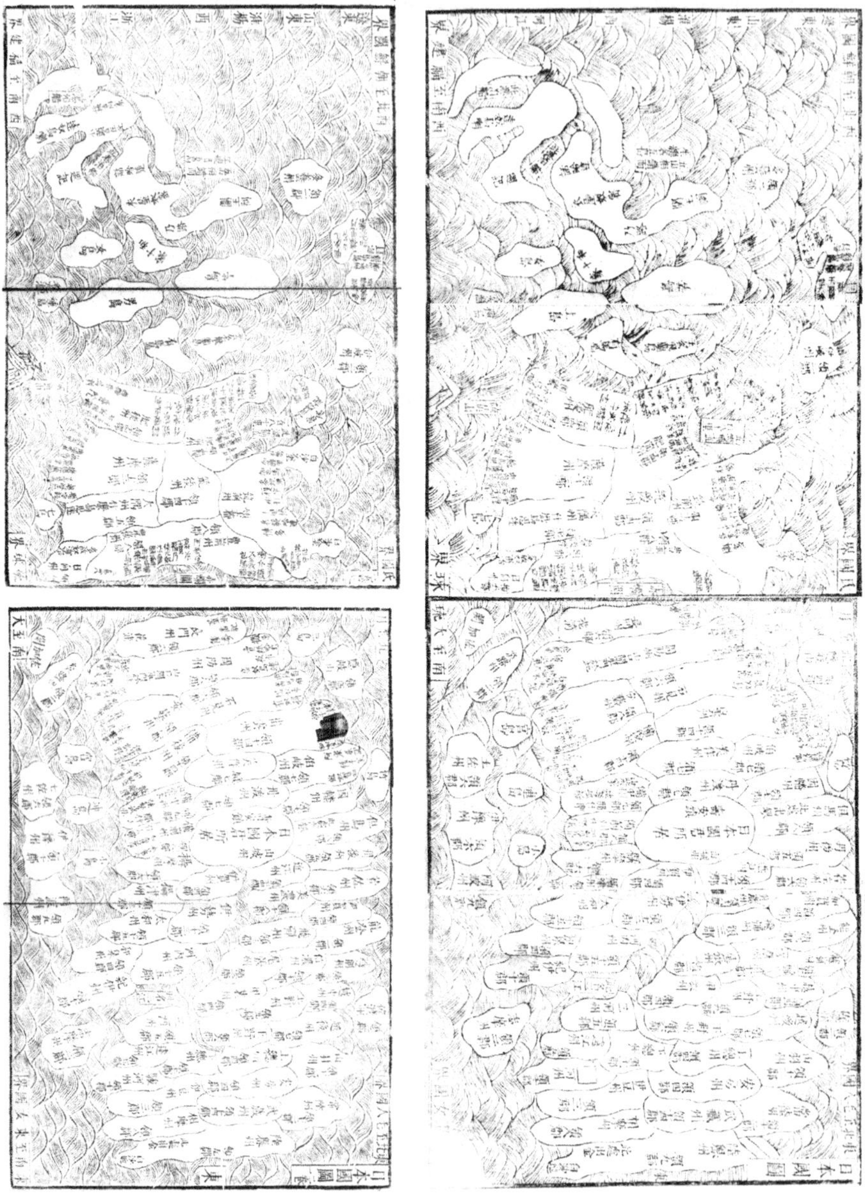

The map at left of Japan was published by Zheng Ruozeng at the beginning of the second chapter of his *Chouhai tubian.*

Courtesy of the Harvard-Yenching Library

The map at right, based on Zheng's, appears in the *Sancai tuhui*; illustrated compendium of the three fields of knowledge), *juan* 13, pp. 8a-9b. This and all other images from the *Sancai tuhui* are taken from an undated Chongzhen edition (1628-1644).

Courtesy of the Harvard-Yenching Library

he has published, of what the maps used by upper ranking military men may have looked like, but we can only guess how the 'state of the art' maps in official hands differed from those that were committed to print. Official coastal maps surviving from the Qing suggest there were probably no great qualitative differences between the types of maps that have survived in print and those that were held for official purposes. If the maps weren't changing, what was? To explore this question we turn now to the genre that did the most to synthesize the cartographic resources that were in circulation by the late Ming: encyclopedias.

Encyclopedias

Leishu, often translated 'encyclopedias,' are collections of material gathered from other works, sometimes supplemented with original material, presenting authoritative texts about various subjects in a systematic way, categorized along particular lines of organization. Late Ming *leishu*, though based on earlier precedents, tied together the developments of administration, military affairs, commerce, religion, and cultural exchange that took place in the latter half of the sixteenth century. I have chosen two of the most important encyclopedias of this time, not only for their popularity and influence, but because they exemplify different forms of syncretism. The *Tushu bian* 圖書編 (Compendium of illustrations and texts) is a prime exemplar of the 'academic' end of the late Ming *leishu* spectrum: wide-ranging, comparative, and inclusive. Text is dominant, though it is profusely illustrated. By contrast, the pictorial aspect is ascendant in the *Sancai tuhui* 三才圖會 (Illustrated compendium of the three fields of knowledge), a more abbreviated work that (surely for this very reason) was reprinted numerous times.

Tushu bian 圖書編 (Compendium of illustrations and texts)[63]

This 127-volume work is a carefully edited amalgam of texts and illustrations taken from more than two hundred highly disparate sources, combined with substantial new material, covering topics ranging from astronomy to parsing the Confucian Classics. Zhang Huang 章潢 (1527-1608) spent about fifteen years compiling it, finishing most of the work around 1577, although substantial portions of the geographical section must have been composed later because they discuss Jesuit materials. It was not until 1613

63 I have used the Chengwen facsimile of the 1613 edition.

that a fellow Nanchang scholar, Wan Shanglie 萬尚烈, brought the project to print by contributing his salary from an administrative post to support its carving on woodblocks.[64]

The *Tushu bian* is organized into three general sections, *Tian dao*, *Di dao*, and *Ren dao*, representing the fields of Heaven, Earth, and Humanity respectively. In this regard it is similar to the *Sancai tuhui* discussed below. In its textual material, however, Zhang's work is more sweeping than the *Sancai tuhui*. Zhang's inclusiveness is cited as the first rule in the *fanli*: 'Illustrations and writings originate from the sages of old or the scholars (*ru*) who expanded upon them. Although it is not possible to treat the importance of each in detail, all that are not offensive to the Way are included; I dare not be exclusive or yield to the majority.'[65]

The *Tushu bian* includes a wealth of cartographic 'landmarks,' maps that are copied or adapted from important works. Originality was not Zhang's claim, yet there is a sense throughout the *Tushu bian* of a personality interested in synthesis and comparison. While he quotes at length from texts with the patina of authority – ranging from the Confucian classics to military memorials from his own dynasty – Zhang proves willing to consider and absorb new and fundamentally different ideas, and to challenge old ones.

Zhang Huang had made the acquaintance of the Jesuit missionary Mateo Ricci, and his discussion of various worldviews demonstrates a willingness to discard certain aspects of tradition in favor of a convincing argument, though he takes pains to show ways in which the earth's form as described by the Jesuits might be compatible with ideas from the Chinese past. His inclusion of Ricci's map is one of the most discussed aspects of the *Tushu bian* in both Western and Chinese scholarship, in which Zhang is heralded as an example of late Ming intellectual openness.[66]

In his prefatory notes, Zhang states that among the illustrations there are some which appear to be true but are in fact false; it is impossible not to use them because they are relevant to the topic at hand, so they appear

64 In the meantime, Zhang worked on the revision of the Nanchang, Jiangxi gazetteer published in Wanli 16, which is illustrated with maps not substantially different from the typical bird's-eye views (Fan Lai 范淶; Zhang Huang 章潢. *Wanli xin xiu Nanchang fuzhi* 萬曆新修南昌府志). Another edition of the *Tushu bian* appeared in 1623 (Goodrich and Tay, 'Chang Huang,' *Dictionary of Ming Biography*, pp. 83-85; Franke, *An Introduction to the Sources of Ming History*, p. 313).

65 '圖書肇自先聖或衍之先儒雖不能一一詳考所自要皆無詭於道者存之不敢杜撰傅會.' Zhang Huang, *Tushu bian*, *fanli* 1a.

66 Cordell Yee has challenged the notion of Ming receptiveness to western cartography by arguing that Jesuit maps had little domestic influence; see Yee, 'Traditional Chinese Cartography and the Myth of Westernization,' especially pp. 171-177.

in the book for reference.[67] To this category belongs a Buddhist world map he includes. Strikingly, Zhang does not simply disregard such perspectives. Instead he uses them to discuss the merits of different views and elaborate his reasons for choosing some over others.

Illustrating the degree to which Jesuit-introduced cosmological perspectives had influenced Zhang's view of the relationship between the heavens and the earth, the first European-style map appears not in the section on the Earth, but in that on the Heavens, along with an image of an armillary sphere.[68] There is a somewhat blurry transition between the first portion of the *Tushu bian*, which deals with cosmology, metaphysics, and the heavens, and the subsequent portion, which turns its focus to the Earth. The nature of the relationship between the heavens and the earth is the theme of several transitional chapters, including not only the Jesuit-inspired illustrations noted above but also a picture in Chapter 28 illustrating the traditional Chinese conception of a square Earth within the round heavens.

The purely geographical section of the *Tushu bian*, a section that takes up almost two thousand leaves, begins with Chapter 29, in which we find two very different maps. The first, the *Ershi ba su fenying gesheng dili zongtu* 二十八宿分應各省地理總圖 (Comprehensive map of the field allocations of the twenty-eight lunar lodges and their relationship to the provinces), mixes cosmology and geomancy with geography. Though Ming territory occupies almost its entirety, its original maker appears to have intended it as a world map; its periphery lists all of the primary tributary states as well as the Western barbarians.[69] After this map we find the *Yudi shanhai quantu* 輿地山海全圖, a Ricci-style 'Complete map of the mountains and seas of the earth.' The prefatory text to this map begins with a quote from Lu Xiangshan about the unfathomed reaches of the world, but contrasts this with the quite fathomable potential to use such maps to determine the distance from the Great Western Sea (i.e., the Atlantic) to Guangdong.[70] The *Diqiu tushuo* 地球圖說 (Explanation of the map of the globe) appended afterwards is a lengthy quote from Ricci's own map preface, as discussed below in Chapter Three.

Late Ming syncretist that he is, Zhang cannot display these contrasting worldviews without including yet another that had some influence in his

67 Zhang Huang, *Tushu bian*, *fanli* p. 1b.

68 Zhang's incorporation of European knowledge is discussed in Chapter Three.

69 Ibid., *juan* 29, pp. 3a-b.

70 Ibid., *juan* 29, p. 33a. The map he illustrates is of course too small and simplified to make such measurements; Zhang's comments must be applied to larger examples of the same type.

time, though he finds it less than appealing to do so. This is a Buddhist Jambudvīpa map, here titled *Sihai Huayi zongtu* 四海華夷總圖 (General map of the Chinese and Barbarians in the Four Seas). Though this map and others like it are discussed in Chapter Two, it is important to note here that although Zhang politely denies its trustworthiness, he lives up to his stated effort to provide breadth by quoting at length from a Buddhist geographical description of Jambudvīpa and the other continents populating Buddhist cosmology, presenting the basic elements of this contrasting worldview. In the geographical context, however, his tendency towards tolerance is embodied more in efforts to show common ground between Jesuit and Confucian ideas.

The next chapter contains maps and essays related to geomantic themes. The *Zhongguo san da ganlong zonglan zhi tu* 中國三大幹龍總覽之圖 (Survey of the three great 'Main Dragons' of the Middle Kingdom) depicts three vaguely horizontal parallel landforms lined up between the Yangtze and the sea, the Yangtze and the Yellow River, and the Yellow River and the Yalu River.[71] A few pages later comes the *Zhongguo dili haiyue jianghe dashi tu* 中國地理海嶽江河大勢圖 (Map of the general form of the Middle Kingdom's seas, marchmounts, and rivers). If one compares this image to the first map of the *Da Ming yitong zhi*, the 1461 gazetteer discussed above, it immediately becomes apparent that this is the same map, but with the black rectangles around place names removed. The inclusion of the five marchmounts, or sacred mountains, is probably the reason for this map's categorization with geomantic materials.

Having outlined some general ideas about the nature of the earth, its form and the forces thought to flow through its surface, Zhang turns to explication of the distant past and the development of the landscape over the course of recorded human history. To do this Zhang turns first to sources that discuss the world as it was in the days discussed in the Classics, later shifting to the administrative changes accompanying the changing of dynasties and the coming and going of invading armies.

The first of these maps is the *Yugong suo zai suishan junchuan tu* 禹貢所載隨山浚川之圖 (Map of 'determining the high mountains and the great rivers' from the *Yugong*), based on a chapter from the *Book of Documents*. Following this are depictions of the *Jiuzhou*, the Nine Regions created by the sage King Yu when he tamed the floodwaters in distant antiquity. Although the title of each of these maps ends with *jiangjie* 彊界, implying fairly defined

71 For a Qing use of this map with a different title, see Henderson, 'Chinese Cosmographical Thought: The High Intellectual Tradition,' p. 221.

Figure 1.6 Map of ancient Yangzhou, one of the Nine Regions, from the *Tushu bian*

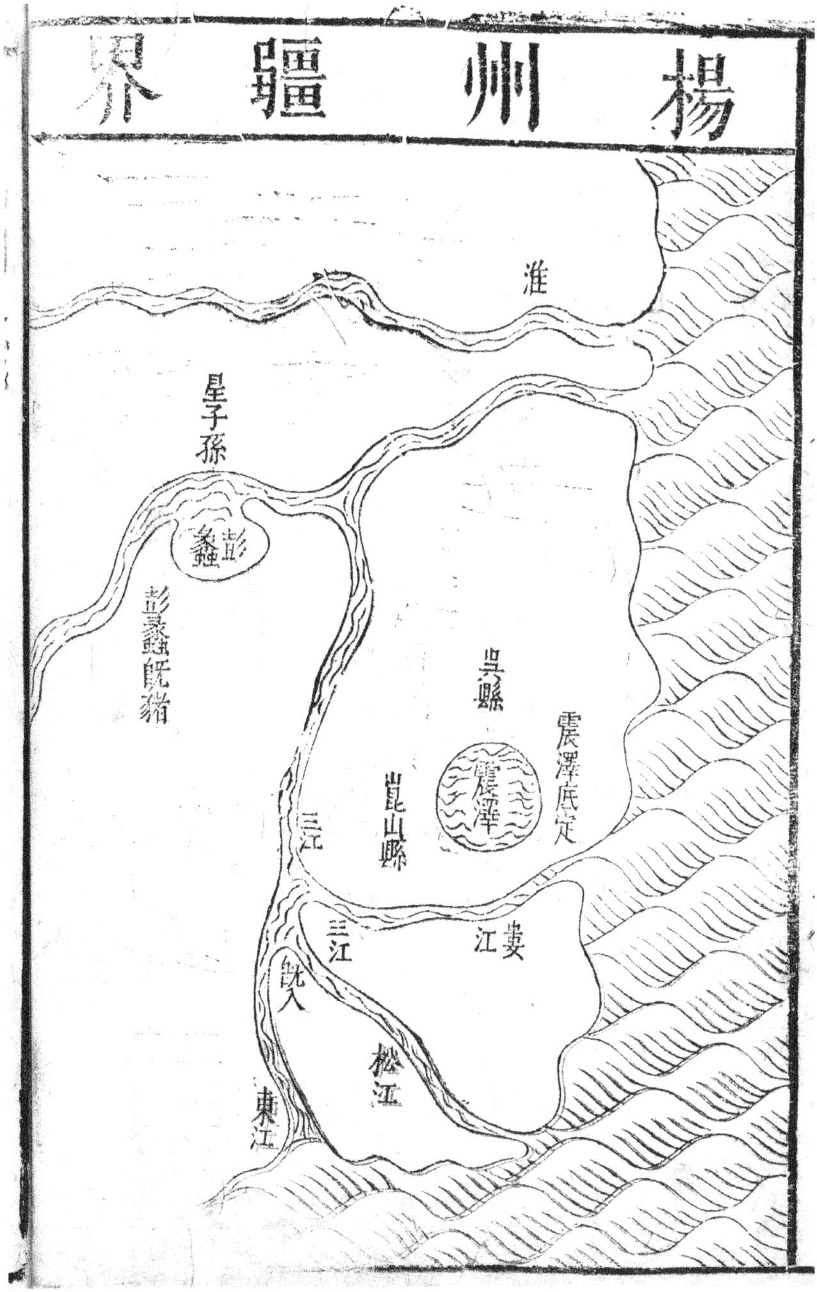

Compendium of Illustrations and Texts, 1613, *juan* 31, p. 16b
Courtesy of the Harvard-Yenching Library

Figure 1.7 Zhifang shi jiuzhou shanze chuanjin zongtu (Official map of the Nine Regions' mountains, marshes, and waterways)

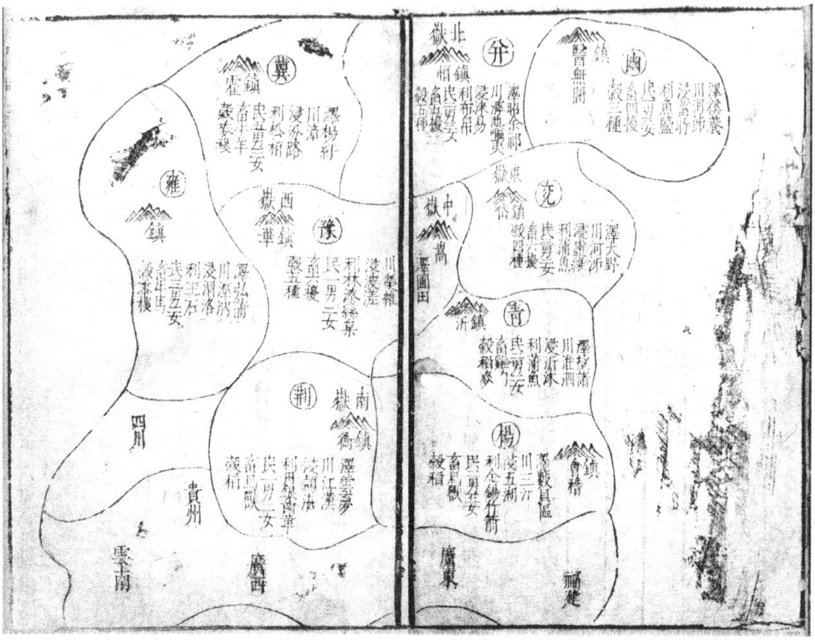

From the *Tushu bian* (*juan* 32, pp. 2a-b)
Courtesy of the Harvard-Yenching Library

regions, none of the maps delineates a boundary, showing instead a very general territorial overview. Given the great antiquity and paucity of records about these alleged regions, the maps show few details other than basic hydrological features. Figure 1:6 shows a reconstruction of Yangzhou.[72]

A different sort of map of the Nine Regions, one that functions as a simple one-page gazetteer, appears in Chapter 32. The *Zhifang shi jiuzhou shanze chuanjin zongtu* 職方氏九州山澤川浸總圖 (Official map of the Nine Regions' mountains, marshes, and waterways) is taken from a commentary on the Rites of Zhou.[73] A close look at the map (Figure 1:7) reveals that the

72 The concluding essay for this section also appears in a Song-era work, Fu Yin 傅寅, *Yugong shuoduan* 禹貢說斷, *juan* 4, p. 59a ff., though it is not certain that this was Zhang's own source. In some cases he cites the origin of the texts he uses but in many cases this information goes unmentioned.

73 The text following it appears under the title *Zhifang shi jiuzhou shanze chuanjin limin xugu tu* 職方氏九州山澤川浸利民畜穀圖 (Official map of the Nine Regions' mountains, marshes, and waterways, useful products, population, domesticated animals and cultivated grains), in the *Zhouli zhuan* 周禮傳 of Wang Yingdian 王應電, *juan* 4 xia, pp. 29b-30b.

most basic level of data in each of these categories is provided in a few words of text in each region.

For example, in Yanzhou we see that its main swamp was the Daye ('Great Wild'), its waterways (divided into two types) the He 河 and the Ji 泲, the Lu 盧 and the Wei 維, its products were rushes and fish, there were three men for every two women, all six of the standard beasts used by humans (pigs, sheep, horses, oxen, dogs and fowl) were domesticated, and four classes of grains were grown. The major mountain in each region, in this case Mount Tai, is indicated visually; above Mount Tai a note indicates that it is the Eastern Marchmont.

While such maps of distant antiquity had been printed since at least the Song to help students of the Classics visualize the locations and characteristics of places mentioned in the texts, by the late Ming this was thought to be a subject of potential interest to any literate person. For the reader who has noticed that the Nine Regions depicted on this map are different from those depicted on the first *Yugong* map, Zhang includes a short essay on the differences between the Nine Regions as given in sources from different periods.[74]

In the next maps Zhang moves into more recent historical times, with one map depicting the arrangement of states during the Spring and Autumn period, and another that tallies the Nine Regions of the *Yugong* with current prefectures, counties, and rivers and mountains. These are followed by many pages of text outlining such topics as the changes in territorial administration of each of the Nine Regions and the evolution of place names. An illustration of the divisions of land quality and tax of the Nine Regions (*Jiuzhou tianfu dengze tu* 九州田賦等則圖) presents these data in a matrix of relative quality rather than the spatial arrangement of the regions (Figure 1:8; compare to the depiction of the same regions seen in Figure 1:7). Each square here denotes the character of agricultural land and the tax base, ranked within nine degrees (upper upper, middle upper, lower upper, upper middle, middle middle, lower middle, etc.). For the student of administrative history, these would have been essential data to keep in mind when comparing the regions.

Earlier in this chapter I have briefly noted the proliferation of printed sheet maps, including a 1555 map published in Fujian titled *Gujin xingsheng zhi tu* 古今形勝之圖 (Map of topographic advantages, past and present). The map that starts off Chapter 34 in the *Tushu bian* strongly resembles this

74 Zhang Huang, *Tushu bian*, *juan* 32, p. 4b.

Figure 1.8 ***Jiuzhou tianfu dengze tu*** **(Illustration of land quality and tax levels of the Nine Regions)**

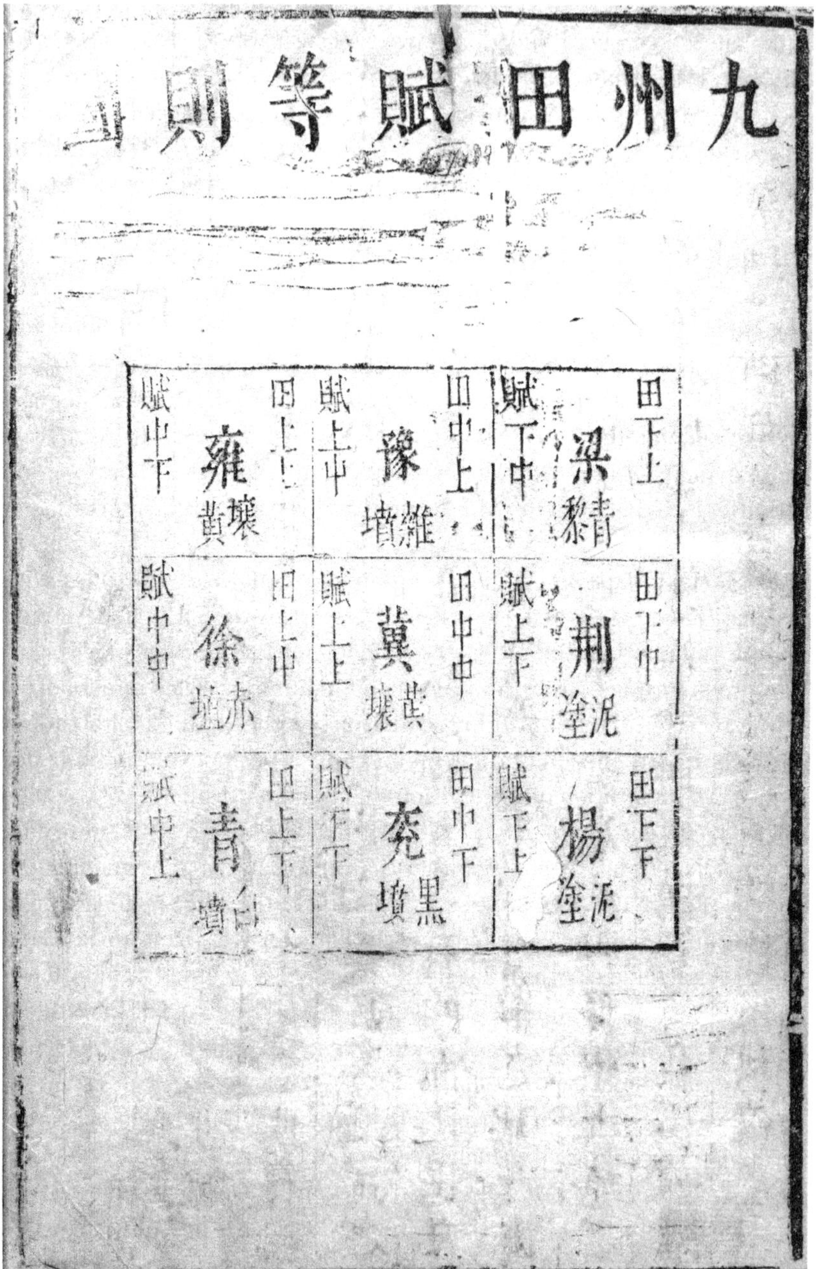

From the *Tushu bian* (*juan* 32, p. 19b)
Courtesy of the Harvard-Yenching Library

Figure 1.9 ***Huayi gujin xingsheng zhi tu*** **(Map of Chinese and Barbarian topographic advantages, past and present)**

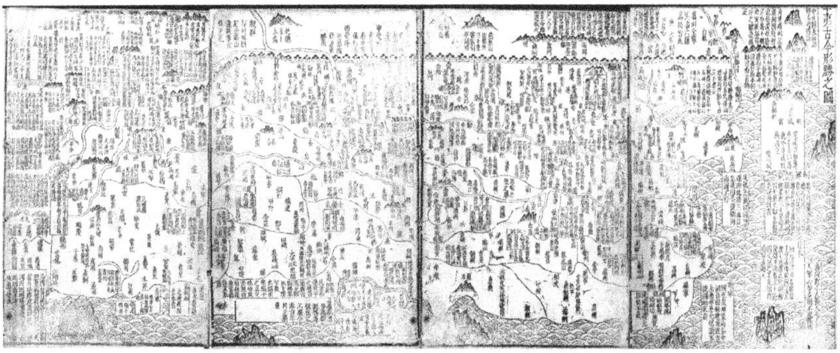

Tushu bian, juan 34, pp. 2a-4b
Courtesy of the Harvard-Yenching Library

progenitor.[75] The version Zhang presents here, titled *Huayi gujin xingsheng zhi tu* 華夷古今形勝之圖 (Map of Chinese and Barbarian topographic advantages, past and present), is a compressed rendering of this map or one very similar to it, squeezed from its original size of over one square meter into an elongated horizontal strip spanning four pages. Horizontally elongating a map such as this and spreading it over several pages was one way of adapting a large sheet map to the printed page.

There were other ways to adapt sheet maps to book format. Earlier in the *Tushu bian*, in the case of Ricci's map, Zhang had simply miniaturized the original at a scale approximating its original ratio of height to width. In this case, he has instead dramatically flattened the original, stretching it across four pages with the Great Wall running along the upper portion and the sea at bottom. It is prefaced with an uncited quotation of the preface from the *Da Ming yitong zhi*, while the following text mentions earlier maps, including a list of imperially sponsored early Ming cartographic projects, and discusses the reactions of emperors to what they saw, especially as they compared the extent of the Ming domains to those of earlier dynasties. The essay mentions two Ming maps by Xu Lun and Gui E, noting that they revealed a world in which the Ming had surpassed the extent of the Song and the Tang, although the Han was still held to have been larger because the Ming did not control the various frontier zones over which the Han held sway.[76]

75 The map spans pp. 3a-4b of *juan* 34. This similarity has also been noted by Richard Smith, 'Mapping China's World: Cultural Cartography in Late Imperial Times,' p. 68.

76 Zhang Huang, *Tushu bian, juan* 34, p. 6b

Figure 1.10a and b The *Tushu bian's* dual view of Zhejiang copied from the *Huangyu kao,* highlighting topography versus administration

Tushu bian, juan 38, pp. 26a-b and 28a-b
Courtesy of the Harvard-Yenching Library

We next move fully into contemporary times with a long series of provincial maps, adapted from the *Huangyu kao* 皇輿考 (Investigation of the Imperial domain).[77] Each province is represented twice, once with a map the *Huangyu kao* has copied from the *Da Ming yitong zhi,* followed by another which shows same province but drawn in a style emphasizing its administrative subdivisions. In the latter, each province is divided into prefectures by simple lines, within which are the names of county seats in cartouches. This dual-view format is thus not Zhang's own invention, but a borrowing from an earlier text that had already made its own fusion of two different perspectives.[78]

The textual material in these sections, which continue province-by-province from North Zhili in Chapter 35 to Guizhou in Chapter 42, is reminiscent of the *Guang yutu* in its combination of basic administrative matter with essays relevant to each province. For instance, the section on Fujian includes a historical essay on the prohibition of contact with evil seafarers (禁通海奸民), and discussion of the *Wokou* troubles that were trailing off around the time Zhang began compiling the *Tushu bian.*[79]

77 Zhang Tianfu's 張天復 *Huangyu kao* 皇輿考 was first published in 1557 in ten *juan*. A 1588 reprint adds two additional *juan* at the end, the maps of which are copied from the second half of the *Guang yutu*. Only these maps use a grid pattern, preserved from Luo's work. This is but one example of a work originally published before the *Guang yutu* was widely available, reissued later to include its influence (probably based on the 1579 edition). A facsimile of the 1588 edition is in the *Xuanlantang congshu* 玄覽堂叢書, chu ji; vols 6-7.

78 Compare Figure 1:10 to the *Huangyu kao* Chapter 6, pp. 2a-3b.

79 Zhang Huang, *Tushu bian, juan* 40, p. 29a.

I have mentioned that the genre of Northern frontier maps depicting the stretch from Liaodong across the northern frontier to Gansu continued to be republished through the Ming dynasty. While they were gradually supplemented by maps dealing with the eastern seaboard and then the northeast, they never disappeared from the catalog of places to be visually comprehended by the officer or layman interested in affairs of state. Zhang's next section incorporates a ten-leaf-long map of this type titled *Jiubian xingsheng* 九邊形勝 (Topographical advantages of the Nine Frontiers), and beginning in Chapter 44, Zhang reproduces the series of frontier maps from the *Guang yutu*, with a few additional close-up maps interspersed (such as a depiction of that all-important pass through the Great Wall, the Shanhai guan).

There is a dramatic difference on the frontier maps as they are reproduced here, however – they are missing the spatial grid that they had in the *Guang yutu*. This erasure has been incomplete. For example, there is still a textual note on the map of Jizhou 薊州 indicating that each square represents 40 *li*, even though there are no squares to be seen. This suggests that the maps might have been made by pasting copies of the original maps facedown on boards, with instructions to the carvers to reproduce them while omitting the grids – only to have a dutiful but oblivious workman trace and carve out bits of text that should have disappeared together with the lines in the background.

The text accompanying depictions of the frontier makes use of a wide range of outside sources, some of them surprisingly old. For example, the text on the *San bian si zhen* 三邊四鎮 (Three frontiers and four commanderies) in Shaanxi is from a memorial that appeared in a work by Ni Yue 倪岳 (*jinshi* 1464).[80] Discussion of the Ming's own frontiers continues through Chapter 50, at which point the *Tushu bian* turns to neighboring countries including Chosŏn and Japan. To represent these places, Zhang follows the lead of the later editions of the *Guang yutu* by incorporating a map of Japan originally published by Zheng Ruozeng, as well as a memorable stylized depiction of the Ryukyus that Zheng took from some other work. The style of the Ryukyu map is so distinctive from the rest of Zheng's works (exemplified by the royal compound taking up the entire island) that it must have been appropriated from some other, perhaps lost source.[81]

80 Zhang Huang, *Tushu bian, juan* 47 p. 4b; Ni Yue 倪岳, *Qingxi mangao* 青谿漫稿, *juan* 33, p. 41b.

81 The map of Japan is illustrated in figure four. The Ryukyu map is followed by text copied from Zheng's *Liuqiu kao* 琉球考 (On the Ryukyus).

Figure 1.11 Map of the border region of Jizhou from the *Tushu bian,* based on a map from the *Guang yutu* but with the grid removed

Tushu bian, juan 44, pp. 19a-b
Courtesy of the Harvard-Yenching Library

Figure 1.12 Map of the Ryukyus

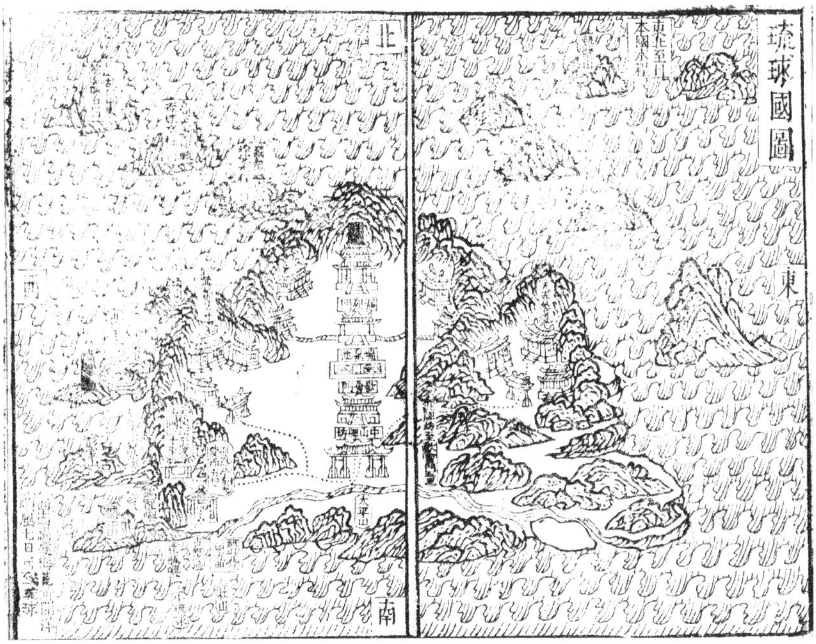

Tushu bian, juan 50, pp. 57a-b
Courtesy of the Harvard-Yenching Library

The following chapter includes among its depictions of more distant foreign lands a map of the southeastern barbarians, copied from the *Guang yutu*, which shows Africa at the far left.[82] On this *Dongnan haiyi tu* 東南海夷圖 (Map of the Barbarians of the Southeast Seas) there is again a note about the distance represented by each square of the grid, even though the grid has not been retained from the original.[83]

After this extensive discussion of security-related issues, illustrated with a host of maps copied from the relevant works of previous decades, Zhang

82 Benjamin Elman writes of this map from the *Guang yutu* that 'Luo's map [...] printed about 1555, included a clear depiction of the Cape of Good Hope and southern Africa, leading some to speculate that Zheng He was the 'Vasco da Gama of China' and had traversed from the Indian to the Atlantic Ocean and back some seventy years before da Gama did it in reverse' (Elman, 'Ming-Qing Border Defence, the Inward Turn of Chinese Cartography, and Qing Expansion in Central Asia in the Eighteenth Century,' p. 32). It appears to have escaped the notice of these 'speculators' that the *Guang yutu* took the map from some version of the c. 1389 *Da Ming hunyi tu*, which would obviously predate Zheng He's voyages! This depiction of Africa, its distorted V shape distinctively similar to those earlier maps, derives instead from Persian or Arab sources imported during the Yuan.

83 *Tushu bian, juan* 51, p. 16b.

turns next to rivers, making a cartographic excursion up the Yellow River that fills fully twenty-five leaves, following it with essays taken from multiple sources, such as an extract on the source of the Yellow River credited to the Yuan dynastic history, and brief notes about it that are attributed to the *Han shu* and the *Shanhai jing*. Changes in the course of the river and ideas about river control are also touched upon, but in snippets of unattributed text.[84] A map of the Grand Canal, following its course south from Beijing with indications of waystations, nearby towns and even the depths of some parts of the river, stretches on for fifty-four leaves. Other rivers around the empire are explored until we reach the sea lanes, for which Zhang relies on Zheng Ruozeng's *Haiyun tushuo* with its map, followed by a text on the use of the compass.[85] A map of the 'myriad-li' Yangtze River (*Wanli Changjiang tu* 萬里長江圖) differs dramatically from the depictions of other rivers, its scenic quality appearing to derive from a painted scroll.[86] There are numerous paintings by this name, and this may well have been taken from such a source; its painterly touches (such as a boatman with a companion floating on the river in the first panel and the dramatic portrayal of the mountains) are only rendered 'cartographic' by the inclusion of toponyms in the occasional rectangle or cartouche.[87] Whatever its source, the variant style of this chart highlights the eclectic nature of the inclusive *Tushu bian*.

The final installment of the geographical section of the *Tushu bian* focuses on mountains. Just as Zhang prefaces the geography section as a whole with a discussion of geomancy, he begins the section on mountains in Chapter 59 with geomantic renderings of the Kunlun Mountains and the five marchmounts, or sacred mountains. One of these images, the *Wu Yue zhenxing tu* 五嶽真形圖 (True forms of the five Marchmounts) was used as a talismanic charm by Daoists, and the subsequent close-ups of various mountains include depictions of some Daoist temple complexes, such as that at Wudan Mountain. Of course, sites important to the literati tradition are not forgotten, such as the birthplace of Confucius or the White Deer Grotto at Lushan, where Zhu Xi had his academy.

The basic organization of Zhang's discussion of mountains is a province-by-province breakdown, with text about each province prefaced by a map

84 The discussion of the changing course of the Yellow River which appears in *juan* 53, pp. 39b-40a is present in the somewhat earlier *Bianjing yiji zhi* 汴京遺蹟志 (Gazetteer of the traces of Bian) by Li Lian 李濂, *juan* 5, pp. 4b-5b.

85 Zhang Huang, *Tushu bian*, beginning *juan* 56, p. 4a.

86 Ibid., *juan* 58, p. 2a.

87 On the links between cartography and other visual arts, see Yee, 'Chinese Cartography Among the Arts: Objectivity, Subjectivity, Representation,' especially pp. 139-153.

showing the famous mountains in each of its districts. The outlines of the maps and the way they are divided on the page strongly suggests that they are based on the provincial maps from the *Guang yutu*, but the details of the rivers have changed, there is no grid, and most dramatically, mountains have sprouted everywhere. The close-ups of specific peaks come from a variety of sources and have varying styles, from sober cartographic depictions to exaggerated, fanciful renderings of lofty, twisting rock formations. Each provincial listing includes texts describing these important mountains, with sketch maps or artistic renderings of the most important.

The *Tushu bian*'s extensive coverage of geography has taken us from competing visions of the world's shape to descriptions of individual provinces, from the states of the distant past to distant corners of the contemporary earth, from the rivers and seas to lofty mountain peaks. But as much as a tour of the world, it has given us a tour of previously published materials on all of these subjects, with a large proportion of its maps still traceable to specific sources despite the lack of citation.

Zhang Huang's method is to intersperse supplementary, or in many cases contradictory, information, to cite authoritative texts and scholars of the past without fearing to dispute their conclusions. Despite the highly derivative nature of the work there is a strong sense of authorial voice in the selection and ordering of the maps and the text surrounding them. Zhang has truly repackaged these old maps for a syncretic late Ming audience. We see much less such initiative in the *Sancai tuhui*, a roughly contemporary encyclopedia with rather different characteristics.

Sancai tuhui 三才圖會 (Illustrated compendium of the three fields of knowledge)[88]

The *Sancai tuhui*, compiled by the Shanghainese scholar-official Wang Qi and his son Wang Siyi, is similarly inclusive but lacks much of the scholarly apparatus of the *Tushu bian*. The 'Sancai' (three types of knowledge) to which its title refers are *Tianwen* (astronomy and astrology), *Dili* (geography), and *Renwu* (people).[89] The first printed edition dates to the late Wanli period,

88 The edition I cite in this discussion is the one with a preface by Zhou Kongjiao 周孔教 dated Wanli 己酉 [1609], reproduced in facsimile by the Shanghai guji chubanshe.

89 Wang, *Sancai tuhui*, 1a. For a general English introduction to the *Sancai tuhui*, see Goodall, *Heaven and Earth*. Goodall's focus is on the work's illustrations of exotic animals, plants, and foreign peoples; he does not address cartography or the textual aspects of the work. The images reproduced here all come from a Chongzhen-era edition held at Harvard, dating to between 1628 and 1644.

with a preface dated 1609.[90] Its profuse illustrations, lists and charts made it a popular work, even abroad; it directly inspired the Japanese *Wa-Kan sansai zue* 和漢三才圖會, discussed in Chapter Five. In the *Sancai tuhui* both images and text are copied from other sources without citation, and errors in transcription are sometimes dramatic.

At first glance the most obvious difference between these books is the higher ratio of illustrations to text in the *Sancai tuhui*. For example, while the *Tushu bian* follows its map of the Yellow River with more than thirty leaves of topical essays and textual descriptions before moving on to the Grand Canal, the *Sancai tuhui* has just three leaves of text. In contrast to the *Tushu bian*, in which Zhang Huang borrows extensively from works like the *Guang yutu* but intersperses these maps with illustrations from other sources and textual interludes, the *Sancai tuhui* sticks to more complete runs of maps copied intact from other sources. The comparative strategy of the *Tushu bian* is markedly less present in the *Sancai tuhui*. The text used in the latter book is often an abbreviated version of similar text used in the *Tushu bian* or earlier books. Granted these shortcomings, its comparative brevity and highly pictorial nature were doubtless reasons for its popularity.

The geographical section begins with the *Shanhai yudi quantu* 山海輿地全圖 (Complete map of the mountains and seas), an adaptation of a European world map that is cruder than the example included in the *Tushu bian*. The captions mention latitudinal bands, but they have been erased from the map itself.[91] As with the *Tushu bian*, this map is accompanied by a long and unattributed quotation from a version of Ricci's map preface, in this case apparently using his 1602 edition. This is immediately followed by text taken from the *Da Ming yitong zhi*. Note that while Zhang Huang quotes from both but puts them in different places, the Wangs stitch these seemingly contending texts together. As I will show in Chapter Three, by placing a textual depiction of Ming unification immediately after Ricci's introduction to the vastness of the earth, Ming unity was situated within an expanded context that now included more distant lands.

From here on, the geographical section turns to views of the empire as established by a series of influential books. The *Huayi yitong tu* 華夷一總圖 (General map of the Hua and Barbarians) and the subsequent maps of provinces all derive from the *Guang yutu*. Figure 1:13 illustrates the derivative and markedly sloppy nature of the *Sancai tuhui* copies, exemplified by the depiction of Huguang. The map is based on the *Guang yutu* depiction we

90 Franke, *An Introduction to the Sources of Ming History*, p. 313.

91 Wang, *Sancai tuhui, dili juan* 1, pp. 2a-b.

Figure 1.13 Huguang province as shown in the Sancai tuhui

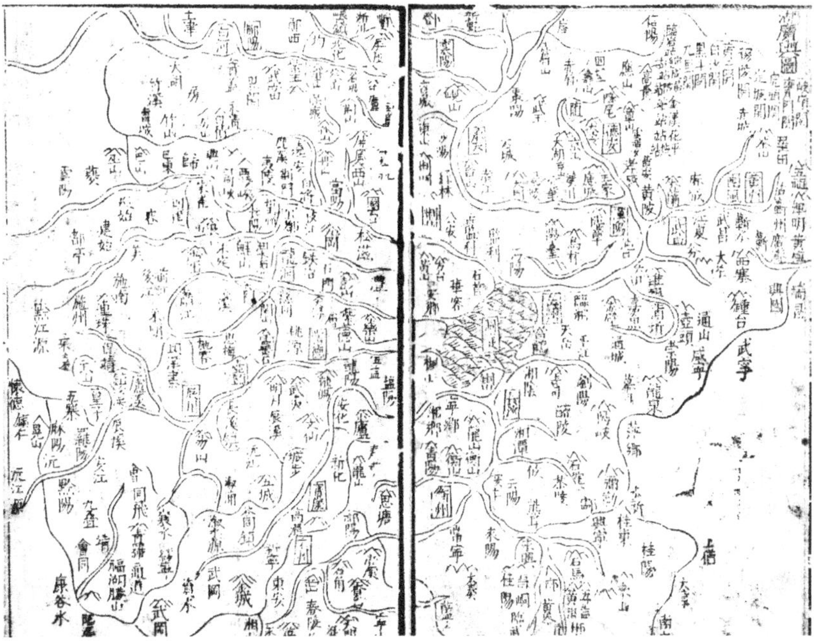

Sancai tuhui, dili juan 2, pp. 13b-14a
Courtesy of the Harvard-Yenching Library

have seen in Figure 1:2, but this version zooms in slightly, bleeds over the provincial borders and is generally much less carefully executed than the *Guang yutu* map. The grid is also removed.

Examples of simplification and omission of detail abound in these provincial maps, from the omission of roads on the *Sancai tuhui* version of the Guizhou map, to the compression of the map of Shaanxi from three pages to two. Furthermore, if we compare the *Sancai tuhui* maps to those of both the first edition of the *Guang yutu* and later editions, it is evident that the *Sancai tuhui* copies a later version, probably that of 1579. The *Sancai tuhui* continues following the model of the *Guang yutu* by next reproducing its series of northern frontier maps.[92] Whereas the provincial maps have had their grid lines removed in the *Sancai tuhui*, these frontier maps retain the grids.

After a series of river maps, including depictions of the Yellow River and the Grand Canal taken from the *Guang yutu*, we reach a long section on maritime defense. Yet again, the entire section is organizationally and

92 Ibid., *juan* 3 of the *dili* section.

cartographically derived from an earlier work, in this case Zheng Ruozeng's *Chouhai tubian*. The organization of the entire series of maps is repeated, province by province, starting from the south, with prefectures individually mapped.

In the subsequent pages, maps of cities mingle with painterly illustrations of the surrounding scenery. The number of essentially artistic renderings of local scenes is far greater than the number seen in the *Tushu bian*. The text accompanying these images comes from a variety of sources; the commentary on Yuhuatai Mountain 雨花臺山[93] is quoted from Chapter 6 of the *Da Ming yitong zhi*, while that for Bagong Mountain 八公山[94] is from Peng Dayi's *Shantang sikao* (Extended investigations of the Mountain Hall).[95]

We shift from close-up depictions of locales back to broad perspectives when we turn to foreign countries, starting with the *Chaoxian guo tu* 朝鮮國圖 (Map of Chosŏn).[96] This square-footed depiction of the Korean peninsula derives from late editions of the *Guang yutu* and is already familiar from its use in the *Tushu bian*; part of the associated text also comes from the *Guang yutu* but the list of sites of interest is different. Among the maps copied from the works of Luo Hongxian and Zheng Ruozeng in this chapter, we see again the illustration of the southwestern oceans that appeared in the *Guang yutu* and the *Tushu bian*, with its crude rendition of Africa. More urgent contemporary concerns come after all of this material, with a map titled *Dongbei yi zhu guo* 東北夷著國 (Various countries of the Barbarians of the Northeast), followed by a list of Jurchen tribes.[97] By this time the *Wokou* threat had been supplanted by the establishment of the worrisome Latter Jin state in the northeast, which would become the Manchu polity that spelled the demise of the Ming dynasty.

A long portion on historical geography begins in Chapter 14 and continues through the beginning of Chapter 16, illustrated with maps taken from the twelfth century *Lidai dili zhizhang tu* 歷代地理指掌圖 (Convenient historical atlas), and accompanied by narrative text that is in many cases copied word for word from that source (though at times heavily abridged). It spans eras from that of Emperor Ku to the Song, but such is its loyalty to the Song source that it makes no effort to add coverage of the later Yuan or Ming dynasties. In some cases, as with the *Songchao Taizong tongyi zhi tu* 宋

93 Ibid., *dili juan* 6, p. 31b

94 Ibid., *dili juan* 6, pp. 44b

95 Peng Dayi 彭大翼, *Shantang sikao* 山堂肆考, *juan* 17, p. 19a.

96 Wang, *Sancai tuhui, dili juan* 13, pp. 2a-b.

97 Ibid., *dili juan* 13, pp. 34a-b

朝太宗統一之圖 (Map of the Song unification under Taizong) the title and text have simply been changed from those used in the *Lidai dili zhizhang tu* to reflect the fact that the Song was no longer in power.[98]

In the original *Lidai dili zhizhang tu,* the text occupies margins to the left and right of the maps, but here the marginal text has been moved to separate pages. There are a number of additional diagrams of land distribution under the rule of the Sage Kings, reflecting continued interest in the well-field system. The Nine Regions are illustrated using the same maps found in the *Tushu bian.* In Chapter 16 we take a sudden turn to agriculture, and with a following section on geomancy the *Sancai tuhui*'s coverage of the Earth comes to a close.

Both the *Tushu bian* and the *Sancai tuhui* retraced the steps of earlier cartographers, reaching as far back as the Song dynasty for both individual maps and lengthy series that are copied directly from earlier atlases. It is taken for granted in both works that in explaining the trinity of Heaven, Earth, and Humanity, it would be impossible to explain the Earth without the use of maps. Perspectives that were once available in separate books by identifiable authors have merged into a single corpus. These compendia retained something of a literati gloss, but as we shall see next, by the early seventeenth century a simple version of some of this material was available in household almanacs, much-simplified encyclopedic works that lay at the most 'popular' end of the spectrum of readership.

Popular miscellanies

Lucille Chia has wryly noted the preference of Jianyang publishers (and presumably their customers) for redactions or overviews of classic historical texts, rather than the works themselves.[99] Abbreviated versions of important books proliferated like Ming versions of Cliff's Notes. Even in the case of route books, where our modern expectations based on using road maps might lead us to expect relatively fresh information, much of the content is cadged

98 The title of this map in the *Lidai dili zhizhangtu,* where it appears on p. 45, is given as 'Taizong huangdi tongyi zhi tu 太宗皇帝統一之圖,' whereas the emperor is cited here as Song chao Taizong 宋朝太宗. The order of maps has been changed in some places; the *Tian xiang fenye tu* 天象分野圖 is used as a break between the historic maps and the maps of the present dynasty in the Song edition of the *Lidai dili zhizhang tu,* but now that all of these maps are of past dynasties, the distinction may not have been considered important, and this item is moved after the Song maps.

99 Chia, *Printing for Profit,* pp. 226-227.

from earlier authoritative texts. Chia figuratively shrugs that 'perhaps the printed word and image even as produced in a cheap commercial edition printed on inferior paper and blighted by nondescript calligraphy and a cramped page layout had an authority and appeal that modern readers can imagine only with difficulty.'[100]

Something similar could be said of the geographical section in many popular miscellanies or household almanacs of the Wanli period. This section in these miscellanies usually includes a list of major transit routes around the empire, including both overland and river routes, as well as a 'census' of the various provinces and their administrative subdivisions, tax burdens, products, and so forth. This is all typical national gazetteer fare, but very skimpily sketched out.

For example, the 'gazetteer' portion of the *Wuche wanbao quanshu* 五車萬寶全書 (Five-cart compendium of a myriad treasures) begins with an essay on the historical capitals of former dynasties that cites the *Di Wang shiji* 帝王世記 of the third century writer Huangfu Mi 皇甫謐.[101] The inspiration to use this quote was no personal revelation, however – the very same passage opens the geographical section of the *Shantang sikao* 山堂肆考 by Peng Dayi 彭大翼, with editions from 1595 and 1619,[102] as well as the *Wuche bajin* 五車拔錦, another household miscellany from 1597,[103] and perhaps other works of the time.

The next portion of the *Wuche wanbao quanshu*'s geography section, prefaced *Yu shu jiu he* 禹疏九河 (Yu untangles the Nine Rivers), abbreviates the wording used in the *Yudi tu zongkao* 輿地圖總考 from the *Tushu bian*.[104] The map that follows is strikingly similar between different miscellanies, although the titles inscribed above the maps can differ.

In our survey of the intersection of cartography and the publishing boom, such books stand at the pinnacle of cartographic penetration into popular works, but at the nadir of quality in both drafting and execution. They demonstrate the extent to which basic geographical orientation and

100 Ibid., pp. 229-230.

101 Xu Qilong 徐企龍, *Wuche wanbao quanshu* 五車萬寶全書, *juan* 2. The title's 'Five carts' suggested the great wealth of knowledge contained within; in the old days one would need five carts to carry so many books.

102 The 1619 edition held by the Harvard Yenching library is inscribed as a reprint of the 1595 edition.

103 Xu Sanyou 徐三友, *Wuche bajin* 五車拔錦, *juan* 2, p. 1a.

104 The relevant passages are, in the *Wanbao quanshu*, *juan* 2, p. 1b, and in the *Tushu bian*, *juan* 34, p. 5a. As for whence Zhang Huang appropriated it, that remains a mystery for the time being.

exposure to a simple map were considered important for readers of such books.

Text and cartography

Scholars of cartography tend to focus on the graphic component of books like those discussed above, for obvious reasons, but this runs the risk of overlooking the significance of the text which these maps are meant to illustrate. As R. A. Skelton has argued, maps tend to be more constant in form than does text. Replication of a map printed in edition after edition does not necessarily indicate lack of change in actual knowledge; in late Ming security-related texts we have examples of textual 'updates' paired with old maps (though text was often copied from old sources as well). Given the degree to which precedent determined the structure of certain types of texts, such as gazetteers, and their inherently conservative nature, such works must be examined in the context of other types of maps and texts to provide a meaningful picture. The maps in a woodblock-printed work cannot be taken to represent the state of the art in cartography.

Cordell Yee has eloquently demonstrated the inseparability of text from maps in the Chinese context.[105] But what of materials that have the functions we expect in modern times of maps, but that are not illustrated at all? In his GIS work with historical cartography, Peter Bol has argued for a vision of spatial organization that emphasizes connective geography – linkages from one place to another – as opposed to bounded geography, the depictions we are now accustomed to in a world where borders appear as lines on charts. In Bol's emphasis it is the linkage between places rather than separation that is of primary interest.[106] This resonates with the practice in local gazetteers of enumerating stations along the roads to neighboring regions, rather than depicting the courses of borders separating them. Textual materials such as route books provide the most vivid case for this notion of linkage rather than boundary.

Many of the activities for which we now take maps for granted have been done throughout most of history without them. Even by the end of the Ming, when maps had permeated literate society, the textual component of route books was far more useful to the actual traveler than the few sparse maps that occasionally appeared in such works. The addition of maps to route

105 Yee, 'Taking the World's Measure: Chinese Maps Between Observation and Text.'
106 See Bol, 'Creating a GIS for the History of China.'

books was a feature of the publishing boom, in which illustration became an expected premium for the armchair traveler, but the maps that were added were so simple that they could not have played much of a role in navigation. While earlier route books tend to be purely textual, like the *Huanyu tongqu* 寰宇通衢 (Major routes of the realm), first published in the Hongwu reign (1368-1398) after having been compiled in the capital by imperial decree, a later guide like the *Yitong lucheng tuji* 一統路程圖記 (Comprehensive illustrated route book) by Huang Bian 黄汴, printed in 1570, uses dotted lines to map out the roads cited in its descriptions.[107] By no stretch of the imagination, however, could such a map have been of practical use while on the road.[108] The increased incidence of maps in published materials should be viewed in tandem with the general increase in woodcut illustration during the late Ming, rather than as a distinctly cartographic phenomenon or a development related to the sciences.

Conclusion

We find that the publishing boom of the mid-to-late Ming fostered the proliferation of cartography, but in a very different context from that in western Europe during the same period. Scholasticism, literati attention to military concerns, and the popularization of elite knowledge underwent tremendous shifts without the technological and conceptual changes that were rapidly transforming maps at the other end of Eurasia.

The real transformation in cartographic consciousness had begun in the Song or even the Tang, when maps were first printed for a substantial audience, including not only those in government but those who hoped to work their way into the governing elite through study. By the end of the Ming, maps were more than ever a part of popular literate culture, but this was a function of quantitative rather than qualitative change. In works as late as the *Sancai tuhui* we find some of the very same maps that were printed for students more than four centuries earlier.

If we accept that there is little technical or conceptual change in the burgeoning of cartographic materials in the late Ming publishing boom,

107 Both of these works are available in facsimile reprints in Volume 166 of the history section (*shi bu* 史部) of the *Siku quanshu cunmu congshu* 四庫全書存目叢書.

108 For more on route books see Brook, 'Guides for Vexed Travelers: Route Books in the Ming and Qing.' Brook also includes route books in the listings of his *Geographical Sources of Ming-Qing History*.

what then *did* change? By the time we reach the final years of the Ming it is no longer easy to point out groundbreaking works; there come to be so many books published in fields associated with cartography that we must narrow our focus to particular genres if we are not to be overwhelmed. What are the most influential *historical* geographies, the works with the most acclaimed perspective on *defense*, the works that reveal the most complete *surveys* of existing schools of thought? Which are targeted at scholars and which are marketed to the mob? Maps appear even in books not specifically about geography. Cartography has saturated literate society. The publishing boom was a venue for the formation and reconfiguration of cartographic genres and categories. As we shall see in Chapter Two, even in an already well-established category such as historical cartography, late Ming saw old techniques applied to new purposes.

Works cited

Primary sources

Bing bu (Board of War). 兵部. *Jiubian tushuo* 九邊圖說. S.n.: Longqing year 3 [1569].

Chen Yuanjing 陳元靚. *Shilin guangji* 事林廣記. S.n., Yuan period.

Chen Zushou 陳組綬. *Huang Ming zhifang ditu* 皇明職方地圖. S.n., Chongzhen 9 [1636].

Fan Lai 范淶, and Zhang Huang 章潢. *Wanli xin xiu Nanchang fuzhi* 萬曆新修南昌府志. Facsimile of Wanli 16 edition. In *Riben cang Zhongguo hanjian difangzhi congkan* 日本藏中國罕見地方志叢刊 (32 volumes), vol. 6. Beijing: Shumu wenxian chubanshe 書目文獻出版社, [1990-1991].

Fu Yin 傅寅. *Yugong shuoduan* 禹貢說斷. Wenyange facsimile edition. [Taibei]: Taiwan shangwu yinshuguan 臺灣商務印書館, [1983].

Huang Bian 黄汴. *Yitong lucheng tuji* 一統路程圖記. Facsimile reprint in *Siku quanshu cunmu congshu* 四庫全書存目叢書, Shi bu 史部 (292 volumes), vol. 166: 481-562. Jinan: Qi Lu shushe chubanshe, 1997.

Huang Jinsheng 黄金聲 et al, *Daoguang Jinhua xianzhi* 道光金華縣志. S.n., Daoguang 3 [1823].

Huanyu tongqu 寰宇通衢. Facsimile reprint in *Siku quanshu cunmu congshu* 四庫全書存目叢書, Shi bu 史部 (292 volumes), vol. 166: 164-237. Jinan: Qi Lu shushe chubanshe, 1997.

Li Lian. 李濂. *Bianjing yiji zhi* 汴京遺蹟志. Facsimile of the Wenyange edition. [Taibei]: Taiwan shangwu yinshuguan 臺灣商務印書館, [1983].

Li Xian 李賢, ed. *Da Ming yitong zhi* 大明一統志. Nei fu, Ming Shuntian 5 [1461].

—. *Da Ming yitong zhi* 大明一統志. Facsimile edition. Xi'an: San Qin chubanshe 三秦出版社, 1990.

Luo Hongxian 羅洪先. *Guang yutu* 廣輿圖. [China]: Han Jun'en, Du Si; Ming Jiajing 45 [1566].

—. *Guang yutu* 廣輿圖. Facsimile reprint of 1799 edition. Taibei: Xuehai chubanshe 學海出版社, 1969.

—. *Guang yutu* 廣輿圖. Facsimile reprint of undated Jiajing edition in *Xuxiu siku quanshu* 續修四庫全書, Shi bu (670 volumes), vol. 586: 411-528. Shanghai: Shanghai guji chubanshe, 2002.

Mao Ruizheng 茅瑞徵. *Dongyi kaolue* 東夷考略. Facsimile edition. Taibei: Guangwen shuju, 1977.

Ni Yue 倪岳. *Qingxi mangao* 青谿漫稿. Facsimile of Wenyange edition. [Taibei]: Taiwan shangwu yinshuguan 臺灣商務印書館, [1983].

Peng Dayi 彭大翼. *Shantang sikao* 山堂肆考. Facsimile of Wenyange edition. [Taibei]: Taiwan shangwu yinshuguan 臺灣商務印書館, [1983].

Wang Qi 王圻, and Wang Siyi 王思義. *Sancai tuhui* 三才圖繪. Facsimile of c. 1609 edition. Shanghai: Shanghai guji chubanshe 上海古籍出版社, 1988.

Wang Yingdian 王應電. *Zhouli zhuan* 周禮傳. Facsimile of Wenyange edition. [Taibei]: Taiwan shangwu yinshuguan 臺灣商務印書館, [1983].

Xu Qilong 徐企龍. *Wuche wanbao quanshu* 五車萬寶全書. Chūgoku nichiyō ruisho shūsei 中国日用類書集成, Vols 8-9. Tōkyō: Kyūko Shoin, Heisei 13 [2001].

Xu Sanyou 徐三友. *Wuche bajin* 五車拔錦. Chūgoku nichiyō ruisho shūsei 中国日用類書集成, vols 1-2. Tōkyō: Kyūko Shoin, Heisei 11 [1999].

Zhang Ao 張鏊. *Jiaoli jiaoping shilue* 交黎剿平事略. Facsimile reprint of 1551 edition in Zhongguo bianjiang shi di congshu chubian 中國邊彊史地叢書初編 (12 volumes), vol. 5: 875-1336. Taibei, [s.n.], 1969.

Zhang Huang 章潢. *Tushu bian* 圖書編. Facsimile of Wanli 41 [1613] edition. Taibei: Chengwen chubanshe 成文出版社, 1971.

Zhang Tianfu 張天復. *Huangyu kao* 皇輿考. S.n., Jiajing 36 [1557].

—. *Huangyu kao* 皇輿考. Facsimile of Ming Wanli 16 [1588] edition. Xuanlantang congshu 玄覽堂叢書, Chu ji, vols 6-7. Taibei: Guoli Zhongyang tushuguan 國立中央圖書館, Zhengzhong shuju yinxing 正中書局印行, 1981.

Zheng Ruozeng 鄭若曾. *Chouhai tubian* 籌海圖編. Publisher Hu Weiji 胡惟極. Tianqi 甲子 [year 4, 1624].

—. *Chouhai tubian* 籌海圖編. Wenyange facsimile edition. [Taibei]: Taiwan shangwu yinshuguan 臺灣商務印書館, [1983].

—. *Riben tuzuan* 日本圖纂. In *Nihon kigo no kenkyū* 日本寄語の研究. Kyoto: Kyōto Daigaku Kokubun Gakkai 京都大學國文學會, Showa 40 [1965].

—. *Zheng Kaiyang zazhu* 鄭開陽雜著. Facsimile of manuscript copy. Nanjing: Guoxue tushuguan Taofeng lou 國學圖書館陶風樓, 壬申 [1932].

Secondary sources

Bol, Peter. 'Creating a GIS for the History of China.' In *Placing History: How Maps, Spatial Data, and GIS Are Changing Historical Scholarship*, edited by Anne Kelly Knowles and Amy Hillier, 25-57. Redlands, CA: ESRI Press, 2007.

Brook, Timothy. 'Communications and Commerce.' In *Cambridge History of China*, edited by D. Twitchett and F. Mote, Vol. 8: 579-707. Cambridge: Cambridge University Press, 1998.

—. 'Edifying Knowledge: The Building of School Libraries in Ming China.' *Late Imperial China* 17, no. 1 (June 1996): 93-119.

—. 'Guides for Vexed Travelers: Route Books in the Ming and Qing.' In three parts: *Ch'ing-shih wen-t'i* 4, no. 5 (June 1981): 32-76; 4, no. 6 (December 1981): 130-140; 4, no. 8 (December 1982): 96-109.

—. 'Mapping Knowledge in the Sixteenth Century: The Gazetteer Cartography of Ye Chunji.' *Gest Library Journal* 7, no. 1 (Spring 1994): 5-32.

—. *Mr Selden's Map of China: Decoding the Secrets of a Vanished Cartographer* (Toronto: House of Anansi Press, 2013).

Cao Wanru, ed. *Atlas of Ancient Maps in China* 中國古代地圖集*: The Ming Dynasty 1368-1644*. Beijing: Wenwu chubanshe 文物出版社 1994.

Chia, Lucille. *Printing for Profit: The Commercial Publishers of Jianyang, Fujian (11th-17th Centuries)*. Cambridge, MA: Harvard University Press, 2002.

Chou Tai-chi. 'Kuei O.' In *Dictionary of Ming Biography,* edited by L. Carrington Goodrich and Chaoying Fang, 756-759. New York: Columbia University Press, 1976.

Chow, Kai-Wing. *Publishing, Culture, and Power in Early Modern China*. Stanford, CA: Stanford University Press, 2004.

D'Elia, Pasquale M. *Fonti ricciane; documenti originali concernenti Matteo Ricci e la storia delle prime relazioni tra l'Europa e la Cina (1579-1615) ed. e commentati de Pasquale M. d'Elia sotto il patrocinio della Reale accademia d'Italia.* [Ed. nazionale delle opere edite e inedite]. Roma, Libreria dello Stato, 1942-1949.

Elman, Benjamin A. 'Comparing and Classifying: Ming Dynasty Compendia and Encyclopedias (*Leishu*).' 'Qu'était-ce qu'écrire une encyclopédie en Chine? / What did it mean to write an Encyclopaedia in China?' edited by Florence Bretelle-Establet and Karine Chemia. Special issue, *Extrême-Orient, Extrême-Occident*, hors série (2007): 131-157.

—. 'Ming-Qing Border Defence, the Inward Turn of Chinese Cartography, and Qing Expansion in Central Asia in the Eighteenth Century.' In *The Chinese State at the Borders,* edited by Diana Lary, 29-56. Vancouver: UBC Press, 2007.

—. *On Their Own Terms: Science in China, 1550-1900*. Cambridge, MA: Harvard University Press, 2005.

Franke, Wolfgang. *An Introduction to the Sources of Ming History.* Kuala Lumpur: University of Malaya Press, 1968.

Fuchs, Walter. *The 'Mongol Atlas' of China by Chu Ssu-Pen and the Kuang-Yü-Tu.* Peiping: The Catholic University Press, 1946.

Goodall, John A. *Heaven and Earth: 120 Album Leaves From a Ming Encyclopedia, San-Ts'ai T'u-hui, 1610.* London: Lund Humphries, 1979.

Goodrich, L. Carrington, and C.N. Tay. 'Chang Huang.' In *Dictionary of Ming Biography,* edited by L. Carrington Goodrich and Chaoying Fang, 83-85. New York: Columbia University Press, 1976.

Henderson, John B. 'Chinese Cosmographical Thought: The High Intellectual Tradition.' In *The History of Cartography, Volume 2, Book 2: Cartography in the Traditional East and Southeast Asian Societies,* edited by J.B. Harley and David Woodward, 203-227. Chicago and London: University of Chicago Press, 1994.

Hsu, Mei-Ling. 'The Han Maps and Early Chinese Cartography.' *Annals of the Association of American Geographers* 68, no. 1 (March 1978): 45-60.

Hu Bangbo. *Cartography in Chinese Administrative Gazetteers of the Song Dynasty (A.D. 960-1279).* PhD dissertation in Geography, University of Wisconsin at Milwaukee, 1994.

Huang, Stanley Y.C. 'Cheng Jo-tseng.' In *Dictionary of Ming Biography,* edited by L. Carrington Goodrich and Chaoying Fang, 204-208. New York: Columbia University Press, 1976.

Kornicki, Peter. *The Book in Japan: A Cultural History from the Beginnings to the Nineteenth Century.* Leiden and Boston: Brill, 1998.

Lewis, Mark Edward. *The Construction of Space in Early China.* Albany, NY: SUNY Press, 2006.

McDermott, Joseph. 'The Ascendance of the Imprint in China.' In *Printing and Book Culture in Late Imperial China,* edited by Cynthia J. Brokaw and Kai-wing Chow, 55-104. Berkeley: University of California Press, 2005.

Mungello, David E. *Curious Land: Jesuit Accommodation and the Origins of Sinology.* Honolulu: University of Hawaii Press, 1989.

Needham, Joseph. 'Geography and Cartography.' In *Science and Civilisation in China, Volume 3: Mathematics and the Sciences of the Heavens and the Earth,* 497-590. Cambridge: Cambridge University Press, 1959.

Purchas, Samuel. *Hakluytus Posthumus, or, Purchas his Pilgrimes: Contayning a History of the World in Sea Voyages and Lande Travells by Englishmen and Others.* Facsimile of 1625 edition. Glasgow: J. MacLehose and Sons, 1905-1907.

Skelton, Raleigh Ashlin. *Maps: A Historical Survey of Their Study and Collecting.* Chicago: Published for the Hermon Dunlap Smith Center for the History of Cartography, the Newberry Library, by the University of Chicago Press, 1975.

Smith, Richard J. 'Mapping China's World: Cultural Cartography in Late Imperial Times.' In *Landscape, Culture, and Power in Chinese Society*, edited by Wen-hsin Yeh, 52-109. Berkeley, CA: Institute of East Asian Studies, 1998.

So, Kwan-wai. *Japanese Piracy in Ming China During the 16th Century*. East Lansing: Michigan State University Press, 1975.

Szcześniak, Boleslaw. 'Matteo Ricci's Maps of China.' *Imago Mundi* 11 (1954): 127-136.

—. 'The Seventeenth Century Maps of China: An Inquiry into the Compilations of European Cartographers.' *Imago Mundi* 13 (1956): 116-136.

Tanaka Takeo 田中健夫. 籌海圖編の成立 [The formation of the *Chouhai tubian*]. *Nihon rekishi*日本歴史 57 (1953): 18-23.

Wang Yong 王庸. *Zhongguo dilixue shi* 中國地理學史. Changsha: Shangwu yinshuguan 商務印書館, [1938].

Yang Ching-yi and L. Carrington Goodrich. 'Mao Jui-cheng.' In *Dictionary of Ming Biography*, edited by L. Carrington Goodrich and Chaoying Fang, 1041-1042. New York: Columbia University Press, 1976.

Yee, Cordell. 'Chinese Cartography Among the Arts: Objectivity, Subjectivity, Representation.' In *The History of Cartography, Volume 2, Book 2: Cartography in the Traditional East and Southeast Asian Societies*, edited by J.B. Harley and David Woodward, 128-169. Chicago and London: University of Chicago Press, 1994.

—. 'Reinterpreting Traditional Chinese Geographical Maps.' In *The History of Cartography, Volume 2, Book 2: Cartography in the Traditional East and Southeast Asian Societies*, edited by J.B. Harley and David Woodward, 35-70. Chicago and London: University of Chicago Press, 1994.

—. 'Taking the world's measure: Chinese maps between observation and text.' In *The History of Cartography, Volume 2, Book 2: Cartography in the Traditional East and Southeast Asian Societies*, edited by J.B. Harley and David Woodward, 96-127. Chicago and London: University of Chicago Press, 1994.

—. 'Traditional Chinese cartography and the myth of Westernization.' In *The History of Cartography, Volume 2, Book 2: Cartography in the Traditional East and Southeast Asian Societies*, edited by J.B. Harley and David Woodward, 170-202. Chicago and London: University of Chicago Press, 1994.

2 Chinese Historical Cartographies

Mapping the Past

Abstract
This chapter focuses on the genre of historical cartography, examining works in a broad range of formats, including maps compiled for administration, for education, and for religious purposes. Tracing developments to the end of the Ming dynasty, it shows how the conditions of the publishing boom fostered the recombination and adaptation of historical maps to new contexts. Because of its utility in contextualizing classical works, this genre was intimately intertwined with the publishing industry that emerged around the national examination system, but historical maps could also buttress, or undermine, administrative and Sinocentric perspectives.

Keywords: Ming, historical cartography, Buddhist cosmography, *Yugong*, *Lidai dili zhizhang tu*, county gazetteers

In this chapter I will examine the development of a particular cartographic form, the historical map. Beginning with some of the earliest surviving examples of the genre, we will see the diversity of its incarnations and the changes it underwent during the late Ming. Taking this diachronic view of a single genre highlights the transformation wrought by the late Ming publishing boom and its textual recombinations.

A formalized preoccupation with the past was a hallmark of literati culture in imperial China, which stemmed from a belief that the words and deeds of the ancients offered important lessons in statecraft as well as morality. For many, the Classics reflected an ancient reality that could be revived through a return to the true meanings of these texts. From this zeal bloomed scholarship into textual interpretation. This was the context in which Chinese historical geography developed, giving rise to a variety of traditions of cartographic representation.

Akin, Alexander, *East Asian Cartographic Print Culture: The Late Ming Publishing Boom and its Trans-Regional Connections*. Taylor & Francis Group, 2021
DOI: 10.5117/9789463726122_CH02

Familiarity with this tradition in China remains scarce among specialists in Western cartography. In a prominent study of historical cartography, Jeremy Black devotes a portion of his first chapter to justifying his omission of China from the rest of the book.[1] He first concedes that the concept of historical cartography appeared rather early in China, citing the examples of Pei Xiu 裴秀 and the Song dynasty's *Lidai dili zhizhang tu* 歷代地理指掌圖 (Convenient historical atlas) by Shui Anli 稅安禮, the two best-known early works, but argues that the genre failed to develop over time:[2]

> In part, this reflected a rather limited interest in the outside world and a very limited knowledge of its history. There was nothing to compare with the role of the Bible and the Classics in European society. These works provided very extensive texts that were spatially specific in what they discussed and that generally depicted events in what were distant, foreign countries for most Europeans. In addition, after their major episode of Indian Ocean exploration in the fifteenth century, the Chinese did not benefit from the massive expansion in cartographic information that the Europeans gained in their explorations from the fifteenth century on.[3]

Black's stated reasons for excluding China deserve to be taken seriously and challenged. In doing so, we may also compare the Chinese context with Benedict Anderson's assessment of the situation in Southeast Asia, where he argues that historical cartography was a tool introduced by the West to naturalize and retroactively justify the territorial divisions of colonizing powers. It is vital to recognize that Song China had developed a notion of cartographic self-image through printed maps long before contact with

1 Black, *Maps and History: Constructing Images of the Past*, pp. 2-4.

2 The *Lidai dili zhizhang tu*, which will be cited frequently below using the shorthand *Zhizhang tu*, was long attributed to the polymath Su Shi 蘇軾 (1037-1101), but is now attributed by most to Shui Anli. A facsimile of a Southern Song edition held at the Tōyō Bunko in Japan was published in 1989. All further references are to this edition: Shui Anli 稅安禮. *Lidai dili zhizhang tu* 歷代地理指掌圖, published as *Songben lidai dili Zhizhang tu* 宋本歷代地理指掌圖 with prefaces by Tan Qixiang and Cao Wanru.

3 Black, *Maps and History*, p. 4. In brief response to Black's comment about European interest in distant countries, the point is partially taken. However, for a scholar in Dadu during the Yuan dynasty, for instance, Hainan was spatially just as distant as was Jerusalem for a medieval German scholar. The range of distance covered is often comparable in many contemporaneous European and Chinese works, but in Chinese cases a much greater proportion of this landscape was under the control of the author's state. This does not in itself mean that the Chinese author should be judged as provincial or narrow-minded. One feature of the standard dynastic history is the survey of foreign countries known to that dynasty.

European colonialism. Literate people from any of several East Asian states could recognize the 'map-as-logo' of their homelands, and given the historical maps discussed in this chapter, might even have been able to fit them into a sequence defined by the local historiographical tradition.[4]

The place of the Chinese Classics in the late premodern geographical consciousness differed in important respects from that of early Greek texts or the Bible in the minds of European Renaissance scholars. As we shall see, the *Yugong*, 禹貢 or Tribute of Yu, was a classic text that provided the fundamental impetus for much early historical cartography, but its focus was internal, and it was only a fraction of the length of the biblical texts from which countless place names and states were drawn by European readers intent on reconstructing the ancient landscape of the Holy Land. Nonetheless, scholarly exposition of the Confucian canon fostered far more development of historical cartography than Black concedes. Furthermore, Chinese understandings of the world were not limited to the Confucian tradition. Buddhist texts, with their descriptions of sites both sacred and secular in India, Central and Southeast Asia, played a role in China that was not entirely dissimilar from the inspiration that religious texts gave to Christian cartographers. After all, Chinese Buddhist pilgrims also ventured into distant lands, creating a particularly rich trove of texts during the Tang period. In many cases the geographical and sociocultural observations made by these travelers, in addition to records made by reluctant ambassadors such as Du Huan (taken prisoner by the Abbasid army at the Battle of Talas in 751), reports of envoys sent by the state, and even commercially motivated compendia of information on foreign lands, were collected and systematized in historical works of the standard bureaucratic tradition. The changes recorded by different visitors over time, as well as information brought by diplomats, aided the construction of the brief historical narratives of distant states that were incorporated into dynastic histories, even if the linkage between an ancient record and more recent observations was never drawn so intimately in these summaries as were the connections between, say, the biblical site of Akko and the Crusader state of Acre founded in that place.[5]

4 For discussion of the 'map-as-logo,' see Anderson, *Imagined Communities*, p. 175. Anderson's arguments about the creation of the geo-body better fit the history of Southeast Asia; Thongchai Winichakul's *Siam Mapped: a History of the Geo-body of a Nation* is perhaps the best application of this perspective.

5 For a discussion of the incorporation of Du Huan's mid-eighth century observations of Iraq and several Central Asian states into mainstream historical works such as the *Tongdian* and the *Xin Tang shu*, see Akin, 'The *Jing Xing Ji* of Du Huan: Notes on the West by a Chinese Prisoner of War.' A classic translation of the trade official Zhao Rugua's early thirteenth century *Zhu Fan*

As we shall see, the assumptions underlying historical cartography in China, and the practical uses it claimed, changed significantly over time and varied in emphasis. For the Neo-Confucian or the bureaucrat, there were lessons to be drawn in the realm of government, while to some Buddhists the shifting lines of the map inspired a meditation on the impermanence of all human endeavor. One of the foundational documents of Song printed cartography, the *Lidai dili zhizhang tu* (hereafter *Zhizhang tu*), postulates what John Dardess has called a 'constant China' in the midst of unceasing administrative transformation and a parade of dynasties.[6] To a Southern Song audience, which had the recent Jin incursions on its mind, this atlas was in part a chronicle of things lost. By the late Ming, we see the mirror image of this concern: In his 1638 preface to the *Gujin yuditu* 古今輿地圖 (Maps of Past and Present) of Wu Guofu 吳國輔, Chen Zilong 陳子龍 suggests that the most immediate and relevant purpose of historical cartography is to apply the lessons of history to prevent a *second* Jurchen invasion.[7] In the twentieth century, when the notion of maps as legal tools for asserting national sovereignty had been fully assimilated in its modern form, China began to use historical maps to press territorial claims against neighboring states including India, the USSR and Vietnam, prompting the Soviet critic Vadim Kassis to describe this practice as 'cartographical aggression.'[8]

The low survival rate of early maps makes it impossible to determine when historical maps might first have been compiled in China. The earliest surviving examples reflect an already somewhat mature tradition. The practice of reconstructing historical geography based on analysis of textual records seems to have formed at an early stage around the *Yugong* and other classical texts, as discussed below. Various examples may be cited from the Han to the Tang, and it was certainly a well-established field of

ji is available in Hirth and Rockhill, *Chau Ju-kua: His Work on the Chinese and Arab Trade in the Twelfth and Thirteenth Centuries, entitled Chu-fan-chï.*

6 Dardess, 'Did the Mongols Matter? Territory, Power and the Intelligentsia in China from the Northern Song to the Early Ming,' pp. 112-113.

7 Chen Zilong 陳子龍, in Wu Guofu 吳國輔, *Gujin yuditu* 古今輿地圖, *xu* 5a. Note that the title of this work is often alternately given as *Jingu yuditu* 今古輿地圖. Different prefaces in this 1638 edition give both variants. Black and white reproductions make it extraordinarily difficult to make out the superimposed maps which were originally done in red, but one map from this work is illustrated in full color in Plate 94 of Cao Wanru's *Atlas of Ancient Maps in China* 中國古代地圖集*: The Ming Dynasty 1368-1644*.

8 'The residents of Zhungnanhai [...] have been indulging for some time in a rather dangerous game of cartography. They have produced a wide range of versions of this destructive passion. World public opinion has been duly suspicious of the geographical "researches" of the Chinese leaders.' Kassis, 'Peking "Cartographers" As They Really Are,' p. 51.

study by the Song. It could be argued that late Song study of the *Yugong* was comparable to the much later resurgence of interest in ancient Greece by scholars in Renaissance Europe – a tradition based on researching the secular achievements of the ancients.

This chapter will examine how analysis of the *Yugong* gave rise to a broad array of visual representations, including a Song historical atlas that combined concepts from the Yugong tradition with later administrative developments. In this period, historical cartography owed much of its popularity to the needs of students of the Classics. From here we will turn to the diversification of historical cartography during the late Ming, taking into consideration not only orthodox perspectives based on administrative or educational needs, but also the fundamentally different views found in Buddhist historical cartography, perspectives that decentered the Middle Kingdom and its imperial heritage.

The *Yugong* and historical cartography

The *Yugong* is among the works collected in the *Shangshu* 尚書, or Book of Documents, one of the Classics traditionally attributed to the editorship of Confucius.[9] As a record of Yu's suppression of the floodwaters that ostensibly ravaged the world in the twenty-first century BCE, the *Yugong* also happens to serve as an outline of topography and regional division. It is the most intensely geographical work to be found in the Confucian classics, providing a dense enumeration of rivers and landscape features, and its brevity (just a few leaves) did not prevent scholars of two millennia from turning out countless pages in their quest to imagine more clearly the lay of the land and the administrative order of its 'Nine Regions' (*jiuzhou* 九州). As the preface to one late Ming work on the subject states, 'The *Yugong* is the ancestor of all geographical treatises, past and present.'[10]

The *Yugong* begins: 'Yu laid out the lands. Going along the mountains, he cut down the trees. He determined the high mountains and the great

9 Though modern scholars differ on the dating of this work, Edward Shaughnessy suggests that the *Yugong* chapter was composed as late as the Qin dynasty (Shaughnessy, 'Shang Shu,' p. 378).

10 '禹貢一書古今地理志之祖.' Ai Nanying 艾南英, *Yugong tuzhu* 禹貢圖註, vol. 3, p. 1a. Kai-wing Chow cites Ai Nianying as an example of the men of broad literary reputation who were sought out by publishers to write prefaces, a late Ming phenomenon combining status in literati culture with commercial motivations (Chow, *Publishing, Culture, and Power in Early Modern China*, p. 77).

rivers' (禹敷土，隨山刊木，奠高山大川).[11] In brief and telegraphic prose we read of the Sage's travels throughout the known world, newly freed from the grip of unprecedented floods. Yu tames rivers, shaping the landscape into forms familiar to later ages, and assesses the taxes of the various territories based on their productivity. In the tradition of activist government, Yu is a pattern-setter, one whose hydrological feats can hardly be replicated by the men of later ages, but whose diligent service to humanity is a model for all. The importance of the *Yugong* text as background for reading the Classics and early histories is reflected in the mnemonic devices created to teach it to students, such as rhyming songs and methods of laying out a map on the fingers of one's hand.[12]

Significantly, much of the text of the *Yugong* is obscure and presents alternate or even conflicting possibilities for spatial arrangement. To further complicate matters, it mixes two types of spatial layout: a relative geospatial description – 'Between the Ji and the He rivers lies Yanzhou' (濟河惟兗州) – and an idealized hierarchy of concentric zones of the ruler's influence. When scholars began representing the *Yugong's* text in map form, two different graphic traditions arose for dealing with these two perspectives. Representations of the rivers, their courses, and the lands between them were drawn in a manner that can unhesitatingly be called cartographic. The zones of influence, however, were depicted in the form of a graph that could not be easily overlaid upon a map. These two understandings could not be superimposed over one another intelligibly; they remained separate but coexistent. In Figure 2:1, we see a presentation of these two distinct forms; at right is a portion of a map showing the cartographer's understanding of the routes of rivers in Yu's era; at left is a graphic depiction of the concentric zones of administration, each zone extending one hundred *li* outward from the zone within.[13]

The text itself offers no key for meshing these two very different representations; indeed, as Needham notes, the rectangular concentric zonal representation emerged independent of anything in the text, which does not actually specify the shape of the zones.[14] This difficult-to-untangle

11 Karlgren, *The Book of Documents*, p. 12. Translations of the *Yugong* are here adapted from Karlgren's publication, which provides parallel Chinese and English text.

12 Such mnemonics based on the hand were common; Lucille Chia illustrates three examples of hand-based memory aids for other fields of study used in various Wanli-era books (Chia, *Printing for Profit*, p. 60).

13 On diagrammatical and schematic representations of space see Henderson, 'Chinese Cosmographical Thought: The High Intellectual Tradition,' especially pp. 204-216.

14 Needham supposes that the rectangular form was chosen for the zones based on the assumption that the earth was square (Needham, 'Geography and Cartography,' p. 501).

Figure 2.1 Typical examples of the two most common forms of representation used in discussions of the ancient text of the *Yugong* (Tribute of Yu)

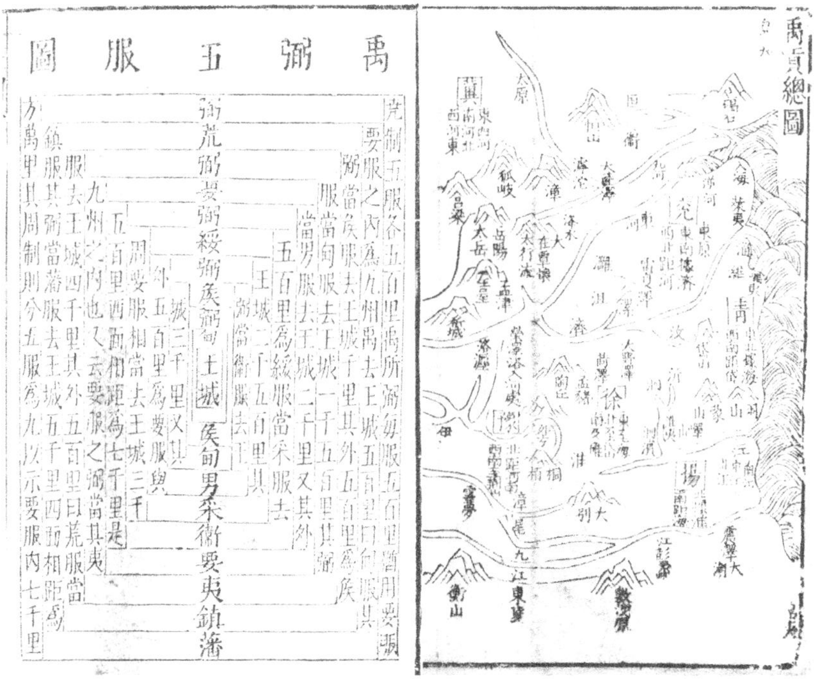

Both are from the *Sancai tuhui*, *juan* 14, pp. 8a and 9a
Courtesy of the Harvard-Yenching Library

relationship, as well as the fact that many of the places described could not be identified with certainty, served all the better to fuel discussion of the text over the centuries. As a constituent portion of the Book of Documents, the *Yugong*'s authenticity was essentially unassailable; in cases where current knowledge seemed to contradict the *Yugong*, later commentators were inclined to believe that they were misinterpreting the text. Their task lay in finding a way to engage its truth with the world as it was known.[15]

15 Sometimes current knowledge was placed alongside that of the *Yugong* without explicitly refuting the latter. For example, Cao Wanru notes that the *Yujitu*'s depiction of the Yangtze River denotes its source textually as being in the Minshan mountains, though it graphically depicts the main source of the river as flowing southwest of the Dadu River. The compilers were thus apparently aware that the source of the Yangtze lay further westward than had formerly been believed, but in Cao's words, 'The only obstacle that prevented them from clearing up the fact was that they dared not challenge the established statement in [the] Confucian Classics.' See Cao Wanru's *Atlas of Ancient Maps in China* 中國古代地圖集: *From the Warring States Period to the Yuan Dynasty*, Notes on the plates, p. 21.

Pei Xiu (224-271), whose cartographic principles are cited in countless gazetteers and whose systematic approach earned him Chavannes's and Needham's coronation as the father of 'scientific cartography,'[16] used the *Yugong* and other records to compose cartographic representations of antiquity and its relationship to the present. Pei's silk maps from the *Yugong diyutu* 禹貢地域圖 (Map of the regions in the Yugong) are long lost, but the preface has been preserved. It reads, in part:

> Referring back to antiquity, I have examined according to the *Yugong* the mountains and lakes, the courses of the rivers, the plateaus and plains, the slopes and marshes, the limits of the nine ancient provinces and the sixteen modern ones, taking account of commanderies and fiefs, prefectures and cities, and not forgetting the names of places where the ancient kingdoms concluded treaties or held meetings; and lastly, inserting the roads, paths and navigable waters, I have made this map in eighteen sheets.[17]

> 今上考禹貢山海川流原隰陂澤古之九州及今之十六州郡各縣邑疆界鄉陬及古國盟會舊名水陸徑路為地圖十八篇。

Rather than representing a single era, Pei Xiu's maps depicted places and events from different moments in history. Whether these were all superimposed on the same maps, or separately indicated in individual vignettes, is unclear. Both of these techniques were commonly used later in the imperial period, the former frequently adapted for decorative wall maps or teaching materials (such as maps highlighting all of the former capitals), the latter form being more suited to books where successive pages could represent installments of a temporal series. In any case, based on the surviving text alone, Hilde De Weerdt describes this work as 'the first to articulate the bidirectional relationship between past and present that is visible in later maps of the empire.'[18]

Texts that survive from the Northern Song highlight the attention paid to the *Yugong* during this time, and the earliest extant maps based on the *Yugong* come from soon after. One reason for interest in the *Yugong* during the Song may have been the contemporary political situation. While the

16 See Needham, 'Geography and Cartography,' p. 538.

17 From Needham's adaptation of the translation of Chavannes and Vacca (Needham 'Geography and Cartography,' p. 540).

18 De Weerdt, 'Maps and Memory,' p. 150.

Yugong listed the idealized Nine Regions of Yu's era, the cold facts of the eleventh century placed northern portions of this territory under the control of the Khitan Liao state. Furthermore, and more importantly, the growth of the exam system was a force behind the proliferation of *Yugong* maps as study aids.[19]

Such a map survives in two examples, early twelfth century stone engravings titled *Yujitu* (Tracks of Yu).[20] These maps, which surely rank among the most significant cartographic artifacts in world history, depict changes in human administration against a river network charted on a regular grid of squares representing 100 *li*. Figure 2:2 illustrates a small portion of the 1136 map. No maps of this era from anywhere else in the world depict such a broad swath of the earth's surface with anything close to such precision. These maps were engraved on stelae so that students or visitors could make rubbings for later convenient reference. Both examples of the *Yujitu* have a small square at the upper left which lists their content as: 'Names of mountains and rivers from the *Yugong*, names of provinces and prefectures from past and present, and mountain and river names and toponyms from past and present.'

On the reverse of the 1136 stele is engraved the *Huayitu* 華夷圖 (Map of Chinese and Barbarian [Lands]), which is based on a Tang-era illustration of historical kingdoms. While the *Yujitu* links the present to the age of Sage Kings, the *Huayitu* covers something of the temporal middle and fills in around the spatial edges, situating the Hua on a broader geographical canvas upon which time has written a constant tale of interaction both internal and external.[21]

After the ascent of the Jurchen Jin in the early twelfth century and the subsequent fall of the Northern Song, scholars who turned to the *Yugong* increasingly lamented the loss of territory to people they saw as barbarians. As Tang Xiaofeng has noted, study of the *Yugong* at this time became linked

19 Note that works with *Yugong* in the title need not be historical in nature; any book dealing with waterways, even contemporary problems of river management, might have these two characters in its title.

20 The title of the 1136 example at Xi'an is written 禹跡圖 while the Zhenjiang example from 1142 uses the variant 迹 for the second character. Reproductions of these maps are published in Cao Wanru's *Atlas of Ancient Maps in China: From the Warring States Period to the Yuan Dynasty*, Plates 54-59 and discussed in the appended text. See also Needham, 'Geography and Cartography,' p. 547.

21 For an illustration and brief discussion of the pairing of these maps, see Yee, 'Reinterpreting Traditional Chinese Geographical Maps' p. 47. For more substantial discussion of the 1136 *Yujitu*, see Akin and Mumford, '"Yu Laid out the Lands": Georeferencing the Chinese *Yujitu* [Map of the Tracks of Yu] of 1136.'

Figure 2.2 ink rubbing of the 1136 *Yujitu* (Tracks of Yu)

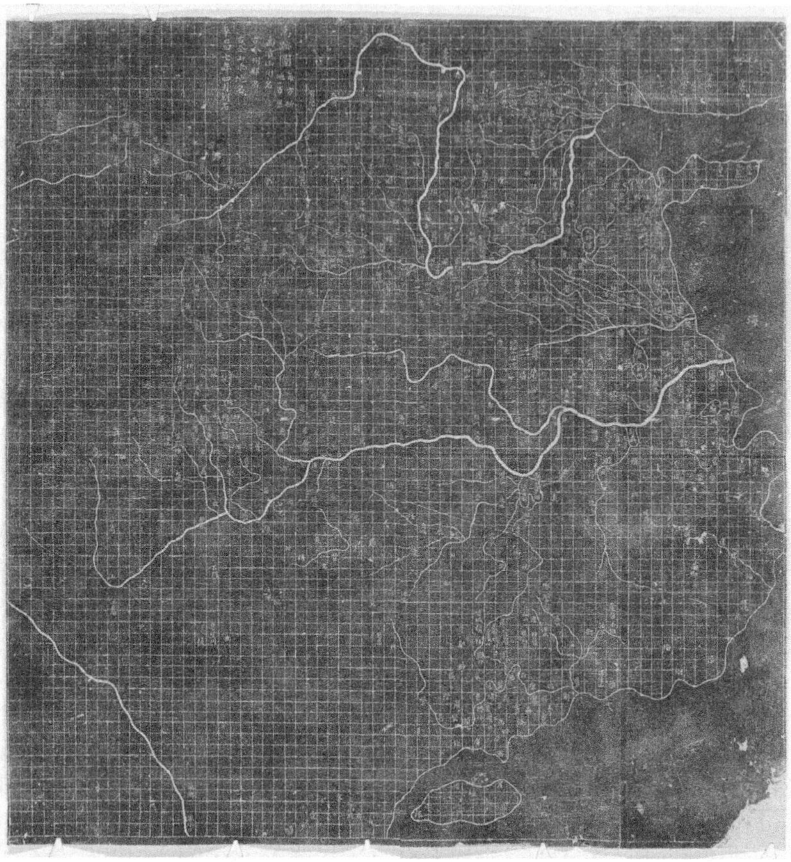

Courtesy of the Harvard-Yenching Library

with an irredentist movement in some literati circles. The Southern Song poet Lu You 陸游 wrote, 'Everything will be empty after death / But still there will be sorrow that the Nine Regions are not unified.'[22]

Because of this political context, it is easy to assume that maps depicting a unified past in this period must have been symbols of irredentism. It should be noted that the 1136 engraving of the *Yujitu* was not commissioned under Song rule, as is assumed by some who look only at the date of 1136. Although (like the 1142 copy) its designs derived ultimately from some earlier and now-lost precedent, in this case chisel was put to stone during the Fuchang

22 Cited in Tang Xiaofeng, *From Dynastic Geography to Historical Geography: A Change in Perspective Towards the Geographical Past of China*, p. 26.

阜昌 reign of Liu Yu 劉豫, who ruled Qi 齊 as something of a Jin puppet from 1130 to 1137. Given the context of its reproduction under a local ruler who held power because of the Jurchen invasion, the 1136 map can hardly have been connected to any Southern Song irredentist movement, unless this was a very subtle move by conservative scholars reluctantly serving Liu Yu.

Comparison of the Nine Regions with the contemporary political situation continued, in different forms, throughout the imperial period. After the Ming consolidation we frequently see a variation of a trope from the *Yugong* – 'To the east [the realm] dips down to the sea, to the west it extends to the flowing sands' (東漸于海西被于流沙)[23] – used in texts lauding Zhu Yuanzhang's establishment of an expansive dynasty. The *Da Ming yitong zhi* (Gazetteer of the Great Ming's unification) states of the Ming's 'unprecedented' breadth that 'to the east it completely [encompasses] everything up to the left bank of the Liao, to the west it reaches the flowing sands' (東盡遼左西極流沙).[24] In such terms was the reunification of the ancient core celebrated under a revived non-'barbarian' dynasty.

The *Yugong lun shanchuan dili tu* 禹貢論山川地理圖 (Treatise on the *Yugong* and Atlas of mountains and rivers) by Cheng Dachang 程大昌 (1123-1195) notes in its preface that there was already an established genre of books that attempted to interpret the *Yugong* either in text or in image. These works included many conflicting interpretations, making it difficult to sift through the errors and create an accurate understanding of the geography of that period and its transformations up to the present.[25] Cheng resolved to undertake this arduous task. His goal was to sort out previous scholars' interpretations, which were often textual rather than mapped, and to represent his findings cartographically, as accurately as was possible in a woodblock printed text. Cheng repeatedly reminds us of his frustration with this medium of printing, noting in his commentary to many maps that details had to be omitted. The best that can be hoped for, it seemed, was for the reader to be able to understand the point of a map when provided with textual context.

Any scholar who wished to represent the *Yugong* in map form was likely to encounter a central problem: Because the text was relational in character,

23 Karlgren, *The Book of Documents*, p. 16 (Chinese), p. 18 (translation).

24 Li Xian 李賢. *Da Ming yitong zhi* 大明一統志, Preface p. 2a. The '遼左' (left of Liao) here refers to the region east of the Liao River; from the southward-facing royal perspective left was to the east. These expanses were even further east than the coastal territories known to the compiler(s) of the *Yugong*.

25 Preface by Peng Chunnian 彭椿年 to Cheng Dachang 程大昌, *Yugong lun, shanchuan dilitu* 禹貢論山川地理圖, pp. 1a-2b.

charting it against the world of the present required stable reference points, features that did not shift over time. The *Tongzhi* 通志, a compilation of historical biographies and other essays by Zheng Qiao 鄭樵 (1104-1162), reveals an awareness of this problem.[26] Zheng notes in his preface to the geographical section that the arrangement of prefectures and counties was temporal, while the forms of mountains and rivers had remained unchanged since great antiquity.[27] Thus, he concluded, the *Yugong* intentionally used natural features, particularly rivers, as its fixed points of reference in order to aid readers of later ages. This emphasis on fixed references that transcend dynastic change and administrative flux fits with Zheng's belief that, as S.Y. Teng puts it, 'History should be treated as a continuous stream [from] ancient to modern [...] it should not be arbitrarily divided into periods corresponding to the dynasties.'[28]

What was the point of these arduous labors of historical reconstruction? The late Ming *Yugong gujin hezhu* 禹貢古今合注 (Combined commentary, ancient and new, on the *Yugong*) by Xia Yunyi 夏允彝 (1596-1645), a scholar from Songjiang prefecture with connections to the late Ming Restoration Society, offers a clue. The preface to this work – written again by the ubiquitous Chen Zilong – states that the greatest problems faced by the realm stem from poverty, which in turn is due to poor management of labor and hydrological resources in the fields.[29] Yu's fame lay in his ability to control floods, so for very practical reasons it was imperative to identify the modern locations of sites mentioned in the *Yugong*, initially for the sake of understanding the ways rivers could change course, and perhaps eventually for replicating the

26 Zheng Qiao 鄭樵, *Tongzhi* 通志. Zheng's *Tongzhi* was later categorized alongside Du You's *Tongdian* 通典 and Ma Duanlin's *Tongkao* 通考 as one of the Three 'Tongs;' it was popularized through multiple printings and an abbreviated version (the *Tongzhi lue*), the latter being a separately circulated collection of twenty short works on specialized subjects such as philology, insects, and of course geography: Zheng Qiao, *Tongzhi lue* 通志略. For discussion of the compilation and contents of the *Tongzhi* see Teng, 'Cheng Chiao,' in Herbert Franke, *Sung Biographies*, pp. 150-153.

27 Zheng Qiao, *Tongzhi, dili juan* 1, *dili xu*. Though the drying up of rivers or changes in course had of course been noted in the past, the assumption that seemingly immovable natural features could serve as eternal frames of reference may have been more severely shaken after the Yellow River dramatically changed course in 1194, followed by further fluctuations which often brought widespread destruction in their wake. Joseph Needham asserts that early Chinese knowledge of geological processes was widespread – see Needham, 'Geology (and Related Sciences),' in *Science and Civilisation in China, Volume 3: Mathematics and the Sciences of the Heavens and the Earth*, pp. 598-604. Some of the passages he selects, however, suggest a belief that the floods of the Xia period and the channels Yu built to control them were the last major hydrological changes on the earth's surface.

28 Teng, 'Cheng Chiao,' p. 149.

29 Chen Zilong, preface to Xia Yunyi 夏允彝, *Yugong gujin hezhu* 禹貢古今合注, p. 1a.

hydrological feats of Yu's generation. Xia's work is a reminder that *Yugong* studies were not only about history for history's sake; the text inspired serious consideration of enormously practical questions.[30] It was not only a remnant of the past but a potential blueprint for the future, an age in which a more perfect alignment of moral and temporal order might be reasserted.[31]

Xia Yunyi's treatise drew its maps from disparate previously published sources, and other late Ming scholars working with historical cartography from the *Yugong* onward also exemplify this late Ming hybridization and synthesis, combining multiple sources and reproducing iconic representations in a single work. For example, a Chongzhen-era edition of Mao Ruizheng's 茅瑞徵 *Yugong huishu* 禹貢匯疏 (Compendium of commentaries on the *Yugong*) combines sources ranging from the historical maps of the *Zhizhang tu* and reconstructions of the courses of ancient rivers, to maps of the 'present' such as the mid-sixteenth century *Guang yutu*, all in order to construct a grand narrative of the evolution of the land and its waterways from ancient times to the present. In a move typical of higher-quality late Ming compendia, the preface briefly evaluates the relative trustworthiness of several texts, and notes which sources were consulted as primary in the compilation of the text.[32] Mao (fl. 1597-1636), a native of Gui'an 歸安, Zhejiang, published some other significant cartographic works, including the *Wanli san da zheng kao* 萬曆三大征考 (Investigation of the three great campaigns of the Wanli Era), about conflicts in Ningxia, Korea and Guizhou, and the *Dongyi kaolue* 東夷考略 (Outline investigation of the Eastern Barbarians), both illustrated with maps copied from earlier books.[33]

In closing this discussion of the *Yugong* and its place in the development of historical cartography, another early work should be mentioned. The *Shanhai jing* 山海經 (Classic of mountains and rivers) served as something of an unorthodox counterpart to the *Yugong* during much of the imperial period. Whereas the *Yugong* is, in Needham's words, 'completely devoid of magic, and even fantasy or legend, apart from the appearance of Yu himself,'[34] the *Shanhai jing* describes an earth arranged in concentric zones, where the familiar lands of the inner regions appear much as in other works, but the countries and inhabitants of the outer zones become increasingly strange.

30 Cordell Yee discusses the *Shuijing* in this context of practical action. See 'Chinese Maps in Political Culture,' pp. 93-94.

31 On this theme see Tang Xiaofeng, *From dynastic geography to historical geography*, p. 51.

32 Mao Ruizheng 茅瑞徵, *Yugong huishu* 禹貢匯疏, *fanli*. For more biographical detail on Mao and a bibliography see Yang and Goodrich, 'Mao Jui-cheng.'

33 See also the discussion of the *Chouhai tubian* in the previous chapter.

34 Needham, 'Geography and Cartography,' p. 501.

Lands peopled by giants and every manner of bizarre creature are its theme, and the commercial publishing industry of the late Ming made it a favorite for illustrated editions.[35] 'Proper' scholarly works on historical cartography tended not to emphasize the *Shanhai jing*, but this did not prevent its service as imaginative fodder for a broad range of other writings.[36]

The *Yugong* was not entirely a 'secular' work, due to its potent Classical authority, but by the Song dynasty, maps of the *Yugong* were placed in the same category as representations of the Warring States and various dynastic periods that were reconstructed from administrative documents and gazetteers. This bridge between the world of the Sages and the more mundane later stages of human existence is clearly illustrated in the *Zhizhangtu*'s historical 'snapshot' format. To the creators of its maps, and to those who used them, the administration of the Sage Kings was just as real as that of the Tang dynasty. The *Zhizhang tu* was one of the earliest and most enduring mass-printed works to insert cartographic renderings of the *Yugong* into a historical series that extended to the author's own dynasty. Since it served as a prototype for later historical atlases and for other works of the late Ming that we will examine below, it merits a brief overview here.

A milestone in historical cartography: The *Lidai dili zhizhang tu* 歷代地理指掌圖 (Convenient historical atlas)

The Song era produced the earliest surviving examples of many types of printed map, and in this panoply of cartographic publications, many had a historical theme. The cartographic collection edited by Cao Wanru, *Atlas*

35 Illustrated versions far predated printing, as one such copy is described in the third century (Fracasso, 'Shan hai ching,' p. 362). Fracasso notes that many later commentators assumed that the text had been compiled from maps, or was designed to be read alongside maps, but no early copies of such maps are known. Creatures from the *Shanhai jing* comprised a large portion of the illustrations in the popular and influential late Ming *Sancai tuhui* 三才圖繪; a picturesque survey of these engravings can be found in Goodall's *Heaven and Earth: 120 Album Leaves from a Ming Encyclopedia, San-ts'ai t'u-hui, 1610*.

36 Certain lands mentioned in the *Shanhai jing*, such as Nü Guo 女國, the Land of Women, were depicted on world maps even at very late stages, but their location was pushed further afield as the radius of geographical knowledge expanded. A counterpart in European cartography would be the kingdom of Prester John, which was pushed further into Asia as knowledge of the Near East increased, and then relocated completely to Africa after expeditions to the Mongols deflated the illusion that this mythical Christian hero was to be found in East Asia. After this, the Christian kingdom of Ethiopia was sometimes referred to as the land of Prester John (Whitfield, *New Found Lands: Maps in the History of Exploration*, pp. 168-169; Woodward, 'Medieval Mappamundi,' p. 333).

of Ancient Maps in China 中國古代地圖集, includes Song maps showing the movement of capitals over time and the distribution of rivers around Chang'an during the Tang. Thus the *Zhizhang tu* is merely one example of a larger development of historical geography during this time, though it was arguably the most influential of its genre; its series of historical snapshots was reprinted for hundreds of years, including in a number of late Ming works.[37]

The preface to the *Zhizhang tu* undertakes three tasks, first outlining the history of cartography, then summarizing the changes of administrative topography and dynastic succession that make a historical atlas necessary, concluding with an overview of the contents of the atlas. It sums up the blessings of historical cartography (having past and present at hand, being able to survey the world without leaving one's door)[38] but leaves unstated the practical use of such an atlas. In fact, it was a potent reference for scholars embarking on the examination track, and we do get the sense that the *Zhizhang tu* was used as a study aid when we note readers' brush marks – the classical equivalent of highlighting – peppering the sole surviving Southern Song edition.

The maps in the *Zhizhang tu* all use essentially the same frame of reference. The Great Wall scrolls along the top, a sandy desert borders the northwest at upper left, the southwest is framed by rivers flowing into Southeast Asia, and the southern and eastern boundaries are defined by coastline. As John Dardess notes, this repeated iteration across the centuries helps to suggest that 'the basic geographical extent of China never changed after [Emperor Ku]; that all that happened in later times was refinement and detail – infilling, renaming, reorganizing.'[39] The comprehensive effect of page after page of identically formatted maps is to impress upon the reader the sense of an eternal space, traversed as it may be by impermanent dynasties with their fickle borders, yet retaining something nevertheless of the grand unity of that era of the Sage Kings when its patterns were established.

The first map in the *Zhizhang tu* is a comprehensive and timeless vision of past and present, domestic, and foreign. It is followed by two pages of

37 A search of the *Siku quanshu* for this text under the various versions of its title (*Dongpo dili zhizhangtu* 東坡地理指掌圖, *Su shi zhizhangtu* 蘇氏指掌圖, etc.) turns up numerous references in other places, including many in unexpected fields such as poetry and ritual. The title phrase that I have translated 'Convenient,' *zhizhang*, could also be translated as 'handy' in more than one sense; it literally indicates pointing at the palm of one's hand.

38 *Zhizhang tu*, pp. 2a-b.

39 Dardess, 'Did the Mongols Matter? Territory, Power and the Intelligentsia in China from the Northern Song to the Early Ming,' pp. 112-113.

text summarizing the general administrative and topographical themes of Chinese history, followed by short blurbs on the Great Wall and various foreign countries, discussing a number of the foreign peoples who will not be addressed cartographically in the remainder of the atlas but whose significance in Chinese history cannot be ignored. Although changes in the names of countries abroad are listed, they are based not on indigenous changes, but on changes in Chinese ways of referring to them. For example, the listing for India states that he Five Indies are equivalent to what was called the country of Shendu in the Han, also referred to as the Land of the Brahmans. Its land measures more than 30,000 *li*, possessing several hundred walled towns, and is located south of the Pamirs.[40] Past administrative changes that were the effect of Han or Tang policy are recorded for the 'Western Regions,' meaning Central Asia. Thus even foreign history develops in the shadow of Hua.

The next map takes mountains and rivers – natural landforms transcending dynastic change – as its main theme, the accompanying text briefly addressing their natural and historical significance. Thus the first two maps have outlined the essential context of this and most later historical atlases: Firstly, this country is ringed by others, its history intertwined with theirs although central to all. Secondly, its historical changes take place on a physical landscape, a palette upon which indelible rivers and mountains play host to the shifting perimeters of human regimes.

The atlas now moves into its historical series, beginning with a representation of the nine regions under Emperor Ku and moving on through the Sage Kings and the Zhou to reach the imperial period. Dynasty-by-dynasty, this sequence reflects a millennium of changing administrative order – with maps illustrating periods of inter-dynastic chaos as well – all the way through the collapse of the Tang. The text alongside each of these maps summarizes major administrative changes, significant border-changing battles, and the absorption of states by their neighbors. The series ends with five maps highlighting the main post-Tang states that preceded the rise of the Song.

After this grand historical tour, before moving to the dynasty still in power at the time of composition we return to the generalized rather than the specific. A chart of the *fenye* system, linking heavenly constellations with regions on the ground, is followed by a diagrammatic representation of the constellations' association with the changing pattern of earthly states. The atlas then presents us with naturalistic depiction of climactic zones bounded by rivers and mountain ranges, first composed by Yixing 一行

40 *Zhizhang tu*, p. 4b.

Figure 2.3 Historical map of the Yuanfeng era (1078-1085), part of the twelfth-century *Lidai dili zhizhang tu* (Convenient historical atlas) as adapted and reprinted in the *Sancai tu hui* encyclopedia

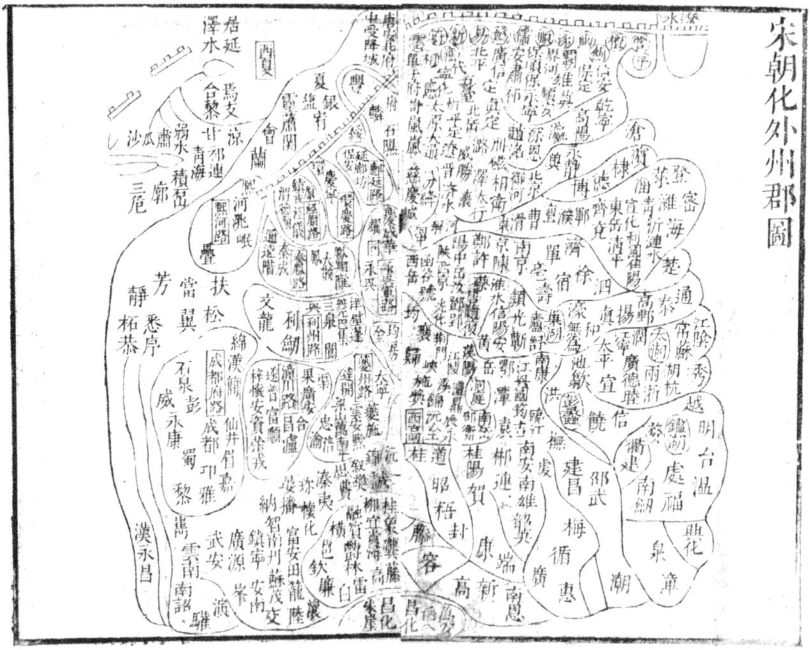

Juan 16 pp. 18a-b
Courtesy of the Harvard-Yenching Library

of the Tang to illustrate the relationship between geophysical terrain and human civilization, and rounds up this lot with a general map of former place names accompanied by lists of dynastic capitals. The charts accompanying this map, which list the capitals of the pre-imperial period at right and the post-Qin imperial period at left, exemplify the collaborative nature of text and image: The chart is handier than the map for quick reference, but the spatial relationship between places is best clarified through cartography.

The foundation and growth of the Song dynasty are the themes of the concluding maps. We see the Song emerge at first as just one of several post-Tang regimes, the maps and text chronicling its absorption of other states. By the Yuanfeng reign (1078-1085), we see the Song in its full glory, albeit marred by the existence of the Khitan regime and the Xi Xia at its fringes. (See a late Ming reprint of this map in Figure 2:3, another example of twelfth-century cartography recirculated through late Ming republication).

Another map depicts the prefectures on the edge of civilization that are in the category of *jimi* 羈縻 (literally 'loose rein'), defined as 'reached by the Teaching although not submitting tribute or tax registers to the Board of Revenue' (雖貢賦版籍不上戶部然聲教所曁).[41] These include the 'lost' sixteen prefectures held by the Liao as well as others in the west and south. The inference is that these remained part of the Chinese cultural sphere although politically out of reach. Saved for last is a map illustrating internal administrative changes, province by province, for the Song dynasty alone.

Although there may be a certain amount of irredentist sentiment expressed in these maps, it must be read between the lines. The maps tracing the establishment and growth of the Song begin by showing the multistate situation existing after the Tang disintegration, the text narrating the assimilation of these short-lived states into the Song polity. The emphasis is thus not on the *loss* of territory, but on the Song's *consolidation*. One might even interpret these pages as implying that the Khitan Liao and the Xi Xia are final holdouts against the expansion of the Song, which might yet incorporate them in the same manner as it has the various kingdoms to the south. In contrast to the nationalist discourse of modern China, no past weakness or humiliation is trumpeted as an inspiration to 'reclaim' these territories. Neither does the historical summary of the Great Wall explicitly state that it marked an eternal border transgressed by the Liao. Rather, the text emphasizes its piecemeal formation in separate segments from the Warring States to the Sui, without mentioning the Liao.[42]

On the other hand, the Liao is treated as a state categorically different from the numerous other kingdoms that arose in the Tang-Song interlude. Although the blow-by-blow narrative of who attacked whom during this period includes the Khitans alongside their contemporaries, the list of dynastic capitals belonging to even small, short-lived kingdoms does not include the Liao (or, for that matter, the Xi Xia).[43] In the commentary to the left of the map depicting the Yuanfeng reign, it is noted that 'at present 16 *zhou* remain under the occupation of the *lu*,' traditionally translated 'caitiff,' suggesting slaves or miscreants (至今十六州尚為虜所據).[44] This appears to be the most directly irredentist statement to be found in the entire work.

The map to which this statement is appended depicts the Khitan and Xi Xia states as bounded units, with their names in dark ovals. The map thus portrays the Song coexisting with two acknowledged states on its borders during the Yuanfeng reign. The use of the insulting term *lu* 虜 and

41 Ibid., p. 47a.
42 Ibid., p. 4a.
43 Ibid., p. 43b.
44 Ibid., p. 46b.

the fact that these states are omitted from the list of capitals remind us of the compiler's displeasure at this fact, but this is still a step removed from the forms of cartographic denial of unwanted circumstances that can be found on many more recent maps. Our use of the term 'irredentism' should be tempered with caution to avoid anachronistic assumptions.[45]

The scope of the *Zhizhang tu*, spanning the earliest years of recorded history up to the eleventh century, made it an ideal study aid that, due to its relative brevity, could be reproduced fairly easily. Although there are occasional small reminders that the extant version is a Southern Song edition rather than the original version,[46] for the most part this (and later editions) faithfully copied the maps and text from the original, maintaining the utility of the atlas not only as a tool for studying pre-Song history, but as an artifact of its time of composition in the late eleventh century. Its distinctive framing of a rectangular map between columns of text was both convenient and, ultimately, iconic. To those who have studied this work, it is often possible to tell at a glance when a map in a later work is derived from it; this visual similarity can frequently be confirmed by examining the text in the later work, which is sometimes copied wholesale. This republication, both in independent editions and as a component of other works, not only made these maps available to a broader audience, it also made the perspective of the *Zhizhang tu* a prevalent way of viewing and representing historical change.[47] As we shall see below, later works like the 1515 *Jinling gujin tukao* 金陵古今圖考 (Illustrated study of Jinling, past and present), in which the Jinling Mountains appear as ramparts of unchanging stone encircling a plain on which humanity's traces shift about in a series of sequenced maps, may owe a debt to its example.[48]

This offers a contrast with the process described by Benedict Anderson in which mechanical reproduction imported by European imperialism brought

45 This is not to deny that scholars opposed to the government's accommodationist policies did use maps of the past as clarion calls for action against the 'barbarians.' Hilde De Weerdt touches on the issue of irredentism and maps in 'The Cultural Logics of Map Reading.'

46 For example, on p. 48a of the *Zhizhang tu* Hangzhou is cited as Lin'an and Jiangning as Jiankang, changes that were not made until 1129 (cited by Tan Qixiang, Preface to *Zhizhang tu*, p. 3).

47 Hilde De Weerdt outlines the publication history and reception of the *Zhizhang tu* (De Weerdt, 'The Cultural Logics of Map Reading,' pp. 244ff.). Regarding Ming reprints of the work, Cao Wanru cites two editions from the Jiajing and Wanli periods (Cao Wanru, Foreword to *Zhizhang tu*, p. 5). As I note below, two major Buddhist works of cosmography from the Song and Ming also quote passages from the text of the *Zhizhang tu*.

48 Chen Yi 陳沂, *Jinling gujin tukao* 金陵古今圖考. In this work, although there is a unidirectional trajectory of development as manmade structures gradually cover more area, the changing titles on each page remind us of the rise and fall of successive dynasties. The diachronic view of a single location reminds us of each dynasty's impermanence.

the cartographic lens to bear on Asian states, introducing a bird's-eye view of their bordered territories for the first time.[49] In the Chinese context we see this formation much earlier, arguably during the Tang but certainly by the Song, with works like the *Zhizhang tu*. Publications of the Yuan and Ming reinforced these images. The shape of the Middle Kingdom, not only in the present but also at various stages of the distant past, became something that could be recognized at a glance; the Nine Regions could be mapped on the hand. Moreover, it was the circulation of Chinese printed maps, rather than the later arrival of European cartography, that helped to spur cartographers on the Korean peninsula and in other neighboring states to respond with their own images.[50]

Mapping the past for the masses: Historical cartography in the Ming publishing boom

The production of historical cartography, closely tied to exegetical studies, had thus emerged by the Song and continued to develop thereafter through reprints, which perpetuated established tropes, as well as through works that used philological tools to critically respond to earlier formulations. The intellectual schisms and temporal concerns that stirred society on a broader level were reflected in views not only of the present, but also of the past. In short, as becomes increasingly clear in the second half of the Ming, there was no single genre of historical cartography, nor any single use to which it was put. As maps of past landscapes proliferated, they also diversified.

The publication and republication of historical cartographic texts in the late Ming was part of a broader interest in the past. Robert E. Hegel suggests that historical fiction became the most popular genre for leisure reading around the mid-sixteenth century.[51] The popularity of both historical and military maps might be viewed in this context; the latter offered a way to 'participate' in conversations on defense from one's own study.

The *Guang yutu* (Expanded atlas), discussed in Chapter One, marked a turning point for Chinese cartography in several ways. Although its primary function was to serve as an atlas of the Ming, highlighting its administrative

49 Anderson, *Imagined Communities*, especially pp. 171-175.

50 The fact that there was such a thing as a visual territorial identity in the Koryŏ, for example, is amply demonstrated by records of that state issuing a form of silver currency in the shape of the country. See Ledyard, 'Cartography in Korea,' p. 240.

51 Hegel, *Reading Illustrated Fiction in Late Imperial China*, p. 29.

and defensive aspects, a concern with history infuses its discussion of foreign countries. The mid-Ming context of alarm about the northern and northeastern frontiers and the pirates on the coast made knowledge of neighboring lands (and their historical relations with successive dynasties, particularly the development of their relations with the Ming) a prerequisite for future action. This applied even to countries that were closely linked in tributary relations. For example, in the *Guang yutu* the Chosŏn dynasty gets a two-page textual summary of its earlier history from the time of Kija to the Ming, although the map is purely contemporary in content.[52] Japan, blamed for the Ming's troubled seas, receives five times as much discussion, partly because of its more complex internal political situation.[53] The map of Japan which appears in the edition edited by Hu Song comes from a work by Zheng Ruozeng 鄭若曾, the famous writer on naval defense.[54]

We see further development in this defensive direction in the 1638 *Gujin yuditu* 古今輿地圖 (Atlas of past and present) by Wu Guofu 吳國輔, but this book adds a new technique to its presentation. Two-color manuscript cartography had long been used to show place names of the past and present side-by-side, but bicolor *printed* cartography was a newer development. The *Gujin yuditu* played this technology to the hilt. Most of the maps in this book take an identical black outline map, with modern nomenclature printed in black, as their foundation layer; this map is then overprinted in red ink with the place names and administrative boundaries of the historical period under discussion. A number of maps related to foreigners differ slightly, placing more details such as place names in black on the fringes. Wherever considered relevant, additional historical notes are printed in the relatively empty far west or in other blank parts of the map.

In his preface to the *Gujin yuditu*, Chen Zilong recalls a time when he was in Yuezhou and had the opportunity to see a copy of a work titled, rather generically, the *Lidai yuditu* 歷代輿地圖 (Historical atlas). Chen writes,

> Its method was to depict a comprehensive map of the present in black, delineating prefectures and counties, and then to present, printed over these maps in red ink, individual maps of the [various eras] from before the Five Emperors up until the beginning of the present dynasty [...]

52 Luo, *Guang yutu, juan* 2, pp. 83b-84a.

53 Ibid., *juan* 2, pp. 97b-102a.

54 For the map's source see note in the *Guang yutu, juan* 2, p. 96b. Wang Yong notes that the maps of the Ryukyus and Japan which are in Hu Song's reprint are new to that edition (Wang Yong, *Zhongguo dili xue shi* 中國地理學史, p. 91).

> Searching for Guide 歸德, one would know that during the Song dynasty this place was the southern capital.[55]
>
> 其法以墨作今一統圖分列郡縣而別以丹書自五帝以來至於國初各為圖。。。稽歸德者知為趙宋之南京。

Extrapolating from these maps, Chen continues, it becomes possible to see the military importance of various frontier garrisons, passes and strategic points. His analysis of the landscape and the tactical lessons of the past led Chen to decide that the Ming held the territorial advantage in the northeast. His hope was that through the military virtuosity of the Ming son of heaven, with his multitude of officials united in action, the empire could be defended on all fronts.[56] Surely the professional or casual interests of the student or scholar were still an important factor in the publication of historical atlases, but the crisis at hand gave new urgency to the learning of history's lessons. This helped foster a market for such books even among those with no personal input in the country's military affairs, who simply wished to keep abreast of current events.

Another bicolor historical atlas from the end of the Ming is the Chongzhen-era *Yueshi yueshu* 閱史約書 (Reviewing history, weighing books) by Wang Guanglu 王光魯, a basic reference to aid the study of the standard histories or other historical texts. Unlike many other historical atlases it does not have pages of text accompanying each map; for the most part the maps must speak for themselves (with some caveats and instructions expressed in the preface, such as directions on distinguishing ancient from modern features and a reminder that the disorder accompanying the collapse of a dynasty is difficult to represent on a single map).[57] A historical narrative and charts are presented separately. This work's color scheme is reversed from the usual, with red being used for the modern maps and black representing the past; perhaps this was done to bring greater emphasis to the historical aspect of each map, as black is easier to read on the page. Its cartographic treatment only continues up to the collapse of the Yuan, leaving the Ming unsullied.

Aside from the military turn, late Ming publications reflect other changes in intellectual interest as well. Historical volumes on specialized topics proliferated, including the preservation of old maps that had nearly been lost. For example, two recompilations of the works of the early Song scholar

55 Chen Zilong, *Gujin yuditu*, *xu* 5a-6a.

56 Ibid., *xu* 8a.

57 Wang Guanglu 王光魯, *Yueshi yueshu* 閱史約書, *fanli* 1a-1b.

Fan Zhongyan 范仲淹 dated Wanli 36 and 37 (1598 and 1599) include a recovered Song-era map of the Xi Xia regime, for which no official dynastic history had ever been published.[58]

Within the genre of historical cartography we find a number of large works assembled from multiple printed sheets to make decorative wall hangings or scrolls. These could, upon a single broad surface, present a wealth of historical information. The *Qiankun wanguo quantu gujin renwu shiji* 乾坤萬國全圖古今人物事蹟 (Universal map of the myriad countries with traces of human events past and present), compiled by Liang Zhou 梁輈 around 1595, lists sources ranging from the *Guang yutu* to Matteo Ricci's world map in its crowded and heavily annotated melding of past with present.

As noted in Chapter One, there were both high and low registers to the very broad genre of thematically organized miscellanies (*leishu*). At the upper end were works like the *Tushu bian*, whose historical maps I will discuss in a moment. At the other end of the spectrum of respectability were the market-oriented household almanacs, encyclopedias of around three dozen brief chapters that included basic information on subjects ranging from comet prognostication to the rules for card games. These shorter works, which were explicitly marketed to all classes, also have something to add to our understanding of the place of historical geography in late sixteenth and early seventeenth century publications, for despite the absence of separate historical maps, historical consciousness informs their geographical content.

The historical maps in the *Tushu bian* are collected from various earlier geographical texts. The map illustrated in Figure 2:4, *Yugong suo zai suishan junchuan zhi tu* 禹貢所載隨山浚川之圖 (Map of 'Determining the high mountains and the great rivers' from the *Yugong*), takes its title from the phrase *suishan junchuan* 隨山浚川, which appears at the beginning of the *Yugong* text.[59] This map is an example of the attempted reconstruction of historical space based on the text of the *Yugong*.

Figure 2:5 illustrates the *Tushu bian's* representation of the dynastic succession of capitals, cast against a grid pattern with a dark band of desert that makes this map instantly recognizable as inspired by (though not precisely copied from) the *Guang yutu*. The grid pattern is more crudely

58 The Wanli-era work was the *Chongjiao Fan Wenzheng gong ji* (重校范文正公文集), according to Cao Wanru's *Atlas of Ancient Maps in China: From the Warring States Period to the Yuan Dynasty, tuban shuoming* p. 8.

59 A map of exactly the same title is recorded as early as Jiading 嘉定 2 (1209), though we do not know whether there is any connection to this particular version.

Figure 2.4 ***Yugong suo zai suishan junchuan zhi tu*** **(Map for 'determining the high mountains and the great rivers' from the *Yugong*)**

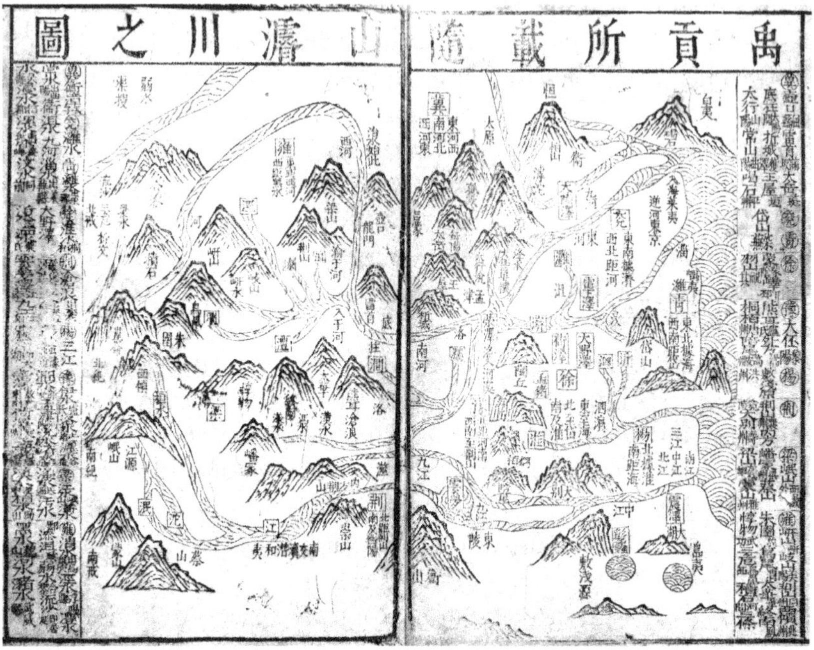

Tushu bian, juan 31, pp. 2b-3a
Courtesy of the Harvard-Yenching Library

applied than in the *Guang yutu,* and at a scale of one thousand *li* to a side for each square. Names of ancient states, former names of capitals, and modern names in rectangles are found along with the marchmounts and a few countries beyond the Ming frontier in this map, which makes no effort to draw administrative boundaries in the fashion of the *Zhizhang tu*. The borders between historical cartography and other categories of map are blurred, as all are presented in a single work that makes their contrasting qualities one of its themes.

Popular almanacs present a different configuration. The geographical section comprises the second chapter of most such works, following a chapter on constellations and heavenly omens (thus pairing heaven and earth). The format of these almanacs is remarkably similar, many of them sharing the same text, which is in turn copied from various earlier books. In all the cases I have examined, the geographical section is preceded by one simple map. In the textual accompaniment to this map, a horizontal line divides the text across the middle of the page; one narrative runs over the midline while another runs below it. One is a listing of historical capitals with their

Figure 2.5 Representation of the dynastic succession of China's capitals

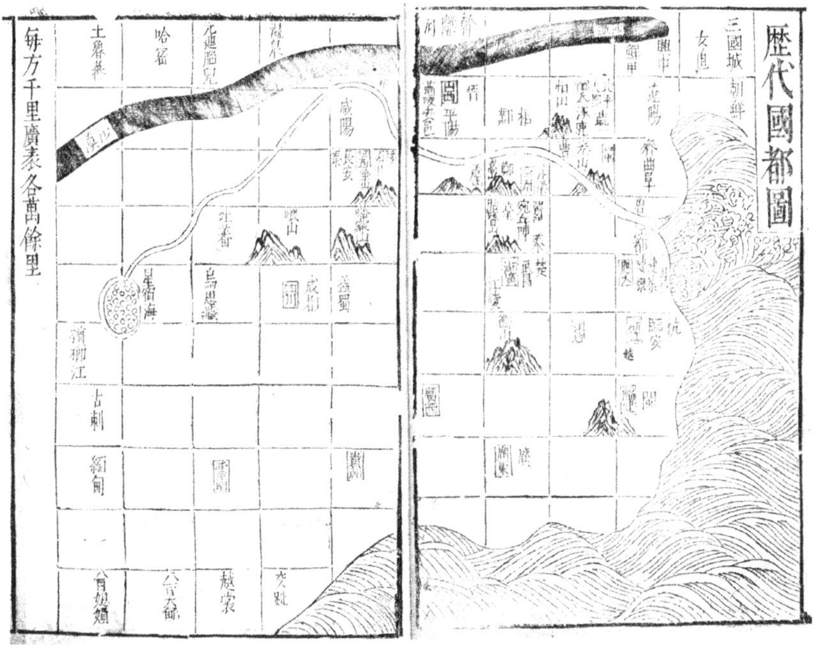

Tushu bian, juan 33: pp. 3b-4a
Courtesy of the Harvard-Yenching Library

modern equivalents, while the other is a simplified administrative history of the successive dynasties, in terms of their organization of territory.

To give an example, the *Wuche wanbao quanshu* (Five-cart compendium of a myriad treasures) with its preface dated Wanli 42 (1614) provides these two different types of information above and below the midline for the reign of Yu. Within the section titled *Lidai guodu* 歷代國都 (Capitals through the ages), the portion on Yu tells us that Yu's capital was at Anyi, now Yudu county in Puzhou county, thus equating the ancient site with its contemporary equivalent. In the parallel section running under the midline, we read that Yao instructed Yu to pacify the waters and the land, dividing all under heaven into nine regions and five zones.[60] Thence begins a simplified narrative of administrative history, which becomes more specific as we enter the dynastic period. Changes in both natural and administrative layout are provided in a cumulative series to inform the reader how the lay of the land in his own era was but the most recent stage in a series of transformations.

60 Xu Qilong 徐企龍, *Wuche wanbao quanshu* 五車萬寶全書, *juan* 2, pp. 1a-1b.

These two angles on historical geography – site-by-site tallying of the past with present geography, and tracing administrative change over time – derive from approaches we have already seen in works of superior scholarship. Almanacs brought this information to anyone who could afford to purchase, rent, or borrow one of these editions. The map appended to this part of the almanac generally lacks any explicitly historical markings, but its historical dimension emerges from the text.[61]

The historical map in gazetteers and other local works

The increased wealth that made the production of these miscellanies popular was part of a broader social change in which the local gazetteer – which by the late Song had already become much more than a mere administrative document – developed further into a community status symbol.[62] By the Ming dynasty the genre of *difang zhi* had become formalized, with a broad repertoire of predictable categories. Still, many individual works contained distinctive characteristics and material. Sometimes this unusual content became a local tradition, as can be seen in a 'lineage' of local gazetteers from Jiaxing Prefecture in northeastern Zhejiang. The gazetteers of this prefecture and the counties within it offer an example of a highly developed local tradition of historical cartography, one that brought the chronological snapshot format of the *Zhizhang tu* to the smaller units of imperial administration.

The beginning of any local gazetteer generally included a historical overview of the changes in administrative organization of that locale. As time passed and new editions of a local gazetteer were compiled, the maps from old editions were sometimes re-engraved and placed together with more recent images as artifacts of interest for nostalgic or comparative purposes. This retention of maps from earlier gazetteers as supplements to the updated version was a common way of showing historical awareness of change in a place.

The development of the maps in gazetteers from Jiaxing prefecture reveals a rare combination of these gazetteer genre traditions with the techniques of the national-level historical atlases discussed earlier. Southward population

61 In this and the other works reprinted in a 14-volume collection of facsimile editions of Chinese miscellanies held in Japanese libraries, *Chūgoku nichiyō ruisho shūsei* 中国日用類書集成 (Tōkyō: Kyōko Shoin, 1999-2001), all share a similar map and much of the same text. This is another reminder that commercial print culture not only has the power to foster comparison and contention, but also to preserve and multiply a standardized view.

62 Bol, 'The "Localist Turn" and "Local Identity" in Later Imperial China.'

shifts in the Song had renewed historical attention to the Wu and Yue states, which had been the subject of older unofficial histories such as the *Wu Yue chunqiu* 吴越春秋 (Spring and Autumns of Wu and Yue) and the *Yuejue shu* 越絶書 (The Glory of Yue, first known in print editions from the Southern Song),[63] as well as the compilation of new works of local history, such as the early Song *Wujun tujing* 吳郡圖經 (Map guide of the commandery of Wu) by Li Zong'e 李宗諤. These minor histories, in addition to classical texts, served as building blocks for the construction of the historical maps in the Jiaxing gazetteers.[64]

Let us examine how these historical depictions evolved over time. The Hongzhi 5 (1492) gazetteer of Jiaxing prefecture, compiled by Liu Yan 柳琰, contains only maps of the prefecture's county seats and their immediate surroundings.[65] The maps are not gathered together in one fascicle, as was becoming common in gazetteers; rather, they are scattered through the book, each map immediately preceding the portion of the text addressing the county depicted. Near the beginning of the gazetteer is a *yange* 沿革 section, providing somewhat detailed records of the various administrative changes known to have taken place in recorded history, from the days when the region played a role in the interaction of the states of Wu and Yue up to the foundation of the Ming, but this record is purely textual in nature.

In the next gazetteer for the prefecture, the 1549 *Jiaxing fu tuji* 嘉興府圖記 compiled by Zhao Ying 趙瀛 and Zhao Wenhua 趙文華, the *tuji* section – now gathered into a single fascicle inserted after the list of contents – has been revolutionized.[66] While individual maps of the subsidiary counties are again included, some of them loosely based on the maps from the earlier

63 The earliest claimed date for a printed edition of the *Wu Yue Chunqiu* is 1224 (Lagerwey, 'Wu Yüeh ch'un ch'iu'), while the *Yuejue shu* was printed in 1212 and 1220 (Schuessler and Loewe, 'Yüeh chüeh shu'). James M. Hargett notes that the latter, ostensibly compiled in 52 CE, has traditionally been identified as the progenitor of the *difangzhi* genre (Hargett, 'Song Dynasty Local Gazetteers and Their Place in The History of Difangzhi Writing,' p. 406).

64 Though in the vast majority of cases historical maps were drawn according to information recorded in textual form, there are a few cases in which textual descriptions are based on maps, as with a description in the *Tongzhi lue* based on the *Tang Kaiyuan shidao tu* (Zheng, *Tongzhi lue*, Chapter 16, *dili lue* 1, beginning p. 23b).

65 Liu Yan 柳琰, compiler, *Jiaxing fuzhi* 嘉興府志.

66 Zhao Ying 趙瀛 and Zhao Wenhua 趙文華, compilers, *Jiaxing fu tuji* 嘉興府圖記. Zhao Wenhua, a native of Ziqi, Zhejiang, memorialized the throne with proposals for coastal defense against the Wokou pirates in his capacity as Minister of Works, leading eventually his appointment as head of armed forces on the southeastern coast. The story of his subsequent fall from grace and the posthumous revelation of his stunning corruption is outlined in Kwan-wai So, 'Chao Wen-hua,' in *Dictionary of Ming Biography*, pp. 132-136. Zhao Wenhua is one of a substantial number of officials charged with addressing the *wokou* threat who published significant cartographic works.

gazetteer,[67] they are now preceded by a twenty-seven-page reconstruction of the natural and administrative changes of the region, shown over a sequence of temporal snapshots from the Warring States period to the Xuande reign of the Ming dynasty (1426-1435). These historical maps illustrate the following temporal views: The Wu-Yue boundary during the Warring States period; the establishment of the Qin; the Han; the Three Kingdoms period; the Jin; the Liang; the Sui; the Tang; the Five Dynasties period; the Northern Song; the Yuan; the Ming before Xuande; and the Ming after Xuande. The cartographic image of each stage is accompanied by a textual description of the changes that transpired during the historical period represented in the map. Very helpfully, the *Jiaxing fu tuji* enumerates the sources used in its compilation; I have presented the complete listing in Appendix 2.

It is possible that the *Zhizhang tu* was an inspiration for this series of maps. The only historical moments mapped in the *Jiaxing fu tuji* sequence that are not present in the *Zhizhang tu* are post-Song developments and a reconstruction of the border between Wu and Yue during the Warring States period, a matter of special local significance. No earlier precedents for locally specific historical cartography in Jiaxing seem to be extant, nor are any cited in the introduction to the historical series, and later gazetteers cite the 1549 edition as the work that began this local tradition. This could well have been a thoroughly original reinterpretation of a formula used in national-level works, now reapplied to a local space.

Half a century later, the Wanli 28 (1600) edition of the Jiaxing prefectural gazetteer reproduces the 1549 *Jiaxing fu tuji* map series in a slightly simpler form, without the commentary attached to each map that was a feature of the earlier work. In its preface, this edition notes briefly that the maps depict the situation at the beginning of each historic interval (as opposed to the middle or chaotic end).[68] The priority given to including such a long historical series in the prefectural gazetteer appears to have faded over time; when we reach Qian Yikai's 1721 edition, the historical section has been winnowed down to thirteen maps. By 1840, there are only two historical maps.[69] By 1878, the historical maps have been excised completely, though their previous existence is mentioned in the preface.[70]

67 The newly festering problem of piracy is also highlighted with a map and short essay discussing fortifications and waterways in the prefecture, which forms part of the coastline (*Jiaxing fu tuji, tuji* section, pp. 37a-38b).

68 Yao Zhong 堯中, *Jiaxing fuzhi* 嘉興府志, *fanli* p. 1.

69 Yu Shangling 于尚齡, *Jiaxing fuzhi* 嘉興府志, *tu* 1b-3a.

70 Hu Yaoguang 許瑤光, Wu Yangxian 吳仰賢, editors. *Jiaxing fuzhi* 嘉興府志, *juan* 1: 1a.

Figure 2.6 Two of the remaining local historical maps included in Qian Yikai's 1721 prefectural gazetteer of Jiaxing, depicting the general layout of the area during the Five Dynasties and the Ming

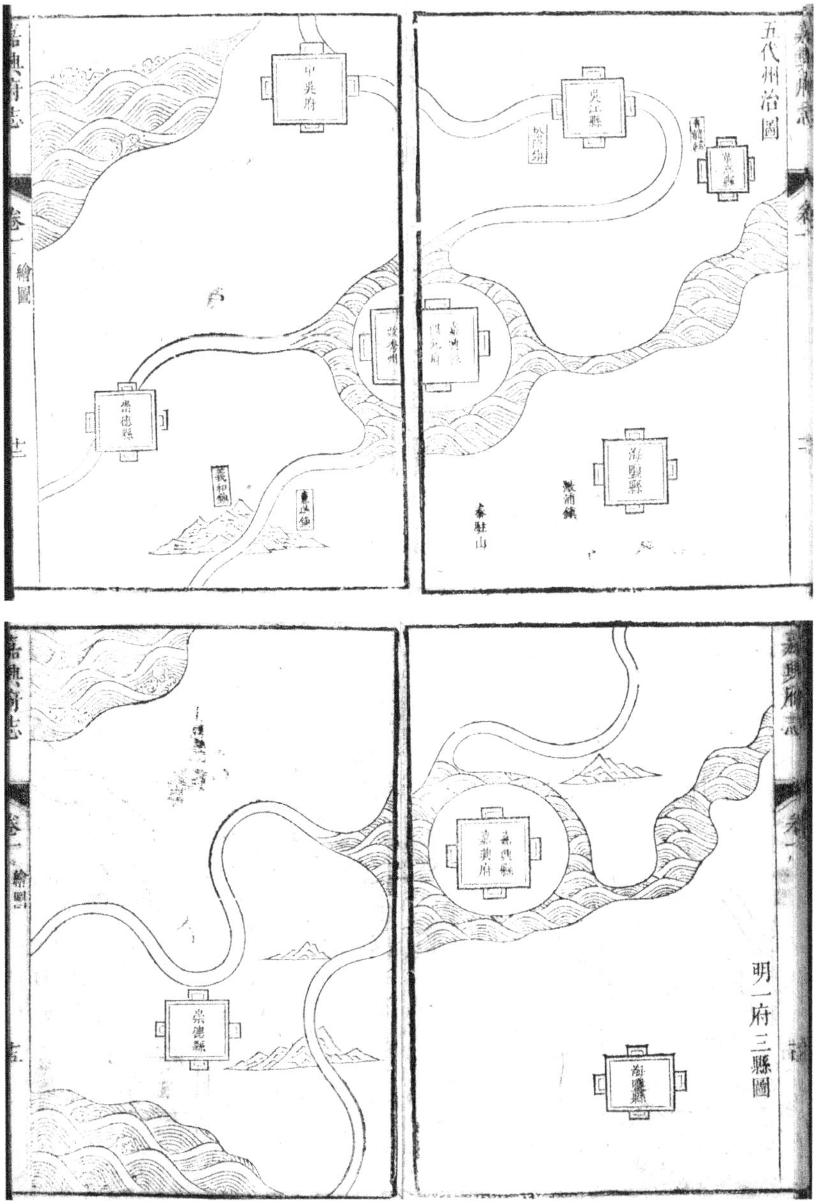

Jiaxing Fu zhi 嘉興府志, *juan* 1, 11-12 and 14-15
Courtesy of the Harvard-Yenching Library

At least three prefectural gazetteers had already been published in Jiaxing before the first county-level gazetteer for Jiaxing, the *Chongzhen Jiaxing xianzhi* 崇禎嘉興縣志, appeared in 1637.[71] This more local gazetteer adopts the tradition set at the prefectural level by including its own set of historical maps, this time focused even more closely on the county level. It does more than simply magnify the county's location as presented on the prefectural gazetteer maps; additional local features are added, and the accompanying text is quite different. In other words, this was not simple imitation; additional research went into the preparation of the county sequence.

In contrast to the Prefectural gazetteers, which dropped their historical maps, the practice of including a historical map series in the County level gazetteer continued even past the turn of the twentieth century; the last imperial edition from 1908 includes nine maps covering the period from the Qin to the Xuande reign of the Ming. The *tushuo* explains that this tradition comes from the *Jiaxing fu tuji*; its use continued here because it was simple and methodical.[72]

A survey of the gazetteers of neighboring counties within Jiaxing Prefecture suggests that the historical series was nowhere as influential as in the seat of Jiaxing itself. Pinghu County's much later 1790 gazetteer makes numerous textual references to the same set of works used in the compilation of the historical map series for Jiaxing, but its maps do not themselves show any influence.[73] Haiyan County's 1622 gazetteer includes some maps, but its historical sequence is shown in a chronological table rather than in map form.[74]

In the case of Jiaxing the vector of influence is from administratively higher to lower works, the Prefectural gazetteer serving as a model for the County level. Only a large-scale survey can reveal whether this was the usual path of transmission for cartographic innovations in local publications. For the lower to mimic the higher would be the hierarchical expectation, but because local gazetteers were sent up the chain of administration as fodder for compiling Prefectural and Provincial gazetteers, influence could potentially have moved in the other direction. Perhaps it will be discovered that it was not so unusual for a county work to influence the *tuji* portion of subsequent prefectural gazetteers.

71 Luo Kai 羅炌, *Chongzhen Jiaxing xianzhi* 崇禎嘉興縣志.

72 Zhao Weiyu 趙惟崙 ed., *Jiaxing xianzhi* 嘉興縣志, *juan* 1, *tushuo* 圖說.

73 Zhang Cheng 張誠 and Wang Heng 王恒, eds, *Pinghu xianzhi* 平湖縣志.

74 Hu Zhenheng 胡震亨, ed., *Haiyan xian tujing* 海鹽縣圖經, *juan* 1, 12a-13a.

In any case, gazetteers were not the only places where historical maps of the locale could be found. Reprinting in books rescued many cartographic images from oblivion, but incision of images on stone was another method by which maps were indelibly preserved. Denizens of cities with particular historical significance sometimes represented their homes retrospectively in such forms. A number of historical representations of individual cities and towns, and even specific building compounds, are known to us today only because they were recorded on stone. One such engraving from the Song, a map carved in 1080 to represent Chang'an as it had appeared in the Tang, was discovered in fragments during the Qing and reassembled.[75]

Nor does this exhaust the variety of forms of historical cartography. Records kept for the benefit of prominent lineages sometimes include maps on a smaller scale. Cao Wanru has published an example carved in stone by members of the Meng clan, one of the most prominent lineages of all; this series of engravings represents the spatial arrangement, changes in location, and expansion over the centuries of the tomb of Mencius.[76] Similar stele carvings of historic temples exist, including those of the complex devoted to Confucius.[77] The *jiapu* (genealogical records) of less prominent families sometimes contain painted or printed maps as well, showing either family compounds or grave sites.

Buddhism, in time and place

In the works discussed above, from the *Yugong* to the County gazetteer, the process of change through history is for the most part described within the limits of the Nine Regions, the lands under the Chinese cultural umbrella. Tang Xiaofeng has argued that Chinese geography did not make comparisons between Chinese territories and other lands, but rather, intensively focused on historical change within the same territory.[78] Although the historical records of foreign states in the dynastic histories complicate the picture slightly, it is fair to assert that distant lands were not subject to the same scrutiny as those areas that were believed to have been bequeathed by the Sage Kings to later dynasties in the Middle Kingdom. Nonetheless, this

75 Images of the stele and a brief introduction can be found in Yan Ping, *China in Ancient and Modern Maps*, pp. 44-45.

76 Cao, *Atlas of Ancient Maps in China* 中國古代地圖集*: The Ming Dynasty 1368-1644*, Plates 110-116.

77 Rubbings of these engravings have been published in *Kongzi shengji tu* 孔子聖蹟圖, pp. 6-10.

78 Tang, *From Dynastic Geography to Historical Geography*, p. 51.

tradition was not the sole perspective available in China; Buddhism offered incentives to look further to the west, for the Buddhist heartland lay beyond the frontiers of the lands described in the *Yugong*.

Chinese Buddhist cartography has received relatively little scholarly attention, a problem attributed by Cordell Yee to the lack of available primary sources but also identified by him as an arena deserving of further inquiry.[79] In contrast to works such as the twelfth-century *Yujitu*, widely reproduced with pride today to demonstrate astonishing technical precocity, few modern reproductions of Buddhist maps exist, perhaps because their adherence to the notion of the continent of Jambudvīpa now seems embarrassing. There also appear to have been far fewer Buddhist maps to begin with, as the bureaucratic/literati tradition created more demand for cartography than did the theological or educational needs of Buddhism. There also exists the possibility that anti-Buddhist campaigns in various periods destroyed much of what might have existed; some Tang Buddhist maps, for example, are known today only from Japanese copies.[80]

A common feature of religious historical cartography in Buddhist and Christian traditions is attention to places cited in formative texts. In this respect it can be compared to religious historical cartography in Europe. Jeremy Black notes that attempts to map out the lands mentioned in the Bible took two main forms: They might attempt to reconstruct the world as it appeared in a given period, or they might depict the world of the present, highlighting ancient sites. An example of the latter category is the Hereford world map, composed c. 1300, which highlights the route of the Exodus and other biblical events upon a map of the known world as it was understood in the late 1200s. Black adds that printing dramatically aided the proliferation of maps depicting religious subjects, which became a common feature of bibles during the Reformation.[81]

The content of maps showing biblical lands was of course based partly upon textual analysis, but it was strongly undergirded by more recent experience of the areas in question. Through both pilgrimage and crusade, the biblical sites of the Near East had long been the focus of European scrutiny. This is one major difference between such works and Buddhist maps from China. When Sui and Tang pilgrims to India visited historical

79 Yee, 'Concluding Remarks,' p. 230. Matthew Mosca devotes some attention to the role of Buddhism in cartography during the Qing period (Mosca, *From Frontier Policy to Foreign Policy: The Question of India and the Transformation of Geopolitics in Qing China*).

80 For a discussion of Japanese historical cartography, which cites the relevance of the Buddhist legacy there, see Wigen, 'Orienting the Past in Early Modern Japan.'

81 Black, *Maps and History*, p. 5.

sites relevant to the legacy of the Buddha and to the spread of his doctrine, the routes of their own voyages became part of the cartographic record. The golden age of pilgrimage to India was in the Tang dynasty, however, and such trips declined dramatically thereafter, partly due to the Muslim conquests of Central Asia and much of northern India. Although Buddhist temples continued to operate in some regions under Muslim rule, travelers from China could not have expected to receive the same welcome and aid from local government that men like Xuanzang received during their travels. As such pilgrimages dwindled in number, maps of the region became static representations of a bygone time. Buddhist maps in East Asia came to comprise two historical layers: the representation of lands from records of the Buddha's life, and records of the later visits of pilgrims to the west. They circulated in manuscript form or as woodblock prints without any adjustment to fit the contemporary situation in Southwest Asia.

The basic shape of the world in the Buddhist tradition was cylindrical, with the earth as we know it forming the top layer of one of the flat ends, circumscribed by a ring of iron mountains, within which lay the salty seas with interspersed continents. One of these continents, the southern landmass of Jambudvīpa, comprised the known world, with India in the south and China to the northeast. Cartographic representations of this worldview naturally differ substantially from the maps of the *Yugong* tradition of Chinese scholarship. Most dramatically, China was not the center of this cosmology, but rather situated at the edge of a much larger landmass, which was but one of several continents. Those who argue that the first world map circulated by Matteo Ricci delivered an unprecedented shock to Chinese literati by moving China away from the center have not examined the tradition of Buddhist cartography in China, in which this perspective was normal.[82]

One of the foundation documents for Buddhist geography in China is the Tang monk Daoxuan's 道宣 *Shijiafangzhi* 释迦方誌 (Sakyamuni gazetteer), a brief work of eight chapters completed around 650 CE, which wrestles with some issues of great cosmological significance to Chinese Buddhists. For example, is it incorrect to call the Great Tang the 'Middle Kingdom' when Buddhism arose elsewhere? Should not the term 'middle' (*zhong* 中) be reserved for India?[83] Daoxuan discusses five issues of contention –

82 Cordell Yee briefly touches upon the Buddhist representation of China as peripheral in 'Traditional Chinese Cartography and the Myth of Westernization,' pp. 173-174.
83 This is the subject of the third section in the *Shijiafangzhi*, entitled 'Zhong bian pian' (中邊篇 Essay on centrality and periphery) (Daoxuan 道宣, *Shijiafangzhi* 释迦方誌, [1983], pp. 7-14).

toponyms that imply centrality, geographical distances, the seasons, waters (lakes and rivers), and people – coming to the conclusion that India must be central. By listing three westward overland routes from the Tang to the Buddhist heartland, this work also served as a building block for later cartographic renderings of pilgrimage routes. While two of these routes closely follow those recorded by Xuanzang, a route passing through Tibet and Nepal is new to this work.[84]

Tang Buddhist maps survive only in later copies. The 1364 *Gotenjiku zu* (Map of the Five Indies) by the priest Jūkai, held in Nara, Japan, depicts numerous places listed in the monk Xuanzang's account of his travels, using a thin red line to trace his route.[85] A similar map from 1736, held at the Hōshōin Temple in Tokyo, is accompanied by a manuscript claiming that it is based on Xuanzang's original copy upon which he himself traced his route in red; the priest Kūkai is alleged to have brought a copy to Japan from China in 806.[86] It is not clear whether the later copyist believed the map to represent an India that still existed in his own time.

The Buddhist Tripitika contains an early encyclopedic work that includes both geographical and cartographic renderings of the world, the *Fozu tongji* 佛祖統紀, collated by Zhipan 志磐, dated Jingding 4 of the Southern Song (1263).[87] The historical aspect here is clearer in the text than in the maps. Chapter 32 begins by enumerating the nine regions of Emperor Ku, continuing through the standard dynastic chronology up to the Song. This portion also gives the maximum extent of Han territory as 9,000 *li* east-west by 13,000 *li* north-south. In a manner similar to the introductory pages in the *Zhizhang tu* – in fact, using words copied from that work, with passages lifted almost fully intact[88] – this is followed by brief vignettes on the Great Wall, rivers, the marchmounts, and then a brief overview of foreign countries represented on the map of China. Going around the perimeter of the map in counterclockwise fashion, this annotation begins with an abbreviated history of the states on the Korean peninsula as well as Japan

84 Ibid., Introduction by Fan Xiangyong 范祥雍, p. 2.

85 Muroga and Unno, 'The Buddhist World Map in Japan and its Contact with European Maps,' p. 49.

86 Ibid., p. 50. See also Unno, 'Cartography in Japan,' pp. 371-376. For citation of similar maps and their role in Korea see Ledyard, 'Cartography in Korea,' pp. 254-256.

87 A facsimile edition of a Ming reprint is available in the *Xuxiu Siku quanshu* 續修四庫全書, Zi bu 子部, Zong jiao lei 宗教類, vol. 1287. Because the original text lacks page numbers I use the page numbers from this edition.

88 See for example the description of the Great Wall in the *Zhizhang tu*, p. 4a, as compared with the *Fozu tongji*, p. 421. The later *Fajie anli tu*, discussed below, shows the same pattern (*Fajie anli tu, juan shang zhi shang* 卷上之上, p. 3a).

and Fusang. Instead of the concern with tributary missions that is accented in the administrative tradition, the coming of men from these places to study Buddhism is carefully noted.[89] Each directional quadrant gets its own mini-history, as we move to the north and read of the succession from Liao to Jin to Mongols on the northern frontier. In an island at the bottom of the map can be found the Arabs (*Dashi* 大食), briefly mentioned in the text alongside Byzantium (*Daqin* 大秦). Discussion of the lands of Buddhist concern comes later.

A map depicting, in general terms, what has just been described follows this text; it is titled *Dong Chendan dili tu* 東震旦地理圖 (Map of Eastern Cīnaṣṭhāna). The use of the name Eastern Cīnaṣṭhāna[90] for China – instead of a dynastic name or the term Zhongguo – harkens back to the Tang discussion of centrality mentioned above. The use of a transliterated Indian name, and its identification as an eastern land, dramatically decentralizes the Middle Kingdom.

Immediately after the map of Eastern Cīnaṣṭhāna is an illustration depicting the western thrust of the Han into central Asia, going as far afield as the Yuezhi confederacy and Parthia. A third map describes three routes from Dunhuang westward based on the *Xiyu tuji* 西域圖記 (Illustrated record of the Western Regions) compiled by Peiju 裴矩 in the Sui. The routes lead to the 'Western Sea' via northern, central, and southern passages. This map is accompanied by brief notes on major states in the west. Confucians are chastised here for their narrow-minded assumption of centrality, which blinds them to the majesty and size of India.[91] The India described here is temporally frozen in the Buddhist past. We find an enumeration of the small states in the Western Regions and India, listed in a matrix of relative position and distance, but the historical sources cited include nothing later than the Tang.

What lessons are we intended to draw from the changes that have transpired over the course of history? In other words, what is the point of recording this history? The *Fozu tongji* comes to a rather different conclusion than any of the works we have seen so far. It notes that the many changes in China since the Zhou simply demonstrate the truth of the Buddhist understanding of the impermanence of human states and constructions.[92]

89 Zhipan 志磐, *Fozu tongji* 佛祖統紀, p. 422.

90 On the derivation of 東震旦 see Soothill and Hodous, *A Dictionary of Chinese Buddhist Terms*, p. 445.

91 Zhipan, *Fozu tongji*, p. 425.

92 Ibid., p. 429.

Figure 2.7 The first map in the *Fajie anli tu* (Illustration of the Establishment of the Dharma Realm [i.e., Structure of the Universe])

Fajie anli tu, juan 上之上 2b
Courtesy of The Library of the University of California, Berkeley

Figure 2.8 *Fajie anli tu* map of the Jambudvīpa continent

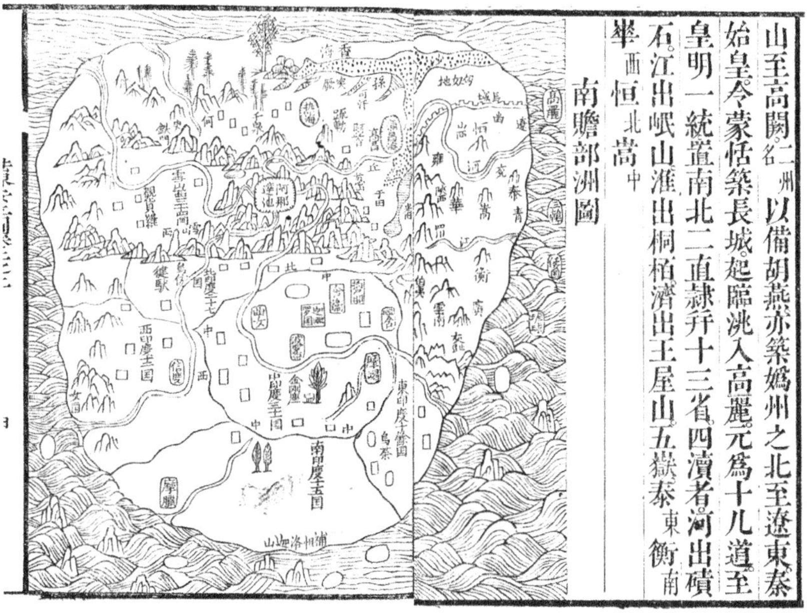

Fajie anli tu, juan 上之上 3b and 4a
Courtesy of The Library of the University of California, Berkeley

A Buddhist cosmological work of the late Ming, the *Fajie anli tu* 法界安立圖 (Illustration of the establishment of the Dharma Realm [i.e., Structure of the Universe]) compiled by Ren Chao 仁潮,[93] represents a continuation of the patterns already apparent in the *Fozu tongji*, again offering a worldview in which China is not central. The first map (shown in Figure 2:7 in an example believed to have been re-engraved in the Qing) again calls China 'Eastern Cīnaṣṭhāna,'[94] and the accompanying geographical/historical text extends from the Sage Kings to the Ming in just fifteen lines.

In contrast, the commentary accompanying the map of the Jambudvīpa continent (Figure 2:8) lists numerous minor states in much the same way

93 Ren Chao 仁潮, *Fajie anli tu* 法界安立圖. This is the Jen Ch'ao of whom Needham writes, 'We cannot identify this man or his work as Herrmann omitted the characters of his name and gave no further details' (Needham, 'Geography and Cartography,' p. 565). The following citations are to the facsimile edition published by the Xin Wenfeng chuban gufen youxian gongsi 新文豐出版股份有限公司, which is not the same edition as that illustrated by Yee in 'Traditional Chinese Cartography and the Myth of Westernization,' p. 174 (note the different arrangement of islands in Yee's copy).

94 Ren Chao, *Fajie anli tu, juan* shang zhi shang 上之上, p. 2b.

Figure 2.9 Jambudvīpa as the southernmost of four continents

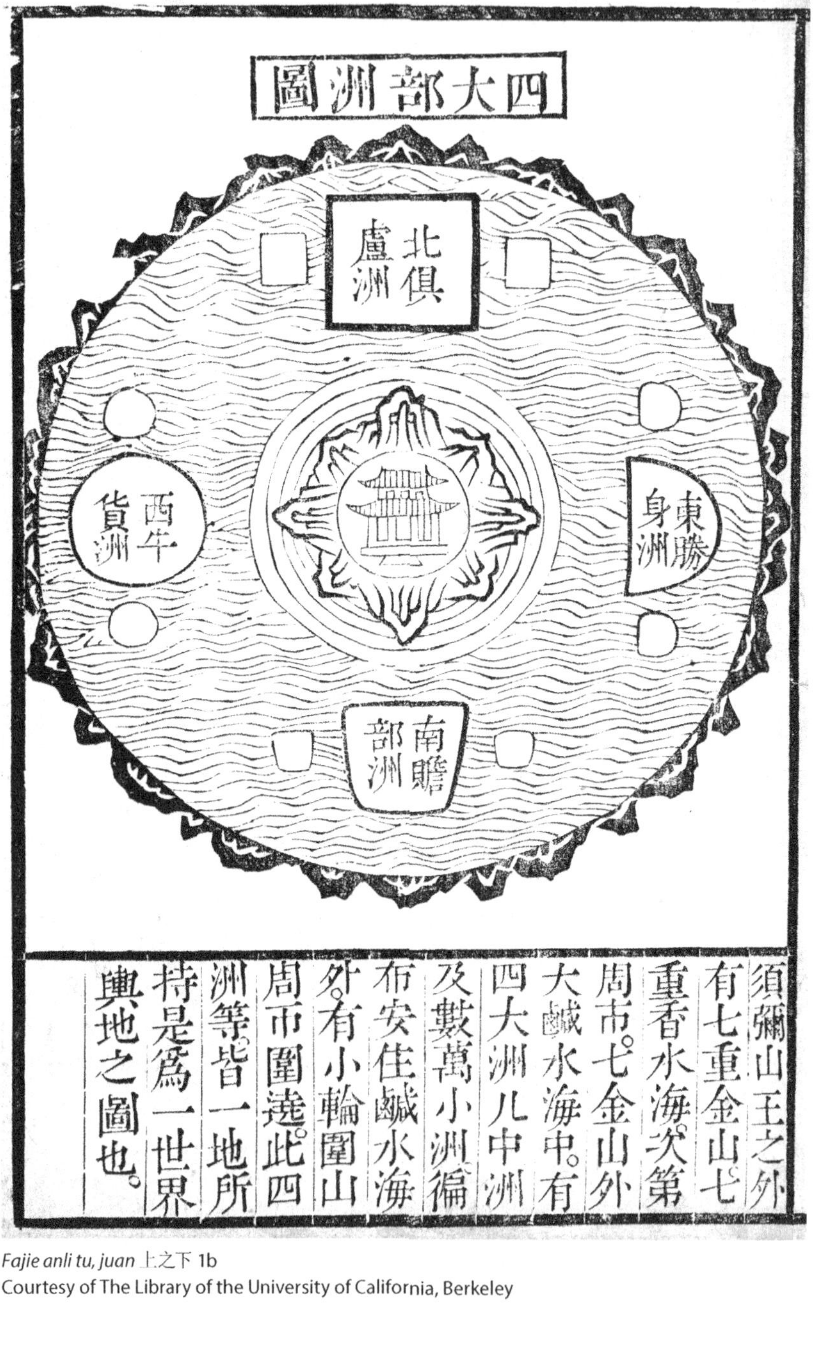

Fajie anli tu, *juan* 上之下 1b
Courtesy of The Library of the University of California, Berkeley

as the *Fozu tongji*. Here the historical component, where it exists, is limited to the relevance of these places as sites of incidents in Buddhist lore. No domestic Indian history of the post-classical Buddhist era appears here, though there are occasional references to later Chinese historical works that mention these places. This map is a reminder that the Jesuits were not the first to 'decenter' China in cartography; Buddhist maps already depicted it as peripheral.

Later illustrations in the *Fajie anli tu* drive home the point that the place of China in the universe is almost infinitesimal. It is a mere corner of Jambudvīpa, which in turn is merely the southernmost of four enormous continents on the face of the earth (Figure 2:9), which in turn comprises merely one of countless worlds (Figure 2:10).

The maps in the *Fajie anli tu* are striking for their anachronisms. The depiction of Eastern Cīnaṣṭhāna shows the long-extinct states of Paekche, Silla and Koryŏ on the Korean peninsula, while the Chosŏn state that actually ruled Korea at the time of publication is named on a cartouche in the sea. The Khitans are still indicated as hovering in the northeast.[95] Although the *Fajie anli tu* is prefaced by a direct rejection of the worldview of the Jesuits, their lands of origin are nowhere to be found on the map.

Ren Chao suggests that Buddhist knowledge is not something these overseas visitors of today, with their art of sailing the seas and their flashy but inconsequential scholarship, can match.[96] This is not simply to say that religious faith is more important than investigation; the terms of investigation themselves as understood by Confucians or Jesuits are rejected. History is understood in terms of grand cycles, each *kalpa* longer than the mind can fathom, just as the universe is shown to be an unbounded matrix of worlds within worlds.

Such Buddhist worldviews were rejected by Zhang Huang, who discusses them in the *Tushu bian*. Having already presented the world in the administrative cartographic tradition and in two projections copied from European maps, Zhang reproduces a map entitled *Sihai huayi zongtu* (Comprehensive map of the Chinese and Barbarians of the Four Seas), identifying it as a map of Jambudvīpa taken from a Buddhist work.[97] Zhang includes this

95 Ibid., *juan* shang zhi shang 上之上, p. 2b. Note also that, in an example of simple carelessness, the label Liaodong 遼東 is applied to the region *west* of the Liao River.

96 Ibid., *juan* shou 卷首, p. 1a. I have paraphrased *huolun dianguang* 火輪電光 as 'flashy,' but the term *huolun* (alātacakra) suggests something that appears real but is in fact unreal, while *dianguang* implies not just the flash of lightning but its transience (Soothill, *Dictionary*, pp. 445, 418).

97 Zhang Huang 章潢, *Tushu bian* 圖書編, *juan* 19, pp. 39a-40b, with explanatory text about Jambudvīpa and the Buddhist worldview continuing to p. 42a.

Figure 2.10 Earth as one of countless worlds

*Fajie anli tu, juan*中之下 1b
Courtesy of The Library of the University of California, Berkeley

map for the reader's comparative reference, but his intellectual sympathies lie elsewhere and he dismisses Buddhist cartography as unconvincing.[98]

Just as medieval European thinkers interpreted biblical references to weave together a retrospective depiction of the ancient territories of the Near East, East Asian Buddhism also gave rise to a religious cosmography endowed with an historical sensibility. Moreover, this tradition was long-lived. From what may have been a position of relative dominance in the Tang period, Buddhist worldviews accentuating the centrality of Mt. Sumeru and the peripheral place of the Chinese cultural zone endured both Confucian criticism and later Jesuit challenges, fading for good only in the Qing.

The historical perspective afforded by these maps differs dramatically from their European cousins, for in Chinese Buddhism there were no crusades to take men into the ancient heartland of their religion and bring into sharp focus the changes that had taken place in the intervening centuries. There were no post-Tang large-scale pilgrimage networks (as vividly described for Europe in Prescott's *Friar Felix at Large*)[99] to help keep the past tallied with a changing present. Buddhist cartography was an ossified image of what a nonbeliever might call a bygone moment.[100] Nonetheless, Buddhist historical maps challenged Sinocentric perspectives in some ways that appear prescient.

Conclusion

From the macrocosm of the state to the individual temple plot, historical cartography placed the present in the context of the past, highlighting the glories or failures that had gone before. It provided impetus for future action,

98 Zhang Huang, *Tushu bian, juan* 19, p. 42a. See also Nakamura Hirosi, 'Old Chinese World Maps Preserved by the Koreans,' pp. 14-15, and Needham, 'Geography and Cartography,' p. 566. As Nakamura Hiroshi has discussed, most of the toponyms on this map for places outside of China derive from the *Xiyou ji* of Xuanzang or the records of other pilgrims, with some mythical islands included for good measure (Nakamura, 'Old Chinese World Maps Preserved by the Koreans,' pp. 13-16).

99 Prescott, *Friar Felix at Large: A Fifteenth-Century Pilgrimage to the Holy Land.* Felix makes numerous scattered references to the use of pilgrim itineraries and competing guidebooks for Jerusalem.

100 It is difficult to gauge what routes there might have been for new knowledge about India, given the relative lack of such updated information in Chinese Buddhist works. There appears to have been interest in such matters, however, as when Oda Nobunaga took the opportunity of a 1569 meeting with the Jesuit Luis Frois to ask what he knew about Indian geography. (See Ayusawa Shintarō, 'Geography and Japanese Knowledge of World Geography,' p. 276.) His primary concern may have been European expansion in India rather than the traces of Buddhism, however.

or served as a tool in debates whose implications might range far beyond the immediate cartographic issue at hand. The *Yugong* inspired scholars to think about the confluence of past and present in spatial terms, and the *Zhizhang tu* established a durable precedent for mapping administrative change.

Some works from the late Ming aplied historical assessments to the military concerns of the period. Works of local history could even bring the perspective of dynastic administrative change down to the level of the individual county. In the gazetteers of Jiaxing, these bird's-eye views represent another iteration of the locale as microcosm of the whole. The application of historicity to Buddhist cartography, meanwhile, represents yet another divergent perspective.

While these trends had origins that can be traced back to the Song, all of them found the opportunity to proliferate in the publishing boom at the end of the Ming. In the next chapter we will examine one of the most dramatic ingredients of this late Ming mix that was truly new to the scene, the alien cartography of the European Jesuit missionaries.

Works cited

Primary sources

Ai Nanying 艾南英. *Yugong tuzhu* 禹貢圖註, Daoguang 11. Facsimile edition in *Lidai Yugong wenxian jicheng* 歷代禹貢文獻集成 (7 volumes), vol. 3: 1017-1062. Xi'an: Xi'an ditu chubanshe, 2006.

Chen Yi 陳沂. *Jinling gujin tukao* 金陵古今圖考. S.n., Zhengde 10 [1515].

Chen Zilong 陳子龍. Preface to Wu Guofu 吳國輔, *Gujin yuditu* 古今輿地圖, Chongzhen 11 (1638). Facsimile edition with title as *Jingu yuditu* 今古輿地圖. In *Siku jinhui shu congkan bubian* 四庫禁燬書叢刋補編, Shi bu 史部 (311 volumes), vol. 26. Beijing: Beijing chubanshe, 2005.

Cheng Dachang 程大昌. *Yugong lun, shanchuan dilitu* 禹貢論山川地理圖. Facsimile of Song edition. Beijing: Zhonghua shuju 中華書局: 1985.

Daoxuan 道宣, with introduction by Fan Xiangyong 范祥雍. *Shijia fangzhi* 釋迦方誌. Beijing: Zhonghua shuju 中華書局, 1983.

Hu Yaoguang 許瑤光, and Wu Yangxian 吳仰賢, eds. *Jiaxing fuzhi* 嘉興府志. Jiaxing: Guangxu 4 [1878].

Hu Zhenheng 胡震亨, ed. *Haiyan xian tujing* 海鹽縣圖經. Haiyan: Tianqi 天啓 2 [1622].

Li Xian. *Da Ming yitong zhi* 大明一統志. Facsimile edition. Xi'an: San Qin chubanshe 三秦出版社, 1990.

Liu Yan 柳琰, compiler. *Jiaxing fuzhi* 嘉興府志, Hongzhi 5 (1493). Facsimile reprint in *Siku quanshu cunmu congshu* 四庫全書存目叢書, Shi bu 史部 (292 volumes), vol. 179: 1-426. Jinan: Qi Lu shushe chubanshe, 1997.

Luo Hongxian 羅洪先. *Guang yutu* 廣輿圖. Facsimile reprint of 1799 edition. Taibei: Xuehai chubanshe 學海出版社, 1969.

Luo Kai 羅炌. *Chongzhen Jiaxing xianzhi* 崇禎嘉興縣志. Facsimile of 1637 edition. Beijing: Shumu wenxian chubanshe 書目文獻出版社, 1991.

Mao Ruizheng 茅瑞徵. *Yugong huishu* 禹貢匯疏. Facsimile reprint in *Lidai Yugong wenxian jicheng* 歷代禹貢文獻集成 (7 volumes), vol. 2: 465-762. Xi'an: Xi'an ditu chubanshe, 2006.

Qian Yikai 錢以塏. *Jiaxing fuzhi* 嘉興府志. Kangxi 60 [1721].

Ren Chao 仁潮. *Fajie anli tu* 法界安立圖. Facsimile edition of undated Qing copy. Taibei: Xin Wenfeng chuban gufen youxian gongsi 新文豐出版股份有限公司, 1977.

Shui Anli 稅安禮. *Lidai dili zhizhang tu* 歷代地理指掌圖. Facsimile edition of Song edition held at the Tōyō Bunko, published as *Songben lidai dili Zhizhang tu* 宋本歷代地理指掌圖 with prefaces by Tan Qixiang and Cao Wanru. Shanghai: Shanghai guji chubanshe 上海古籍出版社, 1989.

Wang Guanglu 王光魯. *Yueshi yueshu* 閱史約書. Facsimile reprint of 1643 edition in *Siku quanshu cunmu congshu* 四庫全書存目叢書, Shi bu 史部 (292 volumes), vol. 32: 741-898. Jinan: Qi Lu shushe chubanshe, 1997.

Xia Yunyi 夏允彝, *Yugong gujin hezhu* 禹貢古今合注. Facsimile reprint in *Lidai Yugong wenxian jicheng* 歷代禹貢文獻集成 (7 volumes), vol. 2: 763-922. Xi'an: Xi'an ditu chubanshe, 2006.

Xu Qilong 徐企龍. *Wuche wanbao quanshu* 五車萬寶全書. Chūgoku nichiyō ruisho shūsei 中国日用類書集成, Vols 8-9. Tōkyō: Kyūko Shoin, Heisei 13 [2001].

Yao Zhong 堯中, ed. *Jiaxing fuzhi* 嘉興府志. Ming Wanli 萬曆 gengzi 庚子 (year 28, 1600).

Yu Shangling 于尚齡. *Jiaxing fuzhi* 嘉興府志, Daoguang year 20 [1840].

Zhang Cheng 張誠, and Wang Heng 王恒, eds. *Pinghu xianzhi* 平湖縣志. Qianlong 55 [1790].

Zhang Huang 章潢. *Tushu bian* 圖書編. Facsimile of Wanli 41 [1613] edition. Taibei: Chengwen chubanshe 成文出版社, 1971.

Zhao Weiyu 趙惟崳, ed. *Jiaxing xianzhi* 嘉興縣志. Guangxu 34 (1908). Facsimile edition in *Zhongguo difangzhi jicheng: Zhejiang fuxianzhi ji* 中國地方志集成浙江府縣志輯, vol. 15. Shanghai: Shanghai shudian, 1993.

Zhao Ying 趙瀛 and Zhao Wenhua 趙文華, eds. *Jiaxing fu tuji* 嘉興府圖記, Jiajing 28 (1549). Facsimile reprint in Siku quanshu cunmu congshu 四庫全書存目叢書, Shi bu 史部 (292 volumes), vol. 191: 302-540. Jinan: Qi Lu shushe chubanshe, 1997.

Zheng Qiao 鄭樵. *Tongzhi* 通志. Facsimile edition. Taibei: Taiwan shangwu yinshuguan 臺灣商務印書館, 1983.

—. *Tongzhi lue* 通志略. Facsimile edition. Shanghai: Zhonghua shuju 中華書局, [1936?].

Zhipan 志磐. *Fozu tongji* 佛祖統紀. Facsimile edition of a Ming reprint, in *Xuxiu Siku quanshu* 續修四庫全書, Zi bu 子部, Zong jiao lei 宗教類, vol. 1287. Shanghai: Shanghai guji chubanshe, [between 1995 and 1999].

Secondary sources

Akin, Alexander. 'The *Jing Xing Ji* of Du Huan: Notes on the West by a Chinese Prisoner of War.' *Harvard Middle Eastern and Islamic Review* 5 (1999-2000): 77-102.

—, with David Mumford. 'Yu Laid out the Lands: Georeferencing the Chinese *Yujitu* [Map of the Tracks of Yu] of 1136.' *Cartography and Geographic Information Science* 39, no. 3 (March 2013): 154-169.

Anderson, Benedict. *Imagined Communities*. Revised edition. New York: Verso, 1991.

Ayusawa Shintarō. 'Geography and Japanese Knowledge of World Geography.' *Monumenta Nipponica* 19, no. 3/4 (1964): 275-294.

Black, Jeremy. *Maps and History: Constructing Images of the Past.* New Haven, CT: Yale University Press, 1997.

Bol, Peter. 'The "Localist Turn" and "Local Identity" in Later Imperial China.' *Late Imperial China* 24, no. 2 (December 2003): 1-50.

Cao Wanru, ed. *Atlas of Ancient Maps in China* 中國古代地圖集*: From the Warring States Period to the Yuan Dynasty.* Beijing: Wenwu chubanshe 文物出版社 1990.

—. *Atlas of Ancient Maps in China* 中國古代地圖集*: The Ming Dynasty 1368-1644*. Beijing: Wenwu chubanshe 文物出版社, 1994.

Chia, Lucille. *Printing for Profit: The Commercial Publishers of Jianyang, Fujian (11th-17th Centuries)*. Cambridge, MA: Harvard University Press, 2002.

Chow, Kai-Wing. *Publishing, Culture, and Power in Early Modern China*. Stanford, CA: Stanford University Press, 2004.

Dardess, John W. 'Did the Mongols Matter? Territory, Power and the Intelligentsia in China from the Northern Song to the Early Ming.' In *The Song-Yuan-Ming Transition in Chinese History*, edited by Paul Jakov Smith and Richard von Glahn, 111-134. Cambridge, MA: Harvard University Asia Center, 2003.

De Weerdt, Hilde. 'The Cultural Logics of Map Reading: Text, Time and Space in Printed Maps of the Song Empire,' In *Knowledge and Text Production in an Age of Print: China, 900-1400*, edited by Lucille Chia and Hilde De Weerdt, 239-270. Leiden: Brill, 2011.

—. "Maps and Memory: Readings of Cartography in Twelfth- and Thirteenth-Century Song China." *Imago Mundi: International Journal for the History of Cartography* 61, no. 2 (2009): 145-167.

Fracasso, Riccardo. 'Shan hai ching.' In *Early Chinese Texts: A Bibliographical Guide*, edited by Michael Loewe, 357-367. Berkeley: Society for the Study of Early China and Institute of East Asian Studies, University of California, 1993.

Goodall, John A. *Heaven and Earth: 120 Album Leaves From a Ming Encyclopedia, San-Ts'ai T'u-hui, 1610*. London: Lund Humphries, 1979.

Hargett, James M. 'Song Dynasty Local Gazetteers and their Place in the History of *Difangzhi* Writing.' *Harvard Journal of Asiatic Studies* 56, no. 2 (1996): 405-442.

Hegel, Robert E. *Reading Illustrated Fiction in Late Imperial China.* Stanford, CA: Stanford University Press, 1998.

Henderson, John B. 'Chinese Cosmographical Thought: The High Intellectual Tradition.' In *The History of Cartography, Volume 2, Book 2: Cartography in the Traditional East and Southeast Asian Societies,* edited by J.B. Harley and David Woodward, 203-227. Chicago and London: University of Chicago Press, 1994.

Hirth, Friedrich and W.W. Rockhill. *Chau Ju-kua: His Work on the Chinese and Arab Trade in the Twelfth and Thirteenth Centuries, Entitled Chu-fan-chï.* St. Petersburg: Printing Office of the Imperial Academy of Sciences, 1911. Reprint, Taibei: Literature House, 1965.

Karlgren, Bernhard. *The Book of Documents.* Reprinted from the Museum of Far Eastern Antiquities, Bulletin 22. Stockholm, 1950.

Kassis, Vadim. 'Peking "Cartographers" As They Really Are.' *Socialism: Theory and Practice* (Supplement), 4 (1981): 51-54.

Kongzi shengji tu 孔子聖蹟圖. Beijing: Zhongguo shudian, 1998.

Lagerwey, John. 'Wu Yüeh ch'un ch'iu.' In *Early Chinese Texts: A Bibliographical Guide*, edited by Michael Loewe, 473-476. Berkeley: Society for the Study of Early China and Institute of East Asian Studies, University of California, 1993.

Ledyard, Gari. 'Cartography in Korea.' In *The History of Cartography, Volume 2, Book 2: Cartography in the Traditional East and Southeast Asian Societies,* edited by J.B. Harley and David Woodward, 235-345. Chicago and London: University of Chicago Press, 1994.

Mosca, Matthew William. *From Frontier Policy to Foreign Policy: The Question of India and the Transformation of Geopolitics in Qing China.* Stanford, CA: Stanford University Press, 2013.

Muroga, Nobuo and Unno Kazutaka, 'The Buddhist World Map in Japan and Its Contact with European Maps.' *Imago Mundi* 16 (1962): 49-69.

Nakamura Hiroshi, 'Old Chinese World Maps Preserved by the Koreans.' *Imago Mundi* 4 (1947): 3-22.

Needham, Joseph. 'Geography and Cartography.' In *Science and Civilisation in China, Volume 3: Mathematics and the Sciences of the Heavens and the Earth,* 497-590. Cambridge: Cambridge University Press, 1959.

—. 'Geology (and Related Sciences).' In *Science and Civilisation in China, Volume 3: Mathematics and the Sciences of the Heavens and the Earth*, 591-623. Cambridge: Cambridge University Press, 1959.

Prescott, Hilda F.M. *Friar Felix at Large: A Fifteenth-Century Pilgrimage to the Holy Land*. Westport, CT: Greenwood Press, 1967.

Schuessler, Axel, and Michael Loewe, 'Yüeh chüeh shu.' In *Early Chinese Texts: A Bibliographical Guide*, edited by Michael Loewe, 473-476. Berkeley: Society for the Study of Early China and Institute of East Asian Studies, University of California, 1993.

Shaughnessy, Edward. 'Shang Shu,' in *Early Chinese Texts: A Bibliographical Guide*, edited by Michael Loewe, 376-389. Berkeley: Society for the Study of Early China and Institute of East Asian Studies, University of California, 1993.

So, Kwan-wai. 'Chao Wen-hua.' In *Dictionary of Ming Biography,* edited by L. Carrington Goodrich and Chaoying Fang, 132-136. New York: Columbia University Press, 1976.

—. *Japanese Piracy in Ming China During the 16th Century*. East Lansing: Michigan State University Press, 1975.

Soothill, William Edward and Lewis Hodous. *A Dictionary of Chinese Buddhist Terms: with Sanskrit and English Equivalents and a Sanskrit-Pali Index*. London: K. Paul, Trench, Trübner & Co., 1937.

Tang Xiaofeng. *From Dynastic Geography to Historical Geography: A Change in Perspective Towards the Geographical Past of China*. Beijing: Commercial Press International, 2000.

Teng, S.Y. 'Cheng Chiao.' In *Sung Biographies,* edited by Herbert Franke, Vol. 1: 146-156. Wiesbaden: Steiner, 1976.

Thongchai Winichakul. *Siam Mapped: A History of the Geo-Body of a Nation*. Honolulu: University of Hawaii Press, 1994.

Unno Kazutaka. 'Cartography in Japan.' In *The History of Cartography, Volume 2, Book 2: Cartography in the Traditional East and Southeast Asian Societies,* edited by J.B. Harley and David Woodward, 346-477. Chicago and London: University of Chicago Press, 1994.

Wang Yong 王庸. *Zhongguo dilixue shi* 中國地理學史. Changsha: Shangwu yinshuguan 商務印書館, [1938].

Whitfield, Peter. *New Found Lands: Maps in the History of Exploration*. London: The British Library, 1998.

Wigen, Kären. 'Orienting the Past in Early Modern Japan.' In *Time in Maps: From the Age of Discovery to Our Digital Era*, edited by Kären Wigen and Caroline Winterer, 37-62. Chicago: University of Chicago Press, 2020.

Woodward, David. 'Medieval Mappamundi.' In *The History of Cartography, Volume 1: Cartography in Prehistoric, Ancient, and Medieval Europe and the Mediterranean,* edited by J.B. Harley and David Woodward, 286-368. Chicago: University of Chicago Press, 1987.

Yan Ping. *China in Ancient and Modern Maps*. London: Sotheby's, 1998.

Yang Ching-yi and L. Carrington Goodrich. 'Mao Jui-cheng.' In *Dictionary of Ming Biography,* edited by L. Carrington Goodrich and Chaoying Fang, 1041-1042. New York: Columbia University Press, 1976.

Yee, Cordell. 'Chinese Maps in Political Culture.' In *The History of Cartography, Volume 2, Book 2: Cartography in the Traditional East and Southeast Asian Societies,* edited by J.B. Harley and David Woodward, 71-95. Chicago and London: University of Chicago Press, 1994.

—. 'Concluding remarks.' In *The History of Cartography, Volume 2, Book 2: Cartography in the Traditional East and Southeast Asian Societies,* edited by J.B. Harley and David Woodward, 228-231. Chicago and London: University of Chicago Press, 1994.

—. 'Reinterpreting Traditional Chinese Geographical Maps.' In *The History of Cartography, Volume 2, Book 2: Cartography in the Traditional East and Southeast Asian Societies,* edited by J.B. Harley and David Woodward, 35-70. Chicago and London: University of Chicago Press, 1994.

—. 'Traditional Chinese cartography and the myth of Westernization.' In *The History of Cartography, Volume 2, Book 2: Cartography in the Traditional East and Southeast Asian Societies,* edited by J.B. Harley and David Woodward, 170-202. Chicago and London: University of Chicago Press, 1994.

3 The Jesuits as Participants in the Late Ming Publishing Boom

Abstract

This chapter takes a Ming-centric approach to the Jesuit cartographic project under Matteo Ricci and his immediate successors, discussing reactions to the missionaries' maps and their citation in works published by Ming scholars. Of all the empires in which the Jesuits set foot, the Ming was the first in which they encountered an already highly developed cartographic culture, leading to an unusually prominent role for cartography in their proselytization. Examining the issues the Jesuits addressed through maps, as well as their methods of production, distribution, and influence, this chapter argues for an understanding of their publications as part of the late Ming publishing boom.

Keywords: late Ming, Jesuit missionaries, Matteo Ricci, *Tushu bian*, *Sancai tuhui*, Chen Zushou

We now turn to one of the few truly novel developments in late Ming cartography, the introduction of a fundamentally different worldview in the late sixteenth and early seventeenth century. In no region of the world other than East Asia did the Catholic missionaries of the Jesuit order make the introduction of translated European maps a tool of evangelism. That they did so in Ming society and in Japan, and indirectly in Chosŏn as well, is due to their serendipitous arrival at a time when cartographic materials were already proliferating in literate society. The Jesuits realized that a discussion was already going on about issues they could answer in a new way, and cartography emerged as a potential channel for opening conversations that they hoped could lead to their more central religious concerns. As the Jesuits became more familiar with their audience, the nature of existing publications already in circulation, and the types of works that would serve as gateways to their broader mission, this familiarity shaped the direction

Akin, Alexander, *East Asian Cartographic Print Culture: The Late Ming Publishing Boom and its Trans-Regional Connections*. Taylor & Francis Group, 2021
DOI: 10.5117/9789463726122_CH03

of their publishing activity under Matteo Ricci (Chinese: Li Madou 利瑪竇) and his successors.

The Jesuits were active participants in the late Ming publishing boom; they introduced strikingly new material to an already burgeoning traffic in maps while still making use of the preexisting infrastructure of Ming publishing. Jesuit publications entered a churning stream of texts from which others could fish out material for recombination and reconstitution in their own books, in much the same way that earlier Ming works provided raw material for cutting and pasting into late Ming compendia. Presenting their arguments in Chinese in works that touched upon, yet maintained a discreet distance from, the Jesuit religious message, Ricci and his successors joined a late Ming conversation that was in full bloom. Ricci voiced surprise at the affordability and ubiquity of printed books in China, and as he came to see, the domestic circulation of published texts by and about the Jesuits played a vital role in spreading familiarity among those who had not yet met them.

The cartography of the Jesuits is one of the most thoroughly explored aspects of mapmaking in the Ming context. While this chapter will revisit some of this background, the main goal is to frame this existing scholarship in a new light, showing how Jesuit compilers of cartographic and geographical works responded to the materials they found in the dominions of the Ming as part of their larger project of gaining access to the literati. This chapter will address the Jesuits' efforts to simultaneously challenge and adopt Chinese ideas, introducing such radical notions as the 'Five Continents Theory' while making concessions such as rearranging the projection to move the Ming territories to the center, or making a calculated appeal to Confucian literati by using cartographic evidence to note Buddhist 'errors.'

The arrival of the Gutenberg press in Macao in 1588 has been described as 'The first IT revolution of the Sixteenth Century,' heralding a quantum leap in the Jesuits' ability to communicate with a larger audience.[1] Despite the presence of this press, however, the Jesuits continued to rely primarily on woodblock printing. Despite the dramatic novelty of the content of Jesuit maps, it is striking that they followed Ming precedents of physical production. As David Mungello has observed, Ricci 'described how this wood block method enabled one to produce small amounts of a pamphlet or book on religious or scientific subjects which the fathers had done using the domestic servants in their mission houses.'[2] Ricci noted that the skilled

1 Üçerler, 'Gutenberg Comes to Japan: The Jesuit Enterprise and the First IT Revolution of the Sixteenth Century.'

2 Mungello, *Curious Land*, p. 53

craftsmen who carved woodblocks were able to produce a page in about the same time it took for a printer in Europe to lay out a page worth of moveable type, and the plates could be kept in storage to produce new print runs whenever needed, as many as fifteen hundred copies in a day.[3] In short, while the cartographic images themselves were novel to their Ming audience, their means of production were not.

We must remember that cartography was but a small part of the Jesuit program. In the hundreds of pages of Ricci's collected letters, only a handful of brief references explicitly mention maps or globes. Local interest in the maps was clearly disproportionate, however; they attracted significant attention from scholars who had little interest in most of the other Jesuit cultural offerings. Ricci would surely have been grateful if his translated catechisms or other religious apologetics had received anything approaching the level of credence or duplication in mainstream Ming texts that was accorded to his maps.

Measuring the extent of Jesuit influence on Chinese ways of thinking about the physical world is a complex task. As Cordell Yee and others have argued, the vast majority of indigenous maps from the late Ming and most of the Qing reflect no influence at all of Jesuit techniques. While a broad range of earlier scholars have emphasized Ming openness to Jesuit influence, some of the field today has shifted to perceive the Jesuits as ignored, or at best, misunderstood by the few who were interested in their work.[4] This more critical scholarship has steadily undermined the notion that Jesuit cartography 'transformed the Chinese picture of the world.'[5] Examination of a variety of sources reveals that the Jesuit influence is more obvious on certain classes of cartographic material than on others. Works that attempted a worldwide or empire-wide focus were far more likely to incorporate certain aspects of Jesuit cartography – or explicitly state reasons for not doing so – than were publications like local gazetteers.[6]

If we attempt to measure influence by searching for local imitation of Western mapmaking techniques, we will find very limited penetration of quantitative surveying science in late Ming cartographic culture. There

3 Trigault, *China in the Sixteenth Century: the Journals of Matthew Ricci, 1583-1610*, p. 21.

4 The most comprehensive such revisionist argument is to be found in Yee, 'Traditional Chinese Cartography and the Myth of Westernization.'

5 Helen Wallis makes this argument for transformation in 'The Influence of Father Ricci on Far Eastern Cartography,' p. 39.

6 The earliest explicit reference to Jesuit cartographic materials that I have found in a county gazetteer occurs in Huang Jinsheng's 1823 *Daoguang Jinhua xianzhi, juan* 1, p. 6b, though there may well be significantly earlier examples.

are some obvious reasons for this. It would have taken several years to build up a base of surveyors with the mathematical background required for high-quality work, even if for some inexplicable reason the Jesuits had put aside their main religious mission to focus on this unlikely goal. Furthermore, the Jesuits worked in a very unstable environment where no permanent institution could be established to impart this knowledge to local students. The casual scholar encountering Jesuit works, even one convinced of the soundness of their models, can hardly be expected to have become conversant enough to compose his own maps based on geodetic measurement without any of these intellectual tools. Unfortunately we will never have any way of knowing with certainty how deeply the geographical ideas introduced by the Jesuits penetrated late Ming society, for insufficient material evidence survives on which to base a reasoned assessment. The appearance of printed maps in mass-produced texts is not a sufficient yardstick for drawing conclusions.

Another way of gauging influence might be to assess the distribution of Jesuit cartography in private and public libraries. For the most part, our information about the circulation of Jesuit works is anecdotal, but catalogs of book collections from the late Ming do record the possession of such works. Nicholas Standaert notes that of the ninety-six Jesuit books recorded in his survey of nine private catalogs, about seventy percent are works related to the sciences rather than religious texts.[7] Of the works containing maps, we find not only Ricci's world map but two examples of a polar projection that he published, now lost but perhaps the same one that is copied in the *Tushu bian*. Giulio Aleni's *Zhifang wai ji* 職方外紀 (Account of countries not listed in the Records Office), a 1623 geographical treatise imbued with Catholic didactic character, also appears as the sole work of Western scholarship in two Fujianese collections.[8] As Kai-wing Chow has reminded us, the presence of a text in a particular collection can too easily be misconstrued as showing the owner's agreement with its contents (or, in large collections, even awareness of its presence).[9] Therefore, while mere ownership tells us something, assessments of cartographic influence should take into consideration the contemporary discussion of texts, quotations of Jesuit textual material related to maps (quotations that often lack precise

7 Standaert, 'Note on the Spread of Jesuit Writings in Late Ming and Early Qing China,' p. 22.

8 Ibid., p. 25. One of the collectors, Chen Di, died before the *Zhifang waiji* was even published, so Standaert surmises that Chen's heirs must have added the text to the collection. This demonstrates quite clearly the risk of drawing assumptions based on the mere presence of a book on a list.

9 Chow, 'Writing for Success,' p. 121.

indication of their sources), and imitations of maps derived from Jesuit sources.

The Jesuit critique of Buddhist beliefs, as inelegant or poorly informed as it was at times, had repercussions in cartography. This stance was partly a play for the sympathies of men from the scholar-official classes; as Gernet notes, 'It was for tactical reasons just as much as out of a natural antipathy felt for a rival religion that the Jesuits proclaimed themselves to be enemies of Buddhism and friends of Confucius.'[10] Be this as it may, Ricci and others were able to point to specific Buddhist beliefs about the structure of the earth and its place in the universe that had already been criticized by native scholars (the Buddhists' location of the Ming in a corner of Southern Jambudvīpa, their speculations about Mount Sumeru, and so on), creating a common cause against the Buddhists while subtly legitimizing the truly new arguments they themselves were putting forth.

As many scholars have been quick to note, the scientific expertise that the Jesuits introduced varied in its depth and was beholden to Church doctrine. Ulrich Libbrecht refers to the Jesuits as 'self-made men of science,' with little formal training but making diligent efforts to keep up with new developments in Europe.[11] Henri Bernard outlines Ricci's education to show how basic it was in comparison to that of many of his colleagues.[12] Liam Matthew Brockey briefly compares the training of Jesuits on the China mission with that of Jesuits assigned elsewhere, and suggests that their famed scientific prowess and its effectiveness in the Ming has been exaggerated, though the missionaries sent to Beijing did include some men accomplished in scientific fields.[13] In any case, it is vital to remember that science was but an ancillary aid to the true Jesuit mission. By exemplifying the power of reason in fathoming the natural sciences, the missionary in Ming China could serve his ultimate goal of proving that Catholicism was itself reasonable. The Jesuit map was meant not only to reveal a broader world than existed in Ming awareness, but also to argue that this world could be comprehended through the gift of intellect imparted by the Creator.

The adaptation of the Jesuit message to the circumstances that men like Matteo Ricci discovered on the ground appears to have been a masterful strategic decision. Recent scholars have emphasized the limits of Jesuit

10 Gernet, *China and the Christian Impact: a Conflict of Cultures*, p. 78.

11 Libbrecht, 'What Kind of Science Did the Jesuits Bring to China?,' p. 221.

12 Bernard, *Matteo Ricci's Scientific Contribution to China*, Chapter Two.

13 Brockey, *Journey to the East: the Jesuit Mission to China, 1579-1724*, Chapter 6, especially pp. 212-217.

influence in the late Ming, but we must ask how much more limited their influence would have been without the maps that often served as the first introduction of these European thinkers to Ming scholars. Examples were printed in various formats, presented as gifts during visits with literati, offered to dignitaries, and carved in stone for people to make their own ink rubbings. As we shall see, they helped pave Matteo Ricci's way to Beijing.

The Jesuits were merely one of several waves of foreign scholars who imported new brands of geographical thought, both cartographic and textual, into early modern China. Although Jesuit contact was indeed novel, it should not be isolated from the complex web of interregional influences that played out in earlier centuries. Before addressing the Jesuits, it makes sense to briefly touch upon Persian and other Muslim influences in earlier periods. We are able to discuss the Jesuits in greater detail than earlier groups, due to the fact that they were more recent and the survival rate of relevant sources is higher. Discussion of Chinese responses to the Jesuits would be incomplete, however, without consideration of these precedents.

Near Eastern influences before the Jesuits

Takahashi Tadashi notes that although the arrival of the Jesuits has been defined as a turning point in chronologies of Chinese cartography, 'The transmission of graticuled maps to China did not begin in the late Ming and early Qing period;' it was an earlier Muslim import.[14] He compares the medieval reintroduction of works by Ptolemy and other classical thinkers on geography to the West, via Muslim scholars, with the roughly contemporaneous eastward flow of such materials to the Yuan.[15] In other words, the Muslim world was a center of ferment and a source of knowledge for both East and West during this period.

There are a number of distinct factors that shaped the development of geographical thought in the Islamic world, some of which are enumerated by Nafis Ahmad, an early Pakistani historian of Muslim geography. In the trackless desert it was perhaps more necessary than elsewhere to observe the location and movements of stars and planets in the sky, while early Arabic oral histories recorded a wealth of detail about the movement of

14 He cites scholars such as Ogawa Takuji 小川琢治 and Wang Yong 王庸 as so defining the Jesuit arrival. Quotation translated from Takahashi Tadashi, 'Tōzen seru chūsei Isurāmu sekaizu,' p. 77.

15 Ibid., p. 80.

tribes from place to place. When these factors were combined with Islam's conquest of the seats of classical learning, its inheritance of the mathematical heritage of both the Greeks and the Hindus, and its combination of trade and conquest (so reminiscent of the later European expansion), it is not surprising that geography – and mensurate geography in particular – rose to such prominence. Perhaps most important in this respect was the Hajj, a journey to Mecca to be undertaken, if possible, by every capable Muslim whose means allowed. The Hajj contributed incalculably to the amalgamation of navigational experience, and Ahmad describes a side effect of the Hajj, the gathering of Muslim scholars from every corner of the known world in one place, as the medieval version of today's international colloquia.[16] Aside from the needs of pilgrimage, the Umayyad and Abbasid postal systems, which traversed at their greatest extent a swath of land extending from the western Mediterranean to parts of central Asia, also required extensive geographical knowledge. At the local level, the orientation of mosques toward Mecca offered an immediate lesson in geography to any worshipper.

Sassanian trade links with Tang China, including trade by sea, predate the Muslim conquest of Persia. It is unclear who precisely ran these shipping lanes, a question of significance to us because it reflects the degree of navigational knowledge held by the pilots. Despite the mention of 'Chinese ships' in Persian records, George Hourani argues that this term – *sufun min al-Sin* or *sufun Siniyah* in Arabic and Persian respectively – refers to ships on the 'China run,' rather than ships owned or piloted by Chinese.[17] Even after the Arab conquest of Persia and its conversion to Islam, the Arabs let Persians maintain control of shipping and navigation for some time. By the 750s, a substantial Arab community had settled beside the Persians in Tang territory, reflecting increasing Arab involvement in long-distance travel and

16 Ahmad, *Muslim Contribution to Geography*, pp. 6-8. The merging of religious fervor with geodetic endeavors is illustrated by al-Biruni's description of his plans for determining the longitude of Ghazna (now Ghazni, Afghanistan), the capital of the Ghaznavid state whose ruler, Mahmud, was his patron. 'My object therefore,' al-Biruni writes, 'is to establish the geographic longitude of a certain city on the earth-globe, that is to say, Ghazna. Hitherto I have only been able to determine the degree of latitude of this city; as for what concerns the longitude, I was not able to establish that properly owing to adverse circumstances. But if I were to plead these obstacles as an excuse for such negligence and were to show myself as therefore blameless, I should have portrayed myself as a denier of God's open and secret favours as well as of the benefits of the Dispenser of Kindness [i.e., the ruler Mahmud] whose hand has brought me unto full prosperity.' For a continuation of the translation by Ahmet Zeki Validi, see his 'Islam and the Science of Geography,' pp. 517-518.

17 Hourani, *Arab Seafaring in the Indian Ocean in Ancient and Early Medieval Times*, pp. 47-50.

trade.[18] The movement of the Abbasid capital to Baghdad brought much greater directness to sea trade with India and China. According to Al-Tabari, the caliph al-Mansur boasted that 'This is the Tigris; there is no obstacle between us and China; everything on the sea can come to us on it.'[19]

This increased contact is chronicled in the *Xin Tang shu* 新唐書, and an eighth century map by Jia Dan 賈耽, now lost, may have shown the Western Regions, for his textual itinerary of a route to Baghdad survives. The itinerary does not mention whose ships followed this route.[20] Unfortunately, little relevant material survives from the Tang. Although sources remain very scarce for later periods, we have somewhat better knowledge of Muslim influence, particularly in cartography, during the Yuan.[21]

In 1267, the Persian astronomer Jamal al-Din (Chinese: 札馬剌丁) brought survey tools, an armillary sphere, and other instruments including a wooden terrestrial globe to the court of Khubilai. The globe is described as using a grid of small squares (小方井) to indicate distances, which suggests that this sort of grid remained somewhat remarkable to observers. Unlike the earlier *Yujitu* and similar works, this grid imposed on a sphere probably represented longitude and latitude rather than surface distances, though it is unclear whether this was understood by many observers.[22] Takahashi Tadashi notes that the color scheme for fresh and salt water used on the *Da Ming hunyi tu* (the 1389 Amalgamated map of the Great Ming) is the same as that described for Jamal al-Din's globe, raising the possibility that the globe was a model for that map.[23] Another source of knowledge about geography and travel routes between the Near East and China became available to the Yuan in 1287, when Muslim sailors in Fujian province were ordered to submit their sailing guides to the central government.[24]

18 Ibid., p. 63.

19 Ibid., p. 64.

20 This itinerary, found in *juan* 43b of the *Xin Tang Shu*, has been translated into English by Hirth and Rockhill (*Chau Ju-Kua*, pp. 10-14).

21 It should be noted that geographical works of the premodern Islamic world consist mostly of text rather than cartographic representation – a feature they hold in common with Chinese works. Analysis of maps alone, which are in both cases relatively 'backward' in comparison to knowledge represented in textual form, will result in gross misunderstandings. The disfavor of printing in the Muslim world until the late nineteenth century, and then only in some places, plays a further role in the 'backward' appearance of these hand-copied maps. Mathematical geodesy was far more developed than the pictorial means to display this knowledge.

22 In his discussion of this passage Fuchs disagrees, however, arguing that this grid network must represent a Chinese-style demarcation of distance rather than a network of latitude and longitude (Fuchs, *The 'Mongol Atlas' of China*, p. 5).

23 Takahashi, 'Tōzen seru chūsei Isurāmu sekaizu,' p. 85.

24 Hyunhee Park, *Mapping the Chinese and Islamic Worlds*, p. 107.

The world maps of Al-Idrisi – born in Ceuta, educated in Cordoba, and active at the Sicilian court of Roger II – were extraordinarily influential in the Muslim world; variants of his world maps reached the Yuan where they were used as source material in the fourteenth century.[25] Textual analysis of place names from the far western part of the *Da Ming hunyi tu* shows that they derive from Arabic names found on copies of the Idrisi-style map, such as the island of *Sanguba* 桑骨八 (Zanzibar) off the African coast.[26] The nature of these names clearly puzzled later copyists, as the toponyms appear in different forms, using different Chinese characters, on various later copies. A version painted in Chosŏn Korea in the early fifteenth century, discussed in Chapter Four, represents the furthest eastward influence of this Islamic world map transmitted during the Mongol period.

European cartography owes more to the work of Muslim geographers than is sometimes acknowledged, from their preservation of the Ptolemaic worldview and understanding of mathematical geodetic measurement to their extensive textual records of travels far afield. In the European context, these seeds, deposited by men such as Adelard of Bath and the specialists at the ecumenical court of Roger II of Sicily, fell upon increasingly fertile soil as maritime empires began to expand into the lands where Muslim traders had long held the upper hand. Thus a powerful economic motivation merged together with a new emphasis on scientific reasoning and theoretical issues, all of which took place upon the grander backdrop of competition between states for the skills of geographers and cartographers. Europe's diversity and political fragmentation meant that even if inquiry was suppressed or an orthodoxy imposed in one place, specialists could often move to the sponsorship of a new court or carry out their work elsewhere.

In the Yuan-Ming transition we see an entirely different set of circumstances. Experts from the Near East introduced materials in an environment where there was not the same social call for their services, and their cultural isolation from the broader population may have played a role in restricting their influence to their official positions. The *Yutu* of Zhu Siben, the most influential map from the Yuan in the sense that it served as a model for the very popular *Guang yutu*, specifically excluded foreign lands about which

25 Ibid., pp. 100-109, for discussion of Yuan-period cartography and its adaptation of information from Muslim sources.

26 For a brief discussion of Idrisi's maps and their Asian impact, and a diagram comparing the Osaka version of the 1402 Korean map with the *Guang yutu*, see Takahashi, 'Chūsei Isurāmu sekaizu,' pp. 53-55. See also the same author, 'Tōzen seru chūsei Isurāmu sekaizu,' pp. 85-89, in which he traces the depiction of the Nile from Ptolemy via Idrisi to the *Hunyi tu* and thence to the *Guang yutu*.

the primary sources 'cannot be trusted,' focusing on the Yuan rather than the broader world or even the broken-off lands of the former Mongol empire.[27] The treatises used by Muslim specialists based at the court may never have been translated for a broader audience.[28] On the other hand, having come to serve the court with somewhat esoteric technical skills, these officers had no good reason to spread their knowledge among the Chinese population. The failure to produce a long-term legacy is demonstrated by the fact that although hundreds of pages of Persian texts were found in the Yuan archives in 1374, among the translators of Persian serving the new Ming court, none could understand their technical aspects.[29]

The influence of Islamic geographical knowledge continued within certain communities after the establishment of the Ming. Distinct from Han society, though enveloped within it, Muslim communities continued to nurture the connections with western regions that had prospered during the Yuan. This legacy played a role in the missions of Zheng He (originally surnamed Ma 馬, the most common Sinicized Muslim surname). Far from being an explorer of unknown lands as he is sometimes popularly portrayed today, Zheng was to some extent following in the footsteps of his father, his grandfather, and numerous others who had previously traveled from China

27 Fuchs translates the relevant portion of Zhu's preface: 'Those who speak of them are unable to say anything definite, while those who say something definite cannot be trusted. Hence, I am compelled to omit them here' (Fuchs, *The 'Mongol Atlas' of China*, p. 8). Note that Fuchs' discussion of the 1402 Chosŏn world map and its Ming counterpart, the *Hunyi tu*, dates to soon after the discovery of the latter, and his conclusions are not definitive. He held that the Ming version was a copy of the Chosŏn version from around 1600, while Cao Wanru dates it to several years before the Chosŏn version.

28 The record of a now-lost text by Shams 贍思, the *Xiguo tujing* (西國圖經, Illustrated treatise on countries of the West), may however be an indication of just such an effort during the Yuan (Fuchs, *The 'Mongol Atlas' of China*, p. 4). Fuchs also notes the influence of Muslim astronomers upon men like Guo Shoujing 郭守敬, who reported the latitudes of twenty-seven locales in 1279 (ibid., p. 6).

29 Bernard, *Ricci's Scientific Contribution to China*, p. 14. Distinctly Islamic cartography produced *within* China is a topic yet to be explored, or even discovered, but it would be useful to know what role the pilgrimage to Mecca played in the formation of the navigational tools available to long-distance merchants, and to the fleets commanded by Zheng He. The only extant maps of Zheng's routes are historical reconstructions or copies of long-lost works, dating from long after the voyages had ended, such as an accordion map in the *Wubei zhi* 武備制 that was reprinted numerous times from the late Ming onward. In the form we see them now, these are perfectly secular maps. Originally published in the Tianqi period, a copy of the *Wubei zhi* was given to the Chongzhen emperor in 1628 (see Yan Ping, *China in Ancient and Modern Maps*, p. 147). It continued to be of interest later; revised or abridged editions produced in the Kangxi period and in 1843 are held in the Harvard-Yenching Library.

to Arabia on the Hajj.[30] Zheng's multiple missions took him to a broader range of territories than any of his ancestors would have visited alone, but the Hajj played a significant part in connecting Yuan and Ming Muslims to the southern seas and western territories. Recently some scholars have attributed certain cognizance of the West, particularly the shape of the African coast, to the missions of Zheng He, even though this knowledge actually appears in Muslim sources that predate him.[31]

The fundamental purpose of the Jesuit introduction of geographical knowledge was quite different from that of their predecessors. For the Jesuits, this was a tactic for attracting the interest and respect of literati who might then go on to consider the metaphysical and Deistic claims of the Jesuits. The Jesuits adopted this tool here because they saw that they had entered a society where cartography was already seen as a topic of substance. With only an admittedly basic grasp of European cartographic theory and practice, the early Jesuits discovered almost by accident that their maps were of great interest, and went on to use woodblock printing to mass-produce and strategically introduce a worldview that inspired both interest and concern in late Ming society. The nature of their influence, as visible in Ming cartographic woodblock publications, is the subject of the rest of this chapter.

Interpretations of Jesuit cartography

In late twentieth century Chinese scholarship, the legacy of the Jesuits was viewed through a changing political lens, with this early contact coming to be depicted as a model for international academic exchange.[32] In the West, too, interest in Jesuit cartography has shifted in tune with revision of

30 Zheng's father and grandfather are both recorded as bearing the name *Hajji* 哈只, a title claimed by Muslims who have completed the Hajj to Mecca (Chang Kuei-sheng, 'Cheng Ho,' p. 194).

31 See, for example, Smith, *Mapping China's World*, p. 67, and Elman, *On Their Own Terms*, p. 125. The *Hunyi tu* reflected this knowledge of the African coast as early as 1389, and was based on even earlier materials.

32 For example, Ji Xianlin 季羡林 outlines China's contributions to the world, which he depicts as constituting an uninterrupted font of benevolence for thousands of years, before listing the two brief periods during which 'useful' things have been obtained from the West by China, namely contact with Buddhism in the Han period, and the Jesuit contact in the late Ming. While both of these superficially seem to be importations of religion, he argues, they were more fundamentally philosophical, artistic, and technical contributions. The path of the future, he concludes, is in showing China's youth how to combine patriotism with internationalism. See

attitudes towards the entire missionary project. Some of the most significant work has been published by fellow Jesuits. The fullest foreign treatment of Matteo Ricci's *Kunyu wanguo quantu*, a world map discussed below, is a massive tome by Pasquale M. d'Elia, published at the Vatican in 1938.[33] Aside from Jesuit scholars, a smaller number of writers knowledgeable about the Jesuit project have been more critical, even characterizing the Jesuits in late Ming as being dishonest in such moves as their hiding of heliocentric theory after its teaching was banned in 1616.[34]

In the Jesuit effort to use science to lure Chinese thinkers into consideration of their religious ideas, some see a tactic that backfired.[35] Those kernels of science that became accepted were rigorously separated from their religious husks; the faith of the Jesuits was found by many literati to be more absurd the more they learned of it.[36] As Jacques Gernet has warned, we must not misjudge Jesuit science as more modern than it was, which would lead us to judge Chinese rejections as purely xenophobic.[37]

Neither should we overlook the variety of perspectives from which various Chinese scholars rejected Jesuit teachings. While most literati who were inclined to dispute the new worldview did so on the basis of its radical challenge to established native traditions, as exemplified by Chen Zushou, Buddhists criticized instead its narrow limits. In 1639, for example, one monk lamented Ricci's failure to understand that there existed an infinite number of worlds.[38] Buddhist cosmology had long been comfortable with extremely large numbers, astronomical distances in both space and time. The crystalline heavenly spheres and close distances of the sun, moon and stars described by the Jesuits seemed preposterous – a Buddhist objection that looks wise now.

Wang Qianjin 汪前进, *Xixue dong chuan di yi shi: Li Madou* 西学东传第一师: 利玛窦, Preface, p. 3.

33 D'Elia, *Il mappamondo cinese del p. Matteo Ricci, S.I.* (*terza edizione, Pechino, 1602*) *conservato presso la Biblioteca vaticana*.

34 For a thorough analysis of this issue, and an interpretation which is rather critical of the Jesuits, see Sivin, 'Copernicus in China.'

35 Jacques Gernet has written prolifically on the intersection of Jesuit and Chinese ideas, focusing much of his work on the vast gap in worldviews and the misinterpretations or false assumptions that could underlie apparent synthesis in the minds of Chinese converts. See *China and the Christian Impact*, 'Christian and Chinese Visions of the World in the Seventeenth Century,' and 'Space and Time: Science and Religion in the Encounter Between China and Europe.'

36 Gernet, *China and the Christian Impact*, pp. 58-60.

37 Gernet, 'Christian and Chinese Visions of the World in the Seventeenth Century,' p. 2.

38 Ibid., p. 10.

Mapping out accommodation: The cartographic strategy of Matteo Ricci

Born in Macerata in 1552, Matteo Ricci was to become an intellectually formidable representative of an alien religion at a troubled moment in Ming society. He arrived in Macau just before his thirtieth birthday, having already spent almost four years in Goa studying and lecturing. He had attracted notice in India by dissenting from what he considered a patronizing attitude toward indigenous converts.[39] His recognition of the need to assimilate not only the language but the cultural bearing of his hosts, and his tolerance for long periods of observation and study as prelude to action, eventually enabled him to break into the circles of polite Ming society where, he believed, lay the path to conversion of the entire country. Ricci's use of maps as a tool for opening the doors of native intellectuals was a byproduct of his astute observation of what was taking place in elite culture.

A number of different 'Ricci maps' appeared over the course of his time in the Ming, the earliest of which are no longer extant. The evolution of his maps represents a maturation not only of Ricci's skill as a mapmaker, but his growing understanding of their utility as ideological tools. From simple beginnings, he worked with converted local informants and craftsmen to create magnificent works that melded dramatic presentation with extensive textual explication of key concepts.[40]

Ricci noted the significance of maps in Ming literati culture soon after his arrival; one of his earliest letters written on September 13, 1584 already mentions it. Ricci describes a map he has obtained of 'toda la China, pintada en cartas planas, a nuestro modo, y despues cada provincia de por si en su carta' ('All of China, painted in flat charts, in our way, and afterwards each province on its own chart'). This has been misinterpreted as a reference to Ricci's first edition of his *Yudi shanhai quantu*.[41] In reality, however, the content of Ricci's *Yudi shanhai quantu* differed completely, being a world

39 Dunne, *Generation of Giants: The Story of the Jesuits in China in the Last Decades of the Ming Dynasty*, pp. 24-25.

40 The interested reader may turn to the comparison of different editions in D'Elia, 'Recent Discoveries and New Studies (1938-60) on the World Map in Chinese of Father Matteo Ricci S.I.' as well as Huang Shijian 黃時鑒 and Gong Yingyan 龔纓晏, *Li Madou shijie ditu yanjiu* 利瑪竇世界地圖研究. John D. Day's 'The search for the Origins of the Chinese Manuscript of Matteo Ricci's Maps' includes a census of all editions, both print and manuscript, known to the author. An anonymous review in *Imago Mundi* vol. 3 (1939), pp. 111-112, lists early Western works on the Ricci maps.

41 Ricci, *Lettere (1580-1609)*, p. 62, n. 4.

map rather than a representation of the Ming and its provinces. Ricci's description refers instead to a Chinese atlas following the format of Luo Hongxian's *Guang yutu*, in which individual maps of the provinces follow a general map of the empire.[42] Ricci's observation that the map was drawn *a nuestro modo*, in 'our' fashion, and his later comments on the beauty of its execution (*lindisima vista*), show his early impressions of the existing cartographic culture with which he came in contact.

In a letter of November 30, 1584, Ricci relates that the prefect of Guangzhou (Wang Pan 王泮) 'had me make a map in the manner of ours of Europe, but with the distances and names of countries in the Chinese language. And he immediately printed it without my reviewing it or thinking that it would be sent to press. He esteems it so much that he keeps the print with him, not wanting anyone to learn about it except those to whom he slowly presents it, the more important persons of China.[43]

This brief statement brings several interesting issues to our attention. In Europe, colorful atlases like that of Ortelius had become popular gifts for noblemen, and it appears that beautiful editions were taken to Macau and elsewhere in the event that they could be presented as gifts to local potentates to ease the travails of the church.[44] It may have been the sight of such a map for presentation that inspired Wang Pan to request a Chinese version. Alternatively, it may have been a colorful map displayed on the wall of a church, as it is noted in Ricci's memoir that 'The fathers had in their hall a map of the world.[45] In any case, Ricci complied with a manuscript copy.

Also, from this brief note it seems that the idea of drawing up a Chinese version was originally the prefect's. Ricci was surprised and a bit dismayed at not having had the chance to clean up his draft transcription before it was committed to woodblock. Nevertheless, this signaled to the Jesuit that a

42 The content of Ricci's *Yudi shanhai quantu* differed completely, being a world map rather than a representation of the Ming and its provinces.

43 Translated in Rienstra, *Jesuit Letters from China 1583-84*, pp. 24-25. The letters included in Rienstra's translation are those that were published in a 1586 letterbook for European audiences. Within 1586 alone, three editions were published in Italian, one in French and one in German, highlighting the level of contemporary interest in the missions to East Asia. For details of their publication see Rienstra's introduction, especially pp. 4-5. The original text of Ricci's letter appears in Ricci, *Lettere*, p. 92. Rienstra's translation into English slightly abridges the original for clarity.

44 In 1600, when Ricci visited the Wanli emperor in Beijing together with Didacus de Pantoja, their gifts included a book recorded in the tribute list as the *Wanguo Tuzhi* 萬國圖志 (Atlas of the Myriad Lands). This may have been one of the copies of Ortelius (either the 1570 edition or the 1598) that are now extant in Beijing (Chen Zhengxiang, *Zhongguo ditu xue shi*, pp. 38-39). The Jesuits preserved a copy of the 1570 edition in their Bei Tang Tushuguan in Beijing (ibid., p. 39).

45 Cited in Hsia, *Matteo Ricci and the Catholic Mission to China*, p. 63.

European-style map of the world, translated into Chinese with explanatory material appended, might serve the Jesuits as a tool for attracting attention and establishing relationships. In a letter from October 20 of the following year, Ricci recalls having later made 'two or three' terrestrial globes as gifts, one of which was given to the prefect.[46] After this the production of globes and maps continued apace, with a letter of November 24, 1585 noting their continued manufacture and distribution.[47]

This story also reveals something about the nature of publishing and the circulation of texts. The prefect had the map engraved and printed not for general distribution, but to make a 'permanent' version over which he would still exercise control, sharing it only with those whom he chose. The proliferation of woodblock publishing in this period meant that an infrastructure of engravers, paper suppliers and other specialists was in place facilitating the printing of anything, from vanity publications of poetry to one-off copies of maps, at short notice. Ricci was soon to take advantage of this domestic printing infrastructure to attain a different goal; not to make a small number of maps to foster exclusive ties with the few allowed to see them, but to draw ever-broader attention to the Jesuits and their message. In a bit of evangelical marketing genius, he would use woodblock printing to create larger, visually arresting world maps extensively annotated with short essays explaining the logic and technique underlying their composition. Such maps, at once works of technical art and polemic tracts, attempted to do things that no Chinese map had ever done before. Tellingly, they also raised the hackles of some prominent Buddhists as well as defenders of the Chinese tradition of imperial administrative cartography.

While the idea of translating the first copy of the map seems to have originated with the prefect of Guangzhou, the concept of using it for advancing the Catholic mission belonged to Ricci, although the circumstances under which these maps first entered the consciousness of local literati are not entirely clear. It is not until 1608, in a commentary on works he made for the palace with the help of Li Zhizao, that Ricci's letters contain any detailed discussion of the strategic use of maps for the Cause.[48] By this time it had been more than two decades since the first European-type world map had been printed in Chinese translation on woodblocks.

The impact of these early printed maps on their readers varied, depending partly on the background of the observers. In an oft-cited quote from Ricci's

46 Ricci, *Lettere*, p. 103.

47 Ibid., p. 116.

48 Ibid., pp. 491-492.

correspondence, we read that 'When they saw the world so large and China in a corner of it, so small to their way of thinking, the more ignorant began to make fun of such a description, but the more intelligent, seeing such an orderly arrangement of parallel lines of latitude and longitude [...] could not resist believing the whole thing true [...] it was printed again and again and all China was flooded with copies.'[49] Regardless of Ricci's hyperbole here, pathbreaking works often left a legacy through imitation and inclusion in later compendia, and it is clear from such evidence that Ricci's maps really did became part of the repository of late Ming cartographic sources from which later works borrowed both images and text.

The many social functions of Ricci's maps are illustrated in his second visit to Nanjing. His reception on this occasion was utterly different from his first visit, when he had been ejected from the city during the xenophobic panic that followed Japan's invasion of Korea. The new welcome accorded to Ricci is attributable at least in small part to the fact that before his second arrival, in 1598, Ricci's maps had traveled before him to pave his way.

In Nanchang he had befriended Zhang Huang, the compiler of the *Tushu bian* described in Chapter One. This afforded Ricci the opportunity to participate in the salon-style discussions that were then popular. It is likely that the version of the world map reproduced in Zhang's *Tushu bian*, discussed below, is based on a Ricci prototype from around this time. In 1598, at the suggestion of Wang Honghui 王弘誨, Ricci left Nanchang for Nanjing *en route* to Beijing, with Wang at his side.[50] Ricci was climbing further up the social ladder with each move, as Wang's sponsorship gave him entry to sophisticated audiences in Nanjing.

Zhao Kehuai 趙可懷, the governor at Nanjing, had by this time already obtained a copy of Ricci's map and had even made copies to give away as gifts. Thus he was delighted to discover that Wang had brought Ricci to town in person. Zhao invited Ricci to separate from the rest of Wang's entourage and pay him a personal visit for a week at his residence outside the city, where he mingled in turn with other members of the governing class.[51]

49 Translated by Columba Cary-Elwes in *China and the Cross*, p. 89.

50 Wang, who served two terms as Minister of Rites in Nanjing (1589-1590, 1595-1599) wanted to apply Jesuit mathematical knowledge to calendrical problems (Goodrich, 'Cattaneo, Lazzaro,' p. 32; Franke, 'Ricci, Matteo,' p. 1139).

51 Dunne, *Generation of Giants*, pp. 53-54. Though no Nanjing edition of Ricci's map remains extant, a preface to a version of it published in book form by Guo Qingluo survives in the latter's collected works. A transcription of the Chinese original, paired with an annotated English translation of the preface, may be found in D'Elia, 'Recent Discoveries,' pp. 98-107. In this preface Guo remarks on the similarity between Ricci's map and the works of the Zhou thinker Zou Yan

Ricci's diary, as edited by Trigault, notes that Zhao asked him to revise the map. Ricci happily complied and revised its text; the new map is said to have surpassed the original both in quality and in quantity printed. Zhao also commissioned the carving of a stone stele with an image of the map, so that inked rubbings could be made.[52]

When Ricci eventually reached Beijing, his presence at the capital facilitated further diffusion of the map as officials heading for the provinces took copies away with them. The breadth of the map's diffusion is illustrated by the fact that even remote Guizhou saw local copies printed by Guo Zichang.[53] Life in the capital exposed Ricci to interested officials such as the eventual convert Li Zhizao, who among countless services to the Jesuit cause had an edition of Ricci's map printed in 1602, the earliest edition still surviving today.[54] Li was perhaps the convert most interested in geography; he even claimed that it was the sight of a Jesuit map that had first drawn his attention to the missionaries.[55]

The Jesuit use of maps to promote their mission extended beyond the Ming frontier. The missionaries had no direct access to Chosŏn territory, but in 1631 João Rodrigues met the Chosŏn diplomat Chŏng Tuwŏn and gave him numerous books including the *Zhifang wai ji* and five copies of Ricci's

鄒衍 (c. 305-240 BCE), who had postulated a much smaller place for the Middle Kingdom in the grand scheme of a world with many continents. A list of other times the map was recorded as having been given as a gift is given in Lu, *Zhongguo ditu xue shi*, p. 173.

52 Trigault, *China in the Sixteenth Century*, p. 331.

53 Guo, a prolific writer and able administrator, had come to know Ricci in Guangdong. In 1600, while serving as governor of Guizhou, he received a copy of Ricci's new cartographic edition of that year and reproduced it in one of his own publications (Fang and Goodrich, 'Kuo Tzu-chang,' pp. 775-777). Guo's Guizhou copy of the map is lost but his preface has been preserved and is presented together with an English translation in D'Elia, 'Recent Discoveries,' pp. 97-98. See also Dunne, *Generation of Giants*, p. 92, and Trigault, *China in the Sixteenth Century*, p. 331.

54 Li Zhizao 李之藻 (*jinshi* 1598, d. 1630), a Hangzhou native, was one of the 'three pillars' of Catholicism in China; he is discussed in several works by Gernet, and a brief biography is provided in *Eminent Chinese of the Ch'ing Period* (Yap and Yang, 'Li Chih-tsao').

55 Willard J. Peterson notes that Ricci's description of Li's background includes mention of his composition of a map in his youth that covered the fifteen provinces of China; the impression made upon Li by Jesuit-translated world maps thus not only speaks to the novelty and interest generated by them, but to the fact that the map was a category already in the field of intellectual endeavor. See Peterson, 'Why did they become Christians?,' p. 137. Li's own preface to the 1602 map starts with the complaint that 'There were no good editions of maps in the old days.' (輿地舊無善版). Li collaborated in the composition of the *Tianxue chuhan*, a multivolume collection of Jesuit works intended for a literati audience, which included a geographical treatise by Aleni. This was one of the few Jesuit works included in the *Siku Quanshu*, though as the *zongmu* indicates, astronomical and geographical material was left in while religious content, especially proofs of the existence of God, were omitted. Ma Biao, 'Johann Adam Schall and Chinese Traditional Philosophy,' p. 105.

world map to circulate when he returned to his country.[56] In Japan the distribution of maps was significantly more developed, though the Jesuits were only one of several sources of Western cartography there in the late sixteenth and early seventeenth century.

Accommodation in Ricci's world maps

Ricci was by no means the first cartographer to challenge a Sinocentric worldview, for Buddhists had posited an Indocentric view much earlier, as well as maps that emphasized the minuteness of the known world at a universal scale. What was really new in the Jesuit maps was their specific depiction of continents on a network of latitudinal and longitudinal lines, presented on a scale that dwarfed the territory of the Ming, and outlining new lands unknown from the tribute rolls. All of this was sufficiently novel to be seen as shocking or scandalous by some literati. Yet Ricci's sensitivity to his audience led him to soften the blow and justify his claims by accommodating the expectations of local literati. In his notes around the edges of the maps, he demonstrated familiarity with classical Chinese texts, cited his personal experience in world travel as evidence, included exercises that the reader could undertake himself to test the validity of his claims, and even tried getting close to Confucians by criticizing Buddhists. This was part of a larger pattern of what Harriet Zurndorfer has called 'Flattery as strategy.'[57]

The small number of extant copies of Ricci's maps highlights the ephemeral nature of printed paper – especially in a context where the content of the print countered common social values. Of Ricci's 1584 *Kunyu wanguo quantu* 坤輿萬國全圖 (Complete map of all the countries of the world), not one example survives, nor of the many revisions undertaken, or copies made by others, at various stages from Ricci's arrival in the south until his stay in Beijing. Only a few miniaturized copies in the *Tushu bian* encyclopedia hint at the types of maps being produced in Chinese translation by the Jesuits at this time. Although Ricci was constantly revising his maps during his lifetime, their proliferation had already escaped his control before he died and many printings took place thereafter. Woodblocks carved in the Ming continued in use even after 1644, the characters for 'Great Ming' being

56 Cooper, *Rodrigues the Interpreter: An Early Jesuit in Japan and China*, pp. 348-349. Chosŏn reactions to Jesuit maps are discussed in the next chapter.

57 Zurndorfer, 'Science Without Modernization: China's First Encounter With Useful And Reliable Knowledge From Europe,' p.11.

excised and replaced with 'Great Qing.'[58] Yet, of the thousands of maps that were apparently printed and at least dozens of wooden globes that were manufactured, only a handful of maps survive, all of late editions. This low survival rate presents a cautionary tale for any modern scholar drawing conclusions based on extant copies.

The 1602 version of Ricci's map, printed in Beijing in large numbers, is the earliest version extant.[59] The basic format of this map is copied largely from Abraham Ortelius' *Theatrum Orbis Terrarum*, but Ricci has moved the Ming territories to the center. The most obvious form of accommodation to Ming expectations, visible at a glance, is this recentering of the map as compared to European originals. Ricci had learned that the Ming must lie at the center, lest the map be dismissed out of hand.

Placing the oval map on a rectangular backdrop left substantial empty space in the corners, which are crammed with textual material, and the margins have been enlarged to fit even more text. This includes brief descriptions of countries, both those new to a Ming audience and those long-familiar, as well as essays on the nine crystalline spheres of heaven and the relative size and distance of the sun and moon.[60] This text has been transcribed and punctuated elsewhere, and much of it has been translated by Lionel Giles.[61] Ricci's main preface, however, which establishes a fundamentally distinct worldview and introduces the author as its far-traveling presenter, has been comparatively overlooked.[62] The full text of this preface may be found in Appendix 3, but let us examine several significant passages.[63]

The preface begins,

> The land and the sea are basically round in shape, and together they form a globe, which is situated in the celestial sphere like the yolk of a

58 Day, 'The Search for the Origins of the Chinese Manuscript Copies of Matteo Ricci's Maps,' pp. 112-113.

59 The monumental 1938 study of this map by D'Elia exhaustively traces its history and analyzes its contents, supplemented by the same author's later 'Recent discoveries and new studies.' Important recent work has been done by Unno Kazutaka (*Tōzai chizubunka kōshōshi kenkyū*) and Huang Shijian with Gong Yingyan (*Li Madou shijie ditu yanjiu*).

60 The moon's distance is given as a highly precise 104,401.81 *li* – about one seventh of its actual distance.

61 Transcriptions with Chinese annotation can be found in Huang Shijian and Gong Yingyan, *Li Madou shijie ditu yanjiu*. For translations, see Giles, 'Translations from the Chinese World Map of Father Ricci,' in *The Geographical Journal*, Vol. 52, No. 6 (Dec., 1918), pp. 367-385, and Vol. 53, No. 1 (Jan., 1919), pp. 19-30.

62 Giles only translates a small portion of this text on p. 373.

63 The translation is based on the 1602 edition sent to the Vatican by Ricci for its archives.

> chicken's egg is suspended in the white. It has been said that the earth is '*fang* (方),' but this refers to its being fixed and immobile, not to a [square] shape.
> The heavens envelop the earth and they mutually correspond. Therefore the heavens have two poles, south and north, and the earth does as well. The heavens are divided into 360 degrees, and the earth is too. In the middle of the heavens lies the [celestial] equator. South from the equator 23.5 degrees lies the Tropic of Capricorn, while 23.5 degrees to the north of the equator lies the Tropic of Cancer. Thus, the Middle Kingdom is north of the Tropic of Cancer.
>
> 地與海本是圓形，而合為一球,居天球之中，誠如雞子黃在青內。有謂地為方者，乃語其定而不移之性,非語其形體也。天既包地，則彼此相應，故天有南北二極，地亦有之。天分三百六十度，地亦同之。天中有赤道，自赤道而南二十三度半為南道，赤道而北二十三度半為北道。按中國在赤道之北。

Ricci plunges directly into the literati tradition in the opening line of his text, comparing the earth to the suspended yolk of an egg in a passage that would remind many readers of a similar proposition by the Han scholar Zhang Heng 張衡 (78-139 CE). With one deft stroke he then attempts to resolve any apparent dissonance with the 'round heavens, square earth' (天圓地方) formulation of traditional cosmology by defining '*fang*' as fixed, rather than square in form. By the time he launches into an explanation of the spherical earth's meridians in the next line, Ricci has already conveyed a basic message to his reader: 'Bear with me for a moment, I know where you're coming from and I'm not saying you're all wrong.' All that was required was some retroactive redefinition of terms, something of a Riccian specialty.

Ricci goes on to address a common misconception about the implications of a spherical earth, and buttresses his claim with an appeal to the authority of personal experience:

> Now, the earth has a thickness (diameter) of 28,636.36 *li*. It is inhabited on all sides, whether above or below. It is a sphere, which has no 'up' or 'down' to begin with. Is there anywhere in the world where one does not look up to see the sky? Absolutely everywhere under heaven, wherever the feet plant themselves is down, and wherever the head points, is up. Those who say that one would point up or down based on location are wrong.

> Now, when I sailed by sea from the West to the Middle Kingdom I arrived at the equator and saw that the two poles, north and south, were both on the horizon with no difference in elevation on either side. Continuing to the south, when I passed the Cape of Good Hope I saw that the south [celestial] pole was 36 degrees above the horizon. Thus the Middle Kingdom and the Cape of Good Hope correspond in latitude, above and below. When I was there I saw the sky above me, I never saw it below me. Thus I say it is worthy of belief that the Earth is round and inhabited on all sides.
>
> 夫地厚二萬八千六百三十六里零百分里之三十六分,上下四旁皆生齒所居,渾淪一球，原無上下。蓋在天之內，何瞻非天，總六合內，凡足所佇即為下，凡首所向即為上；其專以身之所居分上下者，未然也。且予自大西浮海入中國，至晝夜平線，已見南北二極皆在平地，略無高低；道轉而南過大浪山，已見南極出地三十六度，則大浪山與中國上下相為對待矣。而吾彼時只仰天在上，未視之在下也。故謂地形圓而周圍皆生齒者，信然矣。

Ricci speaks from the experience of having traveled around the world himself. While some anti-Jesuit critics later called these claims lies, for many literati the mere presence in the capital of this strikingly foreign and yet undeniably intellectually endowed man proved in itself that something unfathomed lay beyond the horizons of the known world. This was another reason to keep reading.

Ricci goes on to enumerate the continents and their characteristics, the reader simply having to take his word, but in this case again he softens the blow of exoticism by ridiculing the Buddhists and their notion of a Jambudvīpa continent, dismissal of which was already common in Chinese anti-Buddhist writings.

> All countries north of the equator are governed by the north pole and are thus in the northern hemisphere, while those countries south of the equator are governed by the south pole and are thus in the southern hemisphere. Thus the folly of the Buddhists is evident in their claim that the Middle Kingdom is in southern Jambudvīpa, and in their calculations of the altitude of Mount Sumeru.
>
> 凡地在中線以上至北極，則實為北方；凡在中線以下則實為南方焉。釋氏謂中國在南瞻部洲，並計須彌山出入地數，其謬可知也。

With these few words Ricci attempted to bridge some of the gulf between himself and his classically educated readers: They were, after all, members of an anti-Buddhist alliance, were they not?[64]

The use of longitude and latitude to locate places on the map was unprecedented for Ricci's audience, despite its superficial similarity to the well-established tradition of grid-based maps. Ricci goes to considerable lengths to explain how this system works, and how it can be linked to the measurement of distance and time:

> The length of night or day varies according to distance from the equator. I have fashioned at the edge of the map a chart, marked at every 5-degree interval; from the length of the day and night, whether in the east or west, above or below, in any given case you can tell the precise distance from the equator.
>
> 其長晝，長夜，離中線愈遠，則其長愈多。余為式，以記于圖邊，每五度其晝夜長何如，則東西上下隔中線數一，則皆可通用焉。

Not only does Ricci make claims, he also discusses how to verify those claims. If Ricci can observe the changing altitude of the celestial pole after traveling 250 *li* northward, anybody can. A measurement of time differences during the next solar eclipse can test whether it is really the *mao chen* for the Jurchens when in Burma it is the *yin chen*. The merging of such reproducible calculations with cartography was unprecedented in the Ming context, yet at no point does Ricci directly criticize the maps he has found in circulation other than to dispute the Buddhists on the existence of Jambudvīpa.

To what degree did this accomodationist tactic work? Was his audience convinced? Xu Changzhi 徐昌治 was not. A supporter of the Three Teachings movement in the late Ming, he accused the Jesuits of disparaging Buddhism, Daoism *and* Confucianism, the latter by abusing its authority. Ricci's veil of accommodation was all too flimsy as far as Xu was concerned.[65] At the

64 The Jesuits famously switched from the robes of Buddhist monks to Confucian garb when they realized that their original costume would hamper their appeal to the literati. It is interesting to note that these tense relations, which led to relatively few Buddhist converts to Catholicism in the Ming, contrast with their experience in Vietnam. Alexander de Rhodes reported that in the kingdom of Tonkin, where he lived from 1627 to 1630, Buddhist monks were among the best converts and talented catechists; see Hertz, *Rhodes of Viet Nam: The Travels and Missions of Father Alexander de Rhodes in China and Other Kingdoms of the Orient*, p. 65.

65 Dudink, 'The Inventories of the Jesuit House at Nanking Made Up During the Persecution of 1616-1617 (Shen Que, Nangong shudu, 1620),' p. 124.

other extreme, there is evidence that the tactic of accommodation worked so well in some cases that it backfired. Rather than simply being swayed to see Jesuit arguments as compatible with respected tenets of antiquity, some became convinced that they were indeed so similar that they were stolen from the Chinese tradition. The Qing scholar Ruan Yuan 阮元 cited Jamal al-Din's terrestrial globe, presented by the Persian astronomer to the Yuan court in 1257, as a predecessor of Jesuit globes, and thus evidence that the thinking of the westerners was derivative.[66] Ruan also wrote of later work by Adam Schall von Bell that 'The frequent reference to [Chinese] histories and gazettes and the systematic study of works by [Chinese] astronomers and mathematicians make me realize that Western methods are the symposium of significant ideas, both ancient and recent, and not independent creations made by the Westerners.'[67]

It would be legitimate to ask whether modern scholars have exaggerated the significance of Ricci's maps in a world where antiquarian cartography has acquired both an intellectual and a commercial sheen. It is clear, however, that native scholars saw these maps as important too. Half a century later, in 1653, Tan Qian 談遷 listed a map from Ricci as one of the important objects the missionary had given to the emperor.[68]

Once the Jesuits realized that maps of the world were useful for breaking the ice with educated audiences, some of them appear to have carried extra maps among their belongings in case an opportunity arose to distribute them. João Rodrigues carried a version of Ricci's map with him; he describes using it to illustrate the purpose of the Jesuits and the route by which they had come to Asia. At a meeting with a Guangdong official in 1615 he pulled out this map as a conversation aid, and he notes in a letter transcribed by his colleague João da Costa that the official requested a copy of a European globe.[69]

Printing was not the only means by which Jesuit maps became visible to a local audience. It was not unusual for Catholic churches in contemporary Europe to display a world map as part of their decoration, symbolizing both the extent of Creation and the growing influence of Christendom over its farther reaches, and this practice was carried to the Ming. After the church

66 Wong, 'China's Opposition to Western Science During Late Ming and Early Ch'ing,' p. 46.

67 Quoted in translation in Wong, 'China's Opposition to Western Science During Late Ming and Early Ch'ing,' p. 45.

68 Chan, 'Johann Adam Schall in the Pei-yu lu of T'an Ch'ien and the Eyes of His Contemporaries,' p. 276.

69 Costa letter in the *Japonica Sinica* series of the Jesuit archives in Rome, folio 7; cited in Cooper, *Rodrigues the Interpreter*, pp. 291-292.

at Beijing was constructed, a 1635 guidebook to the city, the *Dijing jingwu lue* 帝京景物畧, noted in its brief description of the 'Tianzhu tang' (天主堂, Catholic church) that a painting of Jesus in the main hall depicted him holding a world map in his left hand.[70] It appears that such novel depictions of the world did not pass unnoticed.

After Ricci's death in 1610, his successors in the Jesuit mission continued to build on his record. Giulio Aleni (1582-1649), from Brescia, was a student of Christopher Clavius.[71] Eugenio Menegon attributes Aleni's interest in astronomy and geography to his studies under this master of the age, who was a friend of Galileo and the author of Pope Gregory's calendrical reforms. Having arrived on Ming soil in 1613, Aleni stayed at the Jesuit residences in Nanjing and Beijing before moving to Hangzhou where he endured the anti-Jesuit campaign led by Shen Que 沈㴶. Aleni's close relations with Yang Tingyun, an important protector of the Jesuits at Hangzhou, are reflected in their collaborations.[72] It was in Hangzhou that Aleni composed two of his three works of primary geographic and cartographic importance, the *Zhifang wai ji* (Account of countries not listed in the Records Office) and the *Xixue fan* 西學凡 (Introduction to Western learning), both dated 1623. Aleni made southern Fujian his base of operations after 1625. In Fujian his *Xifang dawen* 西方答問 (Responses to questions on the West) was published in 1637, and his *Kunyu tushuo* 坤輿圖說 (Illustrated discussion of the world) held at the Vatican is also thought to be a Fujian imprint.[73]

Aleni perpetuated Ricci's cartographic contributions by reproducing his map. The most widely circulated version is in the *Zhifang waiji*, Aleni's collaboration with Yang Tingyun that includes a small version of the *Wanguo quantu*.[74] Parts of the extensive verbiage from Ricci's large map,

70 King, 'Note on a late Ming dynasty Chinese description of "Ricci's church" in Beijing,' p. 51. Because of its brief description of the church, this guidebook was banned in Japan after the prohibition of Christian writings. See Ōba Osamu 大庭脩, trans. Fogel, 'Sino-Japanese Relations in the Edo Period (Part Three: The Discovery of Banned Books),' pp. 65-69.

71 Eugenio Menegon notes that Alenis is the name Giulio used to sign his letters, although Aleni is the more commonly encountered version of his name in modern scholarship. See Menegon, 'A Different Country, The Same Heaven: A Preliminary Biography of Giulio Alenis, S.J. (1582-1649),' p. 27.

72 Yang Tingyun 楊廷筠 (1557-1627), Hangzhou native, was one of the 'three pillars' of Catholicism in China; see Wang Chung-min, 'Yang T'ing-yün.'

73 The copy of the *Kunyu tushuo* held by the Vatican Library has been exhibited at the Library of Congress. On the *Xifang dawen* see Mish, 'Creating an Image of Europe for China: Aleni's Hsi-fang ta-wen,' pp. 1-87.

74 A similar map was also published as a single separate sheet, signed by 'Aleni of the Western Seas,' on which the influence of the original page format is obvious. Menegon notes two editions

which could no longer fit on this miniaturized version, were adapted within the text of Aleni's book.[75] While the practice of simplifying large maps to fit them into books was well established in the late Ming, the reproduction of so much peripheral explanatory text was a bit unusual; it was necessitated by the special circumstance of introducing a fundamentally alien worldview. Even the non-Christian Ming scholars who reproduced Ricci's maps in their books copied his preface by way of introduction, as seen below.

The globe made by Manuel Dias and Nicolo Longobardi in 1623 is the earliest extant Chinese globe, though famously preceded by Jamal al-Din's now-lost wooden sphere presented to the court during the Yuan, as well as numerous Jesuit globes mentioned in texts but now lost.[76] Among the least fortunate Jesuit successors of Ricci in the field of cartography were Gabriel de Magalhaes and Lodovico Buglio, who were in Sichuan at the time of Zhang Xianzhong's 張獻忠 bloody rule. They created globes of the heavens and the earth at his pleasure, while witnessing horrible brutality at court.[77] The important work of later influential Jesuits such as Johann Adam Schall von Bell (1592-1666) and Ferdinand Verbiest (1623-1688) dates to the Qing, so I will omit its discussion here.[78]

of this version, one in the late Ming and one issued after 1644 but before Aleni's death in 1649 ('A Different Country,' p. 39).

75 The proselytizing character of this text is especially evident in its descriptions of European countries, in which religion is linked with social harmony and prosperity. Bernard Hung-Kay Luk makes an interesting comparison of Aleni's descriptions of lands in the Indian Ocean with those of Ma Huan and other commentators on Zheng He's voyages; he notes that despite their ethnocentrism, Chinese descriptions of places like Hormuz are fairer than the religiously tinged negative portrayals found in the *Zhifang waiji* ('A Study of Giulio Aleni's Chih-fang wai chi,' pp. 68-69). Luk's article also contains useful observations on the sources for Aleni's work and its later influence, which he argues was minimal among Sinocentric mainstream scholars.

76 Wallis, 'The Influence of Father Ricci on Far Eastern Cartography,' p. 39. An image of this globe may be found as Figure 1 in Li Xiaocong, *Ouzhou shoucang bufen Zhongwen gu ditu xu lu*. Preprinted paper forms that could be cut out, pasted onto a sphere of wood and hand-colored made the creation of simple globes relatively easy.

77 Dunne, *Generation of Giants*, p. 325

78 In 1646 the Shunzhi emperor gave Schall the title 'Tong xuan jiaoshi' (Religious teacher who understands the mysterious). Schall had cemented his reputation by predicting the minute during which an eclipse would begin. See Witek, 'Johann Adam Schall von Bell and the Transition from the Ming to the Ch'ing Dynasty,' p. 118. Although his main contributions were in the calendrical sciences, contemporaries linked Schall with the tradition of Jesuit cartography. For example, Wu Weiye 吳偉業 wrote a poem that described him unfolding a map and showing the distant Western Regions, as discussed in Chan, 'Johann Adam Schall in the Pei-yu lu of T'an Ch'ien and the Eyes of His Contemporaries,' p. 288.

Influence in Ming intellectual circles

Though evidence is scattered and often anecdotal in nature, there are signs of an initial wave of intellectual interest in the cartography and other secular contributions of the Jesuits. Despite the almost immediate opposition from some quarters, for perhaps two decades it was fashionable in many literati circles to show knowledge of these works. This period was followed by a backlash, in part due to the shifting of political winds exemplified by Shen Que's crackdown but also exacerbated by an increasingly strident and polemically charged intellectual atmosphere in the wake of the Donglin movement's suppression and the turmoil of the closing years of the dynasty. By the early Qing, despite the continued presence of Jesuits such as Schall at court, the influence of Jesuit cartography in broader society had essentially dropped off the map, especially as compared to the situation in Japan. As noted by George H.C. Wong, the Ming dynastic history compiled during the Qianlong reign includes only scattered and inaccurate records of Europe in its section on foreign countries, due to deep reluctance to trust the information in Jesuit maps and other treatises.[79]

We observe hints of an initial period of positive reaction not only in the writings of Jesuits and their supporters, which must of course be viewed with skepticism, but also from the words of opponents who looked back on this Jesuit heyday as an era when they were underdogs. Wei Jun 魏濬 (*jinshi* 1604) recalled that Matteo Ricci had been quite successful at deceiving scholars with false teachings.[80] The preface to Chen Zushou's *Huang Ming zhifang ditu* 皇明職方地圖 (discussed below) posits a debate between the author, a staunch defender of the Sinocentric *zhifang* cartographic tradition, and a straw man who prattles on about the Jesuits, Magellan, and other nonsense, representing the all-too-easily influenced intellectual mob of recent years.

Unsurprisingly, the most direct influence of Jesuit maps on Ming scholars is to be found among the associates of the Jesuits. Xu Guangqi 徐光啓 (1562-1633), a convert who took the Christian name Paul, is best known for his contributions to agricultural science, but a copy of one of Ricci's maps first captivated him in 1596.[81] Li Zhizao worked extensively with Ricci in composing his large annotated world maps; it is impossible to separate Ricci's writing from the sheen that Li may have had a hand in applying to Ricci's Chinese prose.

79 Wong, 'China's Opposition to Western Science,' pp. 47-48.

80 Cited ibid., p. 44.

81 On Xu's life and his relationship with both Ricci and the Catholic faith, see Sun Shangyang 孙尚扬, *Li Madou yu Xu Guangqi* 利玛窦与徐光启.

Beyond this circle of intimate contacts, levels of influence varied. Aside from works that either adopted the new geographical vision wholesale or rejected it completely, there were some mapmakers who accepted the existence of the Jesuits' far-off homelands without adopting their method of representation, in some cases suspecting that they were lying about the distances involved and actually lived in islands not too far from the Ming's southern coast. The Jesuit Alexander de Rhodes, in his comparison of European and Chinese cartography, describes a hybrid map representing this perspective:

> Also the enormous size of their country and the abundance of goods they have there has made them so presumptuous they are persuaded that China is everything most beautiful in the whole earth, and they are very much surprised when they see our world maps where their kingdom appears so small as compared with the rest of the world. They do quite otherwise, for on their maps they show the world as a square and put China in the middle [...] and paint the sea underneath, in which they scatter a few small islands: one is Europe, another is Africa, another Japan – concerning which we certainly proved to them that they were far less well-informed than we.[82]

The fact that Europe and Africa appear on these maps at all represents the partial digestion of Jesuit claims. The continued use of a 'square earth' format could either indicate Sinocentric rejection of Jesuit geodesy, or simply the persistence of the derivative printed format of *zhifang*-style maps, in spite of acknowledgement that the tradition had been shaken by the interpolation of new knowledge.

While maps such as this suggest little influence from Jesuit cartography, other contemporaneous publications show a dramatically different perspective. Two of the major late Ming encyclopedias, the *Tushu bian* and the *Sancai tuhui*, incorporate Riccian maps directly into their inventories. Cordell Yee has emphasized the crudeness of the engravings to argue that the Jesuit arguments were little understood.[83] I suggest that the limitations of the small woodblock print, and the common late Ming practice of incorporating derivative copies of larger works, must be taken into account before we judge this question by the appearance of the maps alone. More important is the information the editors of these texts have chosen to place with the

82 Translated by Hertz in *Rhodes of Viet Nam*, p. 29.

83 Yee, 'Traditional Chinese Cartography and the Myth of Westernization,' pp. 175-176.

Figure 3.1 Jesuit-style world map illustrated in the *Tushu bian*

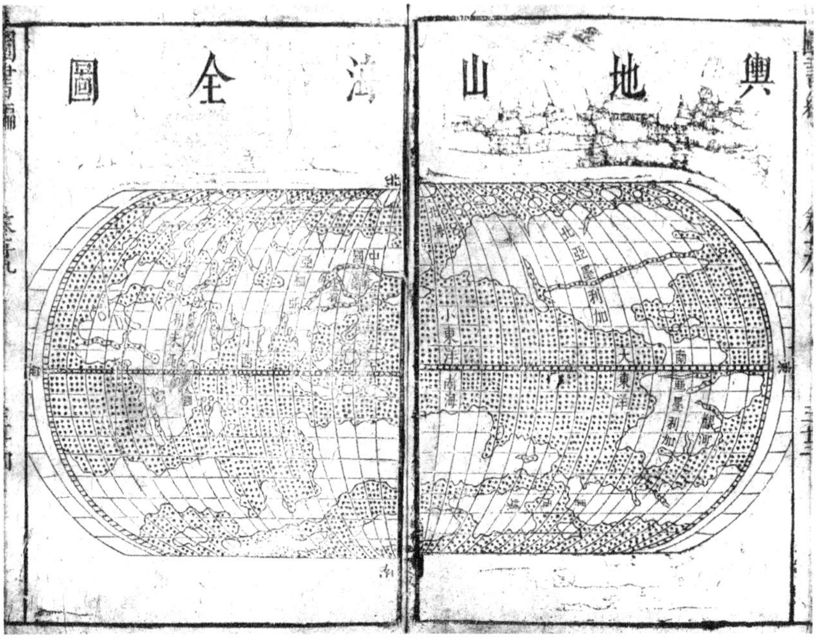

Tushu bian, juan 29, pp. 32b-33a
Courtesy of the Harvard-Yenching Library

maps. Both works copy not only Ricci's maps, but his extensive explanatory text as well. Now we may turn to the question of how the textual context in which these crude copies appear helps to clarify how they were understood.

Citation of Ricci in the *Tushu bian*

Zhang Huang's inclusion of a Jesuit-style map in the *Tushu bian* has drawn the attention of scholars because it is thought to provide the earliest glimpse, albeit condensed and simplified, of one the long-lost primary editions of Ricci's world map (Figure 3:1).[84] Far less attention has been paid to the text accompanying the map, the *Diqiu tushuo* 地球圖説 (Explanation of the map of the globe) (Figure 3:2).[85] The differences between this preface and the earliest extant copy of Ricci's map, the 1602 edition, suggest that

84 Ibid., p. 171.

85 Zhang Huang, *Tushu bian, juan* 29, 34b-35b. The map immediately precedes this description, on pp. 33b-34a.

Figure 3.2 ***Diqiu tushuo*** **(地球圖說, 'Explanation of the map of the globe'), a passage in the *Tushu bian* based on the text of Matteo Ricci's preface**

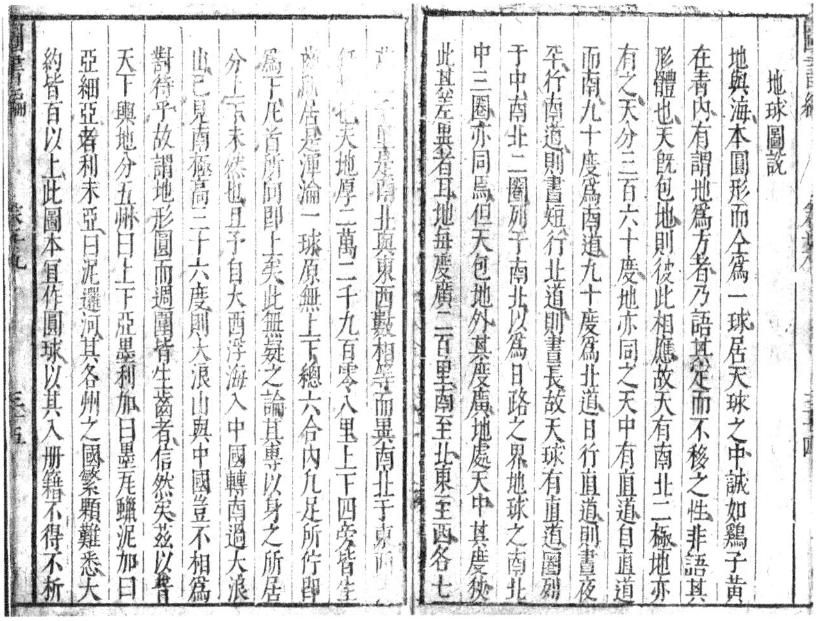

地球圖說
地與海本圓形而合為一球居天球之中誠如鷄子黄
在青內有謂地為方者乃語其定而不移之性非語其
形體也天既包地則彼此相應故天有南北二極地亦
有之天分三百六十度地亦同之天中有直道自直道
而南九十度為南道九十度為北道日行直道則晝夜
平行南道則晝短行北道則晝長故天球有直道圈列
于中南北二圈列于南北以為日路之界地球之南北
中三圈亦同焉但天包地外其度廣地處天中其度狹
此其差異者耳地每度廣二百里南至北東至西各七
里是南北與東西數相等而異南北于東西
也天地厚二萬二千九百零八里上下四旁皆生
齒所居是渾淪一球原無上下總六合內凡足所佇即
為下凡首所向即上矣此無疑之論其專以身之所居
分上下未然也且予自大西浮海入中國轉南過大浪
山已見南極高三十六度則大浪山與中國豈不相為
對待乎故謂地形圓而週圍皆生齒者信然矣茲以[illegible]
天下輿地分五州曰上下亞墨利加曰墨瓦蠟泥加曰
亞細亞者利未亞曰泥還河其各州之國繁夥難悉大
約皆百以上此圖本宜作圓球以其入冊籍不得不析

Tushu bian, juan 29, pp. 34b-35a
Courtesy of the Harvard-Yenching Library

Zhang is quoting extensively, without citation, from an earlier version of Ricci's preface. This would mean that a version of one of the earliest Jesuit cartographic apologia in Chinese has been hiding in plain sight.

It is clear from Zhang's retention of Ricci's authorial voice that the quotation comes from him; the main problem is determining how much of the text is Ricci's and what has been excised or reworded by Zhang. There are several serious errors Ricci would not have made. The fact that Zhang was working with a version of Ricci's preface that predates the extant 1602 version is demonstrated by the more preliminary, less exact conversions of degrees into *li* on the earth's surface. The beginning of the passage, however, is identical:

> The land and the sea are basically round in shape, and together they form a globe, which is situated in the celestial sphere like the yolk of a chicken's egg is suspended in the white. It has been said that the earth is '*fang* (方),' but this refers to its being fixed and immobile, not to a [square] shape. The heavens envelop the earth and they mutually correspond. Therefore the heavens have two poles, south and north, and the earth does as well.

The heavens are divided into 360 degrees, and the earth is too. In the middle of the heavens lies the [celestial] equator.[86] Ninety degrees south of the equator lies the Tropic of Capricorn and at ninety degrees north is the Tropic of Cancer.[87]

When the sun moves over the equator, day and night are equal. When it moves south of the equator, the day is shorter. When it moves north of the equator, the day is longer. Thus the celestial sphere has an equatorial circle in the middle, and southern and northern circles to the south and north, determining the boundary of the sun's movements. Likewise, the terrestrial globe has three corresponding circles in the south, north and middle. However, the heavens are deep and vast as they envelop the earth and extend outward, while the earth, being situated within the heavens, is small and constricted. In this respect they differ.

On the earth each degree equals 200 *li*.[88] Measured either from south to north, or east to west, [the circumference of the globe] is 72,000 *li*. The south-north and east-west measures are identical, so those who suspect that there is a difference between south-north and east-west [distances] have no basis.

Now, the earth has a thickness [diameter] of 22,908 *li*. It is inhabited on all sides, whether above or below. It is a sphere, without an 'up' or 'down;' absolutely everywhere under heaven, wherever the feet plant themselves is down, and where the head points, is up. This is an indubitable theory; it is not the case that one will point up or down depending on one's position. Now, when I sailed by sea from the West to the Middle Kingdom I passed the Cape of Good Hope, where I saw that the south [celestial] pole was 36 degrees above the horizon. Thus, do not the Middle Kingdom and the Cape of Good Hope have corresponding latitudes?[89] Thus I say it is worthy of belief that the Earth is round and inhabited on all sides.

Now the whole earth under heaven is comprised of five large continents: Upper and Lower America, Magellanica, and Asia; Libya is called the Nile River.[90] The countries on each continent are numerous and difficult to distinguish [on the map]. Generally speaking, each [continent] has over

86 Throughout the text 直道 appears in the place of the 赤道 found on Ricci's 1602 map.

87 Note that this should be 23.5 degrees rather than ninety. As we shall see in the *Sancai tuhui* a different but equally colossal blunder appears in this description of the Tropics.

88 Note the shorter distance given here; Ricci's 1602 map indicates 250 *li*. This also results in a different circumference in the following line.

89 That is, one lies at 36 degrees north, the other at 36 degrees south.

90 Note that not only is the order different, but the Americas are not referred to as North and South. On the map the characters for the Nile (泥邏河) appear next to the characters for Libya

> one hundred countries. This map was originally done as a globe, but to put it in a book there was no choice but to transform the sphere into a flat map.
>
> Lines of latitude and longitude are drawn to form squares of ten degrees [to a side]. By means of this it is possible to place each country at its proper meridian, east and west.
>
> The north-south running line of longitude that goes through the Middle Kingdom starts from the Fortunate Islands.[91]
>
> 地與海本圓形，而仝為一球，居天球之中誠如雞子黃在青內。有謂地為方者，乃語其定而不移之性，非語其形體也。天既包地,則彼此相應,故天有南北二極,地亦有之。天分三百六十度,地亦同之。天中有直道,自直道而南九十度為南道,九十度為北道,日行直道,則晝夜平;行南道則晝短;行北道則晝長。故天球有直道圈列于中,南北二圈列于南北,以為日路之界。地球之南北中三圈亦同焉。但天包地外其度廣,地處天中其度狹，此其差異者耳。地每度廣二百里。南至北東至西各七萬二千里。是南北與東西數相等，而異南北于東西者無據也。夫地厚二萬二千九百零八里。上下四旁皆生齒所居。是渾淪一球，原無上下，總六合内凡足所伫即為下，凡首所向即上矣。此無疑之論其專以身之所居分上下未然也。且予自大西浮海入中國轉南過大浪山,已見南極高三十六度。則大浪山與中國豈不相為對待乎。故謂地形圓而週圍皆生齒者信然矣。玆以普天下輿地分五州：曰上下亞墨利加，曰墨瓦蠟泥加，曰亞細亞者，利未亞，曰泥邏河。其各州之國繁夥難悉，大約皆百以上。此圖本宜作圓球，以其入冊籍不得不析圓為平。其經緯線畫每十度為一方,以分置各國于其所東西線數。自中國起南北線數自福島起也。

To conclude this passage Zhang adds a note in his own editorial voice: 'This map was drawn by a far westerner, who claims to have personally experienced what it depicts' (此圖即太西所畫,彼謂皆其所親歷者).[92]

The earlier version of Ricci's preface may have had far fewer details than that on the 1602 map, to judge by Zhang's shorter quoted version, but it is also apparent that many of those details it did contain have been excised. For example, the latitude of the Cape of Good Hope is compared with that of the Ming, which has not yet been introduced; the system of longitude is barely touched on at all, though the citation of the Fortunate Islands suggests

(Africa), so the provision of this 'alternate name' may be Zhang's attempt at clarification based on what he saw on the map.

91 I believe this refers to the role of the Fortunate Islands as marking the Prime Meridian.

92 Zhang, *Tushu bian*, *juan* 29, p. 35b.

that Ricci's original probably described their location in a manner similar to that found in the 1602 version. The name of Europe is entirely omitted.

Zhang's use of this text highlights the difficulty of evaluating the extent to which he was persuaded by or even comprehended the Jesuit geographical worldview. The error in transcription of the latitudes of the Tropics of Cancer and Capricorn suggests that this aspect, at least, was not fully understood, but it could have been a copyist's error; the text was, after all, committed to woodblock after Zhang's death. Even if we cannot determine how deeply Zhang was persuaded, it is evident that the Jesuit maps had become a part of the Ming geographical conversation, one of the threads to be woven together in a scholarly treatment.

Citation in the *Sancai tuhui*

In comparison to the scholarly *Tushu bian*, the more popular *Sancai tuhui* is deficient in various ways, including thoroughness and basic editorial quality. These shortcomings are no less obvious in the adaptation of Ricci's map and preface. Unlike the quotation in Zhang's text, the Wangs do not tell the reader that Ricci's words are those of a foreign scholar; the copied text is simply sutured into the narrative. The way this text is incorporated, even 'naturalized' in the context of the *Sancai tuhui*, might be interpreted as granting it a greater level of credence. This is not a description of the world by a foreigner; it is simply a description of the world. A number of textual errors do complicate the picture, as features that are textually described do not appear on the map, and the transcriber of the preface's text omits a long string of characters, rendering a portion superficially readable but logically incoherent. Would these mistakes have escaped the notice of an editor who thoroughly understood the text?

Looking at the small world map occupying pages 2a and 2b of the first chapter, which is a simplified version of Ricci's *Kunyu wanguo quantu* (Figure 3:3), we notice at once that the grid of latitude and longitude is omitted although the text at lower right mentions its presence; this is but one serious editorial shortcoming. We might be tempted to conclude that the Wangs grasped the basics of 'Five Continent' theory but were at a loss when it came to the details of latitude and longitude, even though Ricci's explanation of the function of the meridians and how to calculate distance based on them appears in the accompanying text. There is no way to be certain whether this omission was due to confusion, or simply the limitations of cheaply produced small-size woodblock prints.

Figure 3.3 World map from the *Sancai tuhui*

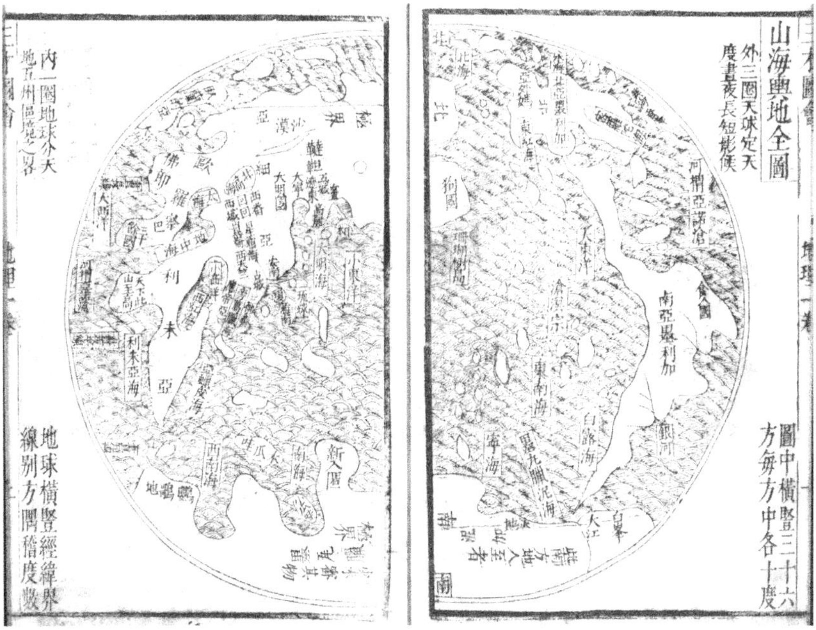

Dili juan 1, pp. 1b-2a
Courtesy of the Harvard-Yenching Library

The explanation of the map comprises the first part of the 'Geography' section in the *Sancai tuhui*.[93] The passage starts off with the familiar passage, 'The land and the sea are basically round in shape, and together they form a globe, which is situated in the celestial sphere like the yolk of a chicken's egg is suspended in the white. It has been said that the earth is '*fang* (方),' but this refers to its being fixed and immobile, not to a [square] shape' (地與海本是圓形，而仝為一球，居天球之中，如鷄子黄在青內。有謂地為方者，乃語其定而不移之性，非語其形體也). It is clear that the Wangs are copying from a later version of the original, however, due to changes such as the citation of North and South America as two separate continents.

I have translated the passage in its entirety in Appendix 4, but several points are worth noting here. A serious error has been introduced into the passage on the Tropics, due to a copyist's failure to transcribe a line. Ricci's original states that 'South from the equator 23.5 degrees lies the Tropic of Capricorn, while 23.5 degrees to the north of the equator lies the Tropic of

93 This portion of the text can be found in the 1988 Shanghai guji chubanshe facsimile edition, pp. 92-94.

Cancer. Thus, the Middle Kingdom is north of the Tropic of Cancer.' The text given here states 'South from the equator 23.5 degrees lies [omitted text] the Tropic of Cancer' (自赤道而南二十三度半為北道). A passive reader might not even notice that this tropic actually lies North of the equator, and the Tropic of Capricorn has been omitted. The fact that such a mistake went into print could be cited as evidence that the entire system was not understood. On the other hand, the process of moving from draft manuscript copy to engraved woodblock presented opportunities for such errors to creep in without attracting notice.

Other changes reflect depersonalization of the text; for example, Ricci's line about having visited the southern hemisphere and still seeing the sky above and his feet below has been removed. This helps to naturalize this text in its new context divorced from Ricci's map. Later, a note about differentiating the continents with color has been removed, as the book was printed in black ink and not colored by hand.

Let us examine how this text is knitted together with the passage that immediately follows it. Having read an overview of the earth's form, we zoom in from the macroscopic view to the particular territory of the Middle Kingdom and its successive dynasties, a transition that further illuminates the 'naturalization' of the Jesuit text. In this passage, we find a more orthodox interpretation of the historical context and the purpose of cartography:[94]

> Since antiquity, the first task of rulers has been to unify *Tianxia*, and after that to make it flourish.[95] The organization of territory in the period before Yi and Nong[96] is unknown today, that time being so remote. In various annals it can be seen that the Yellow Emperor drew boundaries and delineated territories, creating states of 100 *li* and a myriad *qu* [subsidiary regions]. Di Ku created the Nine Regions under which the myriad states were subsumed and governed. There were floods in the time of Yao and *Tianxia* was divided into twelve territories. This led Yu to stabilize the waters and the land, reestablishing the Nine Regions, and he created the five zones of submission. Yu continued the florescence of Yao and Shun.

94 See ibid., pp. 94-96, for this portion. The text is copied from the *Da Ming yitong zhi*'s map preface, where it begins on p. 2a. It is omitted from the version of the *Da Ming yitong zhi* preserved in the *Siku quanshu*. At the time when the Wangs produced the *Sancai tuhui*, the passage had been more recently copied as the beginning of *juan* 34 in the *Tushu bian*.

95 I leave *Tianxia* in Chinese, because I think that in the context of Ricci's broader worldview which appears earlier in the text, it can no longer be taken to mean all under heaven, but more likely something closer to 'the realm.'

96 That is, before Tai Hao and Yan Di in the third millennium BCE.

At the time of the meeting at Tu Shan, a myriad states came bearing jade and silk.[97] Over the course of four hundred years these states absorbed one another so that at the time when Shang received the mandate, there were some three thousand remaining, though *Tianxia* was still partitioned into the Nine Regions. By the time the Zhou subdued the Shang, there were still 1800 states and Tianxia was divided into nine domains (*ji* 畿). In the time of Cheng Wang they were again called regions (*zhou* 州). After this the various lords began to swallow one another up, and the situation deteriorated until the Warring States when *Tianxia* was divided into the Qin and six other states. The lords arranged *Tianxia* into 40 *jun* (郡). Following the Qin system, the Han established *jun* and states (*guo* 國). Wu Di expelled the Hu and opened up Yue, broadening the land in every direction and dividing it into thirteen regions (*zhou*). The *jun* all had governors (*ci shi* 刺史). At the time of the Three Kingdoms, when the Jin arose it started [a process of] unification and established nineteen *zhou*. Before long the north and south split, until the Sui reunified them and all the *jun* were converted to *zhou*.

When the Tang replaced the Sui, the *zhou* grew increasingly numerous. In the beginning of the Zhenguan period [627-650], *Tianxia* was divided into ten circuits. During the Kaiyuan period [713-742] circuits were added to make a total of fifteen. When the Song replaced the Five Dynasties it abolished all of the illegitimate holdings [i.e., the lands of local rulers] but *Tianxia* was not yet divided into fifteen *lu* [this happened in 997, a number of years after the dynasty was founded]. During the Xuanhe period [1119-1126], *lu* were added to total twenty-six. When the Yuan governed *Huaxia* with barbarians, it established a central province to govern the central *lu*, and outside established provinces numbering ten to govern the rest of the *lu* in *Tianxia*.

Now, although the territory [of the Yuan] was greater than before, it did not append the islands and barbarian territories of the southeast, [something which happened] only with our Imperial Ming's glorious reception of the heavenly mandate to unite the Chinese and the barbarians. The broadness of the borders is such that in the east they extend to the 'left' Liao, in the west they extend to the shifting sands; in the south they stretch beyond the seas, and in the north they abut the desert. From the four directions and the eight wastes, there are none who fail to come to court [with tribute]. The system of land distribution is thus: [aside from] the capital and imperial district, prefectures, and sub-prefectures, as well as Zhili and the Six Bureaus,

97 This was a meeting that Yu is said to have summoned local lords to attend before he died.

Tianxia is divided into thirteen provinces: Shanxi, Shandong, Henan, Shaanxi, Zhejiang, Jiangxi, Huguang, Sichuan, Fujian, Guangdong, Guangxi, Yunnan, and Guizhou, which govern their own prefectures and counties. Regional Military Commissions, Guards and Battalions are scattered among them for defense. In total there are 149 prefectures, 128 sub-prefectures, and 1,105 counties. In the border areas the Regional Military Commissions, Guards and Battalions as well as [various types of aboriginal officials]. Those among the four barbarians who have taken official enfeoffments and grasped the rites of tribute have been recorded one after another in the annals.
Now, in the past the officials of the Zhou had annals for surveying affairs, and for scrutinizing the land they had maps. Therefore maps are made again today. Arranged before the maps of the two imperial domains and those of the individual provinces, there is a general map of ***Tianxia***. Gazing at it, the territory's vastness will be fully before one's eyes, and it will be sufficient to prove for a myriad generations the supremacy over past and present of our Imperial Ming's unity.

自古帝王之御世者必一統天下而後為盛。羲農以上疆理之制，世遠莫之詳矣。其見諸載籍者，謂黃帝畫野分州，得百里之國萬區。帝嚳創制九州，統領萬國。堯遭洪水，天下分絶為十二州，使禹平治水土，更制九州列五服。禹繼唐虞之盛，塗山之會，執玉帛者萬國，而四百年間，遞相兼併。逮商受命，其能存者纔三千餘國，亦為九州，分統天下。及周克商，尚有千八百國，而分天下為九畿。至成王時，仍曰九州。厥後諸侯相吞，列國耗盡，陵夷至於戰國，天下分而為七。秦並六國，罷侯置守，分天下為四十郡。漢因秦制，加置郡國。武帝攘胡開越，四履彌廣，分天下為十三州郡，皆置刺史。既而三國鼎峙。至晉始合為一置州凡十有九。未幾，南北分裂，至隋復合為一，盡廢郡為州。唐承隋後，置州愈多。貞觀初，分天下為十道，開元中又增至十五道。宋承五季，削平偏據，至道末，分天下為十五路，宣和中，又增至二十六路。元氏以夷狄入主華夏，内立中書省一以領。雖過於前，而東南島夷則未盡附，惟我皇明誕膺天命統一華夷。幅員之廣東盡遼左，西極流沙，南越海表，北抵沙漠。四極八荒靡不來庭。而疆理之制則以京畿府州直隸六部，天下分爲十三步政司，曰山西，曰山東，曰河南，曰陝西，曰浙江，曰江西，曰湖廣，曰四川，曰福建，曰廣東，曰廣西，曰雲南，曰貴州，以統諸府州縣。而都司衛所則錯置於其間，以為防禦。總之為府一百四十九，為州一百二十八，位縣一千一百五，而邊陲之地，都司衛所及宣慰，招討，宣撫，安撫等司，與夫四夷受官封，執臣禮者，皆以次具載於志焉。顧昔周官詔觀事則有志，詔地事則有圖。故今復為圖分置於兩畿各布政司之前，又為天下總圖於首。披圖而觀，庶天下疆域之大了然在目，而我皇明一統之盛冠乎古今者垂之萬世有足徵云。

The pairing of this passage with text from Matteo Ricci's map has the effect of 'nesting' the orthodox worldview within an expanded awareness of the lands beyond. The broader world is sculpted in general physiographic terms, after which the focus moves in to the reader's own land, a territory that is laden with more historical and cultural significance and that has taken its current form through many stages of administrative change.

For the rest of the geographical portion of the *Sancai tuhui*, the frame of reference focuses on the Ming and its adjacent territories, following established prototypes without either contradicting or citing Ricci's clearly articulated vision of the world as a whole. The only place where a clash might have occurred is with the presentation of a map of the southwestern seas copied from the *Guangyutu*, which includes a distorted image of Africa. This map was by then so far removed from its original context and so little understood (there is no accompanying commentary at all) that the compilers may have been completely unaware that the heart-shaped piece of land depicted there was the same as the continent of Libya on Ricci's map. The Persian-derived names on that earlier map would have presented no obvious clues as to the identity of those places on a Jesuit-derived work, even though both portray the Nile River.

Jesuit maps would not have reached the compilers of these two encyclopedias if not for the flourishing woodblock print industry of the late Ming, which both inspired the Jesuit production of maps as evangelical tools, and facilitated their circulation far beyond the range of individual manuscript copies. By reproducing these maps, Zhang and the Wangs further perpetuated the images introduced by the Jesuits, even if the degree to which they influenced later thinkers is debatable. A Japanese encyclopedia called the *Wa-Kan sansai zue* 和漢三才圖會, styled after the *Sancai tuhui*, borrowed a similar Riccian map. This map and its textual explanation, which are discussed in Chapter Five, further highlight the trans-regional influence of late Ming compendia.

Rejection of Jesuit cartography

Modern scholarship has understandably taken an interest in the local influence of cartography introduced from Europe. The interest shown by some Ming figures has a flip side, however, as evinced by the anti-Jesuit furor inspired by the efforts of these foreign missionaries to preach doctrines that offended certain local sensibilities, both religious and scholastic.

Shen Que, then acting Minister of the Board of Rites, initiated the suppression of the Jesuits in Nanjing in 1616, and he documented his attempted purge.[98] An abbreviated collection of his records was published in 1640 by Xu Changzhi 徐昌治 under the title *Poxie ji* 破邪集 (Record of the smashing of the heterodox). During Shen's crackdown the Jesuit house in Nanjing was demolished and its structural materials used to restore a shrine to Lord Huang (a deified Jianwen loyalist), while an inventory was taken of its books before the Western works, including maps and globes, were burned in front of Shen's residence.[99]

The status of Jesuit cartography among the literati seems to have suffered a severe blow around the time of the crackdown, though it is difficult to weigh the significance of the various factors involved. A changing political scene had resulted in less openness to foreign novelties. Literati may also have become more familiar with the underlying religious purpose of the mission, or their greater acquaintance with Jesuit theological beliefs may have led them to find Jesuit arguments less convincing in other realms. In any case, harsh responses to the Jesuit map seem to become more common by the 1630s, and incorporation in new non-Jesuit works less frequent (though existing works continued to be reprinted). One critic, Chen Zushou 陳組綬 (*jinshi* 1635), looked back on the initial period of welcome as a time when the mainstream of Ming scholarship was deluded.

Chen's *Huang Ming zhifang ditu* 皇明職方地圖 (Administrative atlas of the Imperial Ming), published in 1636, makes a stridently polemical case against the geographical claims of the Jesuits and those Ming scholars who found them persuasive.[100] In his preface Chen struggles with a hypothetical critic in a fictional debate. The critic questions his lack of reference to the theories of Western scholars or the experiences of navigators like Magellan,

98 Shen Que, appointed vice minister of Rites in Nanjing in 1615, led opposition to Jesuit activities on numerous grounds. Local Buddhists joined forces in his campaign (Dudink, 'The Inventories of the Jesuit House at Nanking Made Up During the Persecution of 1616-1617,' p. 124). For a short biography see Lienche Tu Fang, 'Shen Ch'ueh.' For a detailed study of the suppression of Christian activities see Edward Thomas Kelly's *The Anti-Christian Persecution of 1616-1617 in Nanjing*.

99 For a detailed analysis of the *Poxie ji*'s composition and a listing of the inventories that it omits, see Dudink, 'Inventories.' The inventory of confiscated items incidentally gives us some sense of the cartographic materials that the Jesuits kept on hand. Though some objects, mostly religious implements, were smuggled out to safety in the middle of the night on August 30, 1616, making this list incomplete, it does include terrestrial and stone globes, a folding screen with Ricci's *Wanguo quantu*, seventeen unspecified maps, and two copies of a 1600 version of Ricci's map, the *Shanhai tu*, now lost (Dudink, 'Inventories,' pp. 126, 137, 139). The entire Chinese-language library amassed by the Jesuits was confiscated, while the western works were burned.

100 Reprints of this work appeared at least until 1736, but it was then banned and excluded from the *Siku quanshu*, perhaps for its anti-Manchu material related to defense. Copies circulated in manuscript form, exemplified by one from Japan held at the Harvard-Yenching Library.

lamenting Chen's failure to see the broader world. Chen responds that all under Heaven is in the hands of the Ming emperor, who has inherited it along with 3,600 years of history – a history replete with geographical knowledge about the Four Seas and the Nine Regions that has remained unchallenged for centuries. To discard this tradition now would be a betrayal; 'I have heard of Xia changing the Yi barbarians, but never of the Yi changing the Xia' (吾聞用夏變夷者未聞變于夷者也).[101]

In his world map Chen depicts lands that were recorded in old dynastic histories but absent from the Jesuit worldview. For example, we find the term Fulin 拂林, commonly used in Tang works to refer to Byzantium, as if the distant west were still frozen in the configuration it held during the period of the Middle Kingdom's greatest westward contacts. This is a pointed rejection of Jesuit claims.

Chen turns for a model to the path breaking *Guang yutu*, a geographical text that predated the appearance of the Jesuits, which we have discussed in Chapter One. Chen's first map is an overview of the world as one large continent, followed by maps of the various Ming provinces, then frontier regions, then sections on rivers and canals, and ocean defense. He includes maps of Korea and Vietnam, the northwestern deserts and Western Regions, a list of entry routes from foreign lands, and countries paying tribute (including Japan). The Sinocentric character of Chen's work is a revival of the classically educated literatus's default perception of the world, and serves as an explicit call to return to that perspective.

In Chen's preface he notes the age-old reliance on maps within the Board of War (*Zhifang* 職方), thus connecting himself with an ancient tradition. Aware that his more cartographically conscious readers will immediately perceive the similarities between his work and the *Guang yutu*, he enumerates differences between his work and that of Luo Hongxian. Luo's work did not address territories beyond the Ming, and it was precisely Luo's failure to situate the Ming within the broader world that left the door open for the Jesuits to 'Roll out globes to sanctify their claims minimizing the Middle Kingdom and magnify the [territories of the] barbarians of the four directions' (披地球以神其說小中國而大四夷也).

Chen continues in this vein, 'Alas! This is the error of the *Guang yutu*. It does not examine the universe of the Emperors and Kings of antiquity. It does not examine the Classics of the ancient Sages and Worthies. It does not examine the territories of the Rong and Yi tribes of antiquity. It does

101 Chen, *Huang Ming zhifang ditu*, Preface, p. 5b. All of the Chen quotes cited in the next four pages come from this preface.

not examine the near and far, center and frontiers of antiquity. It doesn't look into the places where the three Huang, the five Di, the thousand sages and myriad worthies emerged, or their movements over land and water. By these unexamined words, it minimizes our Middle Kingdom. This is very lamentable' (嗟乎此廣輿之過也，不稽之古先帝王之宇，不稽之古先聖賢之典，不稽之古先戎夷之域，不稽之古先地里之遠邇，古先地里之中邊，不稽之古先三皇五帝千聖萬賢之出處經涉，以此無稽之言得小吾中國是大可痛也). Chen is resolved to correct the record by restoring the classical Sinocentric version of the world's physical organization.

Chen realizes that he could be accused of simply copying some of Luo Hongxian's maps, but he replies to his theoretical accuser, 'Maps depict the land, how can I dare make them differ?' (圖所以圖乎地豈敢有異). His new version is, he claims, still an improvement because Luo's maps could be divided but not merged together; they were acceptable for mountains but not for rivers. This new work depicts the mountain chains spread out clearly, and represents the rivers with their sources individually depicted.

Chen's hypothetical interlocutor next charges him with being prejudiced and behind the times, for the majority decides what is right or wrong. If Chen neglects the position of the majority, the majority will not accept his work. Nonsense, Chen responds, for today's right and wrong are not the right and wrong of antiquity. He depicts himself as a valiant voice of tradition standing against the tide of contemporary public opinion. It is difficult to determine whether Chen was using this underdog ploy as a mere rhetorical device, or whether a significant number of contemporaries truly looked upon him as a backward crank.

Chen portrays himself as a something of a lone sentinel, but he is not without role models in antiquity or friends in the present. His book is very much a product of the Ming, as his list of references shows. If not for Zuo Gong traveling around Xia, the Spring and Autumn Annals would not have been written. By the same token, the Ming dynasty's *Huangyu ji*, the *Da Ming yitong zhi*, the *Da Ming guan zhi*, and the *Guang yukao* all inspired Chen's own work. Chen also lists numerous contemporary friends and colleagues who allowed him to examine maps and texts in their collections.

We next encounter Chen's most direct rejection of the materials imported by the Jesuits. His critic mentions Magellan's circumnavigation of the earth, during which he passed below the equator, traversing more than 300,000 *li*: 'And yet you do not expand your *zhifang* on this basis. I fear that those who [believe in] latitude and longitude will call you narrow-minded, saying that you fail to see the great size of the world' (子不以此廣之職方,恐經緯者推步將以子為管窺蠡測不見天地之大也).

Not so, Chen replies. The world is now in the hands of the Ming emperors, who inherited it from the three Huang, five Di, three Wang: 3,600 years of unbroken transmission. How can the Five Continents theory abide? This would be to take the word of the Jesuits over the Four Seas and Nine Regions known since imperial antiquity. 'I dare not acquaint myself with such an idea. Now, in the Warring States there was one Zou Yan, who expounded that the world had nine great regions.[102] But the Zhou did not add this to its *zhifang*. The Xia had Shu Hai, according to whom the world was 28,000 *li* from west to east and 25,000 *li* from north to south. Yet the Xia did not put this into the *Yugong*.' (我不敢知然。則周有鄒衍衍天下有大九州而周不以加之職方，夏有豎亥步天下之東西二萬八千里南北二萬五千里，而夏不以入之禹貢). The Jesuit theories, he implies, are but another fad to be disregarded by upholders of the noble tradition.

Chen's work represents a backlash against European worldviews, but one that could only happen after Jesuit influence had become familiar to his audience. A milder form of rejection appears in some large-format sheet maps, such as the one described by Alexander de Rhodes above, or the 1593 *Qiankun wanguo quantu gujin renwu shiji* 乾坤萬國全圖古今人物事蹟 (Complete map of the myriad counties between Heaven and Earth, with human vestiges and records of events ancient and modern), which places European place names at the edges of a *zhifang*-style square map of the Ming.[103] Such a hybrid reflects acceptance of an outside world differing from the classical view, without adopting the newly introduced form of visual representation.

European 'echoes'

It is worth noting that maps produced during the late Ming publishing boom also stirred interest and had repercussions in the cartography of distant Europe. Nicolas D'Abbeville Sanson's 1656 map of China, among others, is based on a copy of the *Guang yutu* sent to Rome by Matteo Ricci.[104] A number

102 In the third-century BCE, Zou Yan proposed that there were nine great continents, with the known world of his time existing on one of them.

103 Elman, *On Their Own Terms*, pp. 128-129. Richard Smith has also briefly discussed this map in a paper given at Harvard University's Fairbank Center in 1996, 'Mapping China's World: Cultural Cartography in Late Imperial Times.'

104 Szcześniak, 'Matteo Ricci's Maps of China,' pp. 127-136. On Jesuit transmission of Ming and Qing geographical knowledge to Europe, focusing on the later Jesuits, see also Foss, 'A Western Interpretation of China: Jesuit Cartography.' The works of other geographers who transmitted knowledge to Europe in the late Ming and early Qing, such as Michele Ruggieri, Alvaro Semedo,

of European maps of China were certainly based on the 1655 Amsterdam edition of Martino Martini's *Novus Atlas Sinensis*, which was in turn largely based on the *Guang yutu*, as Martini acknowledges in his preface.[105] These maps served as fodder for yet other cartographers.

A seventeenth-century oil painting of Martini held at Trent depicts him pointing to a map; tellingly, it is a Chinese-style map with the Great Wall spanning its northern fringe and the Heishui flowing down the western edge toward the south.[106] It was upon such maps that Martini based much of his own work. In contrast to the other Jesuit figures cited in this chapter, Martini can be described as more of an introducer of China to Europe than a conveyor of European cartography to China.

In a letter of 1616, João Rodrigues states that he has obtained a Yuan-period geographical text that indicates the distances from various points in East Asia to India. He asserts that it will be of great use in correcting errors in European maps.[107] Unfortunately, it is no longer known what became of this document, or even precisely what it was.

Not only did the broad circulation of maps in the Ming spur the Jesuits to produce their translated world maps, which are themselves certainly landmarks in the history of world cartography, but works like Luo Hongxian's *Guang yutu* and large sheet-maps helped European cartographers revise their own efforts until sufficient surveys had been made along the coasts and inland regions to supersede them. In the Ming, for the first time, the Jesuits had encountered a civilization where mapping was not only a well-established practice, but where local maps showed vastly more information than European explorers had yet ascertained about the region. Thus the Jesuit role was, for a time, that of a bidirectional conduit.

Conclusion

If the Jesuits succeeded in bringing their worldview to the attention of a significant number of Chinese, as I have argued, why does their impact

and Philippe Couplet, are introduced in Szcześniak's 'Seventeenth Century Maps: An Inquiry Into the Compilations of European Cartographers.' Of course Ricci did not only send indigenous sources back to Europe; he used them in compiling his own maps for the Ming audience. On his use of text by Ma Duanlin, see Ch'en, 'A Possible Source for Ricci's Notices on Regions near China,' and d'Elia's discussion in 'Recent Discoveries,' pp. 116-118.

105 Li Xiaocong, *Ouzhou shoucang bufen Zhongwen gu ditu xu lu*, p. 90.

106 This painting is illustrated in Mungello, *The Forgotten Christians of Hangzhou*, p. 36.

107 Cooper, *Rodrigues the Interpreter*, pp. 283-284.

appear as limited as Cordell Yee has suggested? While a combination of cultural pride, suspicion of foreigners, and respect for precedent undeniably played a part, it is clear that some literati in the late Ming were willing to consider geographical ideas imported from the West, and some even studied the detailed methodologies that make advanced cartography possible. Nevertheless, there are several important reasons why so little clear Jesuit legacy appears in the long-term cartographic record.

One factor was the Ming state's limited interest in Western surveying methods. In 1612 João Rodrigues reported a court request that a Jesuit take readings of latitude at various places around the empire, a task for which Diego de Pantoia was chosen.[108] Extensive surveys did not take place until the high Qing, however. Nor was there very much opportunity to impart the necessary technical knowledge to a significant number of students who might have pursued survey-based cartography on their own. While a map is simply a two-dimensional image that can be copied, cartography as a practice requires a broad set of skills used for creating or fine-tuning this image in the first place, ranging from theoretical foundations to the practical procedures of measurement. Such skills were not part of the standard curriculum for students preparing for the Ming bureaucracy, and the Jesuits' attempts to establish their own educational institutions encountered numerous difficulties because of the missionaries' precarious position in Ming society. In contrast to the situation in Japan, the Jesuits were not able to establish formal academies in Ming China.[109] Even if they had been able to do so, it would certainly have been religion that occupied most hours of instruction. Without the necessary technical skills, even the most enthusiastic student could have done little more than imitate existing maps.

Large-scale surveying is expensive, for aside from the basic cost of equipment, the survey party is removed from other work and must make arrangements for food and lodging all along its route. The traces of ink on paper that comprise a well-surveyed map conceal chests full of silver – the expenses of their production – that had to move in the background to make this work possible. The more careful the work of preparation, the higher the

108 Ibid., pp. 275-276. Such practical applications were not always welcome in the region; an example from Cochinchina illustrates how geomantic beliefs could prevail against efforts to apply new survey methods. Alexander de Rhodes relates that the king of Cochinchina was anxious about the visits of Jesuit priests to the home of his aunt, for he feared that their mathematical talents would be manifested in the selection of a geomantically perfect grave site for her, leading her descendants to take the kingdom away from his own (Hertz, *Rhodes of Viet Nam*, p. 113).
109 Üçerler, 'The Jesuits in East Asia in the Early Modern Age: A New "Areopagus" and the "Re-invention" of Christianity,' pp. 32-34.

cost, such that a great many hours' labor could go into producing a couple of square inches on the finished product. Although Ricci was working from existing sources, and did not have to perform any surveys in order to produce his elaborate 1602 world map published in Beijing, it nonetheless took him a full year to fit this work in between his other obligations. When he was asked to produce a version for the emperor in 1608, which he realized would be his last chance to impress the ruler with the virtues of Christianity, it still took him a month to have the new blocks carved.[110]

Some of the Jesuits made every effort to take readings along their route when traveling on other business, as in the case of Ricci's first trip from Nanjing to Beijing, during which he and Lazzaro Cattaneo collected latitudinal data and distances between towns.[111] Though spotty observations such as these could be used to adjust existing maps, they were insufficient to create entirely new maps. The expense of conducting a comprehensive survey of even one province could only have been undertaken with extensive financial backing, most likely coming from the state, or the expectation of profit from the publication and sale of the resulting maps. Such state backing did not arrive until the following dynasty. As for profit to be expected from selling maps, there could be no such confidence in the Ming context. Not only would an independent surveyor immediately raise suspicion, but there was also little if any recourse in cases of unauthorized reprinting. An example of such illicit profiteering, which took place right under the noses of Ricci and Li Zhizao, is recorded in Beijing. The printers entrusted with committing their map to woodblock plates thought that the work would sell well, and they carved a second set of plates at the same time, from which they printed their own versions for the market after the authorized plates had been given to Li Zhizao. The plates remained in use until they were destroyed when the house in which they were stored collapsed in 1607. The copies that Ricci gave out bore his seal, which was lacking on the clandestine edition, of which at least one example survives.[112]

The Jesuit production of maps was effectively subsidized, and a great many copies were given away for free because they served a purpose for the grander mission, but no system of patronage existed to support independent cartographers. It was quite feasible, however, to copy maps for which the hard work had already been done. Thus we see reprints of the Jesuit worldview, the entire earth's grand design, but no new maps of smaller locales produced using Western survey techniques. Local maps in gazetteers were usually

110 Day, 'The Search for the Origins,' p. 96.

111 Dunne, *Generation of Giants*, p. 55.

112 See D'Elia, 'Recent Discoveries,' pp. 114-116.

copied from earlier works or redrawn in much the same way as an artist might look at a sketch portrait and then endeavor to draw the same face from a different angle, without seeing the original face. For the surveying of fields during tax assessments, there were longstanding methods in place that served the purpose well enough. It is not until the nineteenth century that dramatic changes in local surveying appeared.[113]

Once the Jesuit worldview had been circulated and established as another resource in the pool, it was reprinted and positioned alongside other worldviews. In a work like the *Sancai tuhui* each of these representations is rendered with approximately equivalent effort (or lack thereof). There is nothing so uniquely crude about the reproduction of Jesuit representations in books like this as to suggest that they were less accepted than other views. The crude reproduction of Jesuit maps is a general phenomenon of cartography in late Ming books, rather than a display of ignorance about Jesuit-introduced materials in particular. The 'incompetent' rendering of copied Jesuit maps does not prove ignorance or rejection of the basic model. The Riccian text accompanying these reprinted maps, in fact, hints at the converse, as does Chen Zushou's self-depiction as an underdog standing against a stiff barbarian tide.

Ricci's policy of using the sciences as a way of drawing respect and preparing the literati for exposure to Christianity was temporarily proscribed after his death by Valentim Carvalho in 1614. He wanted the missionaries to preach the faith alone. Zhang Cuo has suggested that the unique personality and talents of Ricci himself, including his personal charm and magnetism, were major factors in the acceptance and influence of Jesuit ideas, factors that were not commonly observed among his successors.[114] Shen Que's anti-Jesuit campaign presented yet another layer of difficulty.

All of this being said, the role of ideology cannot be ignored when seeking to explain the fact that Western methods did not have a broader effect on world maps in China, or any effect at all on local maps. In Japan, which was at the fringes of the Sinocentric world vision, Western cartography rose to much greater prominence and did not disappear even after the expulsion of Christian missionaries. Meanwhile Chosŏn, which had bought into the Sinocentric vision more passionately than any other tributary state, rejected Jesuit cartography for decades. These comparative examples suggest that the imperial ideology of the Ming and its tribute system, even in its waning

113 See Amelung, 'New Maps for the Modernizing State: Western Cartographic Knowledge and Its Application in 19th and 20th Century China.'

114 Zhang Cuo 張錯, *Li Madou ru Hua ji qita* 利瑪竇入華及其他, pp. 35-41.

days, was one of the factors dissuading local scholars from seeking ways to 'overcome' the barriers cited above. Perceiving itself as the most powerful state in the world and the most culturally superior, with a selection of cartographic systems and tools that had served manifold purposes since at least the Song dynasty, why should its scholars strive to fulfill the expectations that some might retroactively place upon them hundreds of years later?

Harriet Zurndorfer has noted that 'The Ming and Qing governments treated the Jesuit missionaries like they did all foreign 'technicians,' as minions to serve the court and to support the astronomical, military, and geographical needs of the regime.'[115] Even as they accepted the limitations of their situation, however, the Jesuits differed from their Muslim predecessors by pursuing an evangelical strategy in broader society, under the cover of protection afforded by sympathetic literati. Their cartography was a tool in this strategy, but only one of many. Rather than asking why the Chinese 'failed' to follow the Jesuit lead, we should note how remarkable it was that this small sideline became one of the intellectual hallmarks for which the Jesuits were most respected in Ming society.

Works cited

Primary sources

Chen Zushou 陳組綬. *Huang Ming zhifang ditu* 皇明職方地圖. S.n., Chongzhen 9 [1636].

Huang Jinsheng 黄金聲, ed. *Daoguang Jinhua xianzhi* 道光金華縣志. S.n.: Daoguang 3 [1823].

Wang Qi 王圻, and Wang Siyi 王思義. *Sancai tuhui* 三才圖繪. Facsimile of c. 1609 edition. Shanghai: Shanghai guji chubanshe 上海古籍出版社, 1988.

Zhang Huang 章潢. *Tushu bian* 圖書編. Facsimile of Wanli 41 [1613] edition. Taibei: Chengwen chubanshe 成文出版社, Minguo 60 [1971].

Secondary sources

Ahmad, Nafis. *Muslim Contribution to Geography*. Lahore: M. Ashraf, [1947].

Amelung, Iwo. 'New Maps for the Modernizing State: Western Cartographic Knowledge and Its Application in 19th and 20th Century China.' In *Graphics*

115 Zurndorfer, 'Science Without Modernization: China's First Encounter With Useful And Reliable Knowledge From Europe,' p. 27.

and Text in the Production of Technical Knowledge in China, the Warp and the Weft, edited by Francesca Bray, Georges Métailié and Vera Dorofeeva-Lichtman, 685-726. Leiden: Brill, 2007.

Bernard, Henri. *Matteo Ricci's Scientific Contribution to China*. Translated by Edward Chalmers Werner. Peiping: H. Vetch, 1935.

Boxer, Charles Ralph. *The Christian Century in Japan: 1549-1650*. Berkeley: University of California Press, 1951.

Brockey, Liam Matthew. *Journey to the East: the Jesuit Mission to China, 1579-1724*. Cambridge, MA: Harvard University Press, 2007.

Cary-Elwes, Columba. *China and the Cross: A Survey of Missionary History*. New York: P.J. Kennedy and Sons, 1957.

Chan, Albert. 'Johann Adam Schall in the Pei-yu lu of T'an Ch'ien and the Eyes of His Contemporaries.' In *Western Learning and Christianity in China: The Contribution and Impact of Johann Adam Schall von Bell, 1592-1666*, edited by Roman Malek, 273-302. Sankt Augustin, Germany: China-Zentrum and Monumenta Serica Institute; Nettetal, Germany: Steyler Verlag, 1998.

Chang Kuei-sheng. 'Cheng Ho.' In *Dictionary of Ming Biography*, edited by L. Carrington Goodrich and Chaoying Fang, 194-200. New York: Columbia University Press, 1976.

Ch'en, Kenneth. 'A Possible Source for Ricci's Notices on Regions Near China.' *T'oung Pao* 34 (1938): 179-90.

Chen Zhengxiang 陳正祥. *Zhongguo ditu xue shi* 中國地圖學史. Hong Kong: Shangwu yinshuguan, Xianggang fen guan, 1979.

Chow, Kai-Wing. 'Writing for Success: Printing, Examinations, and Intellectual Change in Late Ming China.' *Late Imperial China* 17, no. 1 (June 1996): 120-157.

Cooper, Michael. *Rodrigues the Interpreter: An Early Jesuit in Japan and China*. New York and Tokyo: Weatherhill, 1974.

Day, John D. 'The Search for the Origins of the Chinese Manuscript Copies of Matteo Ricci's Maps.' *Imago Mundi* 47 (1995): 94-117.

D'Elia, Pasquale M. *Il mappamondo cinese del p. Matteo Ricci, S.I. (terza edizione, Pechino, 1602) conservato presso la Biblioteca vaticana. Commentato, tradotto e annotato dal P. Pasquale M. D'Elia, S.I.; con XXX tavole geografiche e 16 illustrazioni fuori testo*. Vatican City: Biblioteca Apostolica Vaticana, 1938.

—. 'Recent Discoveries and New Studies (1938-60) on the World Map in Chinese of Father Matteo Ricci S.I.' *Monumenta Serica* 20 (1961): 82-164.

Dudink, Adrian. 'The Inventories of the Jesuit House at Nanking Made Up During the Persecution of 1616-1617 (Shen Que, Nangong shudu, 1620).' In *Western Humanistic Culture Presented to China by Jesuit Missionaries (XVII-XVIII centuries), Proceedings of the Conference Held in Rome, October 25-27, 1993*, edited by Federico Masini, 121-123. Rome: Institutum Historicum S.I., 1996.

Dunne, George H. *Generation of Giants: The Story of the Jesuits in China in the Last Decades of the Ming Dynasty.* Notre Dame, IN: University of Notre Dame Press, 1962.

Elman, Benjamin A. *On Their Own Terms: Science in China, 1550-1900.* Cambridge, MA: Harvard University Press, 2005.

Fang, Chaoying and L. Carrington Goodrich. 'Kuo Tzu-chang.' In *Dictionary of Ming Biography,* edited by L. Carrington Goodrich and Chaoying Fang, 775-777. New York: Columbia University Press, 1976.

Fang, Lienche Tu. 'Shen Ch'ueh.' In *Dictionary of Ming Biography,* edited by L. Carrington Goodrich and Chaoying Fang, 1177-1179. New York: Columbia University Press, 1976.

Foss, Theodore N. 'A Western Interpretation of China: Jesuit Cartography.' In *East Meets West: The Jesuits in China, 1582-1773,* edited by Charles E. Ronan and Bonnie B.C. Oh, 209-251. Chicago: Loyola University Press, 1988.

Franke, Wolfgang. 'Ricci, Matteo.' In *Dictionary of Ming Biography,* edited by L. Carrington Goodrich and Chaoying Fang, 1137-1144. New York: Columbia University Press, 1976.

Fuchs, Walter. *The 'Mongol Atlas' of China by Chu Ssu-Pen and the Kuang-Yü-Tu.* Peiping: The Catholic University Press, 1946.

Gernet, Jacques. *China and the Christian Impact: A Conflict of Cultures.* Translated by Janet Lloyd. Cambridge and New York: Cambridge University Press, 1985.

—. 'Christian and Chinese Visions of the World in the Seventeenth Century.' *Chinese Science* 4 (1980): 1-17.

—. 'Space and Time: Science and Religion in the Encounter Between China and Europe.' *Chinese Science* 11 (1993-1994): 93-102.

Giles, Lionel. 'Translations from the Chinese World Map of Father Ricci.' *The Geographical Journal* 52, no. 6 (December 1918): 367-385, and 53, no. 1 (January 1919): 19-30.

Goodrich, L. Carrington. 'Cattaneo, Lazzaro.' In *Dictionary of Ming Biography,* edited by L. Carrington Goodrich and Chaoying Fang, 31-33. New York: Columbia University Press, 1976.

Hertz, Solange. *Rhodes of Viet Nam: The Travels and Missions of Father Alexander de Rhodes in China and Other Kingdoms of the Orient.* Westminster, MD: The Newman Press, 1966.

Hirth, Friedrich, and W.W. Rockhill. *Chau Ju-kua: His Work on the Chinese and Arab Trade in the Twelfth and Thirteenth Centuries, Entitled Chu-fan-chï.* St. Petersburg: Printing Office of the Imperial Academy of Sciences, 1911. Reprint, Taibei: Literature House, 1965.

Hourani, George F. *Arab Seafaring in the Indian Ocean in Ancient and Early Medieval Times.* Revised and expanded by John Carswell. Princeton, NJ: Princeton University Press, 1995.

Hsia, R. Po-chia. *Matteo Ricci and the Catholic Mission to China: A Short History With Documents.* Indianapolis: Hackett Publishing Company, 2016.

Huang Shijian 黃時鑒, and Gong Yingyan 龔纓晏. *Li Madou shijie ditu yanjiu* 利瑪竇世界地圖研究. Shanghai: Shanghai guji chubanshe, 2004.

Kelly, Edward Thomas. *The Anti-Christian Persecution of 1616-1617 in Nanjing.* PhD Thesis, Columbia University, 1971.

King, Gail. 'Note on a Late Ming Dynasty Chinese Description of "Ricci's Church" in Beijing.' In *Sino-Western Cultural Relations Journal* 20 (1998): 49-51.

Li Xiaocong. *Ouzhou shoucang bufen Zhongwen gu ditu xu lu = A Descriptive Catalogue of Pre-1900 Chinese Maps Seen in Europe.* Beijing: Guoji wenhua chuban gongsi, Xinhua shudian jingxiao, 1996.

Libbrecht, Ullrich. 'What Kind of Science Did the Jesuits Bring to China?' In *Western Humanistic Culture presented to China by Jesuit Missionaries (XVII-XVIII centuries), Proceedings of the Conference held in Rome, October 25-27, 1993,* edited by Federico Masini, 221-234. Rome: Institutum Historicum S.I., 1996.

Lu Liangzhi 卢良志. *Zhongguo ditu xue shi* 中国地图学史. Beijing: Cehui chubanshe, 1984.

Luk, Bernard Hung-Kay. 'A Study of Giulio Aleni's Chih-fang wai chi.' *Bulletin of the School of Oriental and African Studies, University of London* 40, no. 1 (1977): 58-84.

Ma Biao. 'Johann Adam Schall and Chinese Traditional Philosophy.' In *Western Learning and Christianity in China: The Contribution and Impact of Johann Adam Schall von Bell, 1592-1666,* edited by Roman Malek. Sankt Augustin, Germany: China-Zentrum and Monumenta Serica Institute; Nettetal, Germany: Steyler Verlag, 1998.

Menegon, Eugenio. 'A Different Country, The Same Heaven: A Preliminary Biography of Giulio Alenis, S.J. (1582-1649).' *Sino-Western Cultural Relations Journal* 15 (1993): 27-51.

Mish, John. 'Creating an Image of Europe for China: Aleni's Hsi-fang ta-wen.' *Monumenta Serica* 23 (1964): 1-87.

Mungello, David E. *Curious Land: Jesuit Accommodation and the Origins of Sinology.* Honolulu: University of Hawaii Press, 1989.

—. *The Forgotten Christians of Hangzhou.* Honolulu: University of Hawaii Press, 1994.

Ōba Osamu 大庭脩; translated by Joshua A. Fogel. 'Sino-Japanese Relations in the Edo Period (Part Three: The Discovery of Banned Books).' *Sino-Japanese Studies* 9, no. 1 (October 1996): 56-74.

Park, Hyunhee. *Mapping the Chinese and Islamic Worlds: Cross-Cultural Exchange in Pre-Modern Asia.* Cambridge: Cambridge University Press, 2012.

Peterson, Willard. 'Why Did They Become Christians?' In *East Meets West: The Jesuits in China, 1582-1773,* edited by Charles E. Ronan and Bonnie B.C. Oh, 129-152. Chicago: Loyola University Press, 1988.

Ricci, Matteo. *Lettere (1580-1609)*. Macerata: Quodlibet, 2001.

Rienstra, M. Howard, ed. and trans. *Jesuit Letters from China 1583-84*. Minneapolis: University of Minnesota Press, 1986.

Sivin, Nathan. 'Copernicus in China.' In *Science in Ancient China: Researches and Reflections*, 1-53. Aldershot: Variorum, 1995.

Smith, Richard J. 'Mapping China's World: Cultural Cartography in Late Imperial Times.' In *Landscape, Culture, and Power in Chinese Society*, edited by Wen-hsin Yeh, 52-109. Berkeley, CA: Institute of East Asian Studies, 1998.

Standaert, Nicolas. 'Note on the Spread of Jesuit Writings in Late Ming and Early Qing China.' *China Mission Studies (1550-1800) Bulletin* 1 (1985): 22-32.

Sun Shangyang 孙尚扬. *Li Madou yu Xu Guangqi* 利玛窦与徐光启. Beijing: Xinhua chubanshe, 1993.

Szcześniak, Boleslaw. 'Matteo Ricci's Maps of China.' *Imago Mundi* 11 (1954): 127-136.

—. 'The Seventeenth Century Maps of China: An Inquiry into the Compilations of European Cartographers.' *Imago Mundi* 13 (1956): 116-136.

Takahashi Tadashi 高橋正. 'Tōzen seru chūsei Isurāmu sekaizu.' *Ryūkoku Daigaku ronshū* 374 (1963): 77-95.

—. 'Chūsei Isurāmu sekaizu' 中世イスラーム世界図. In *Chizu no shisō* 地図の思想, edited by Hasegawa Kōji 長谷川孝治, 53-55. Tōkyō: Asakura Shoten, 2005.

Trigault, Nicolas. *China in the Sixteenth Century: The Journals of Matthew Ricci, 1583-1610; Translated from the Latin by Louis J. Gallagher.* New York, Random House, 1953.

Üçerler, M. Antoni J. 'Gutenberg Comes to Japan: The Jesuit Enterprise and the First IT Revolution of the Sixteenth Century.' Accessed August 2014. http://www.usfca.edu/ricci/events/Ucerler.pdf.

—. 'The Jesuits in East Asia in the Early Modern Age: A New "Areopagus" and the "Re-invention" of Christianity.' In *The Jesuits and Globalization: Historical Legacies and Contemporary Challenges*, edited by Thomas Banchoff and José Casanova, 27-48. Washington, DC: Georgetown University Press, 2016.

Unno Kazutaka. *Tōzai chizu bunka kōshōshi kenkyū* 東西地図文化交渉史研究 [Monographs on the history of cartographical exchange between the East and the West]. Osaka: Seibundō, 2003.

Validi, Ahmet Zeki. 'Islam and the Science of Geography.' *Islamic Culture, the Hyderabad Quarterly Review* 8 (October 1934): 511-527.

Wallis, Helen. 'The Influence of Father Ricci on Far Eastern Cartography.' *Imago Mundi* 19 (1965): 38-45.

Wang Chung-min. 'Yang T'ing-yün.' In *Eminent Chinese of the Ch'ing Period*, edited by Arthur W. Hummel, 894-895. Washington, DC: United States Government Printing Office, 1943-1944.

Wang Qianjin 汪前进. *Xixue dong chuan di yi shi: Li Madou* 西学东传第一师: 利玛窦. Beijing: Kexue chubanshe, 2000.

Witek, John W. 'Johann Adam Schall von Bell and the Transition from the Ming to the Ch'ing Dynasty.' In *Western Learning and Christianity in China: The Contribution and Impact of Johann Adam Schall von Bell, 1592-1666*, edited by Roman Malek, 109-124. Sankt Augustin, Germany: China-Zentrum and Monumenta Serica Institute; Nettetal, Germany: Steyler Verlag, 1998.

Wong, George H.C. 'China's Opposition to Western Science During Late Ming and Early Ch'ing.' *Isis* 54, no. 1 (March 1963): 29-49.

Yan Ping. *China in Ancient and Modern Maps*. London: Sotheby's, 1998.

Yap, Paul Teh-lu, and J.C. Yang. 'Li Chih-tsao.' In *Eminent Chinese of the Ch'ing Period*, edited by Arthur W. Hummel, 452-454. Washington, DC: United States Government Printing Office, 1943-1944.

Yee, Cordell. 'Traditional Chinese Cartography and the Myth of Westernization.' In *The History of Cartography, Volume 2, Book 2: Cartography in the Traditional East and Southeast Asian Societies*, edited by J.B. Harley and David Woodward, 170-202. Chicago and London: University of Chicago Press, 1994.

Zhang Cuo 張錯. *Li Madou ru Hua ji qita* 利瑪竇入華及其他. Hong Kong: Xianggang chengshi daxue chubanshe, 2002.

Zurndorfer, Harriet T. 'Science Without Modernization: China's First Encounter With Useful And Reliable Knowledge From Europe.' Paper for the Fourth Global Economic History Network Workshop. Accessed December 28, 2020. https://www.lse.ac.uk/Economic-History/Assets/Documents/Research/GEHN/GEHNConferences/conf4/Conf4-HZurndorfer.pdf

4 Chosŏn Cartography in a Trans-regional Context

Abstract
This chapter examines cartographic exchange and influence between Ming China and the neighboring Chosŏn state. The *Tongguk yŏji sŭngnam* (Newly Augmented Geographical Survey of the Land of the East) was directly inspired by the *Da Ming yitong zhi* (Gazetteer of the Great Ming's Unification). Long before the emergence of a substantial commercial publishing industry, the maps of the *Sŭngnam* were widely copied in manuscript atlases. The chaotic richness of later Ming works tended to be rejected in Chosŏn, however, in favor of adherence to orthodox precedent. The chapter also discusses the importation of materials into Ming China, examining one such work in detail – the *Chaoxian tushuo* (Annotated maps of Chosŏn), printed in 1600.

Keywords: Ming, Korea, cartography, Chosŏn, *Sinjŭng tongguk yŏji sŭngnam*, *Chaoxian tushuo*

The role of classical Chinese as a written language shared by scholars throughout East Asia meant that books were shared as well. Sometimes they were bestowed as official gifts, sometimes purchased by visiting diplomats, but on a larger scale – albeit one that is difficult to measure – they moved through commercial trade. Such trade sometimes took place directly over the ocean between the mainland and Japan, and sometimes via the Korean peninsula. Ōba Osamu has emphasized the interconnectedness of East Asia in his studies of the 'Book Road,' an intellectual reconfiguration of the 'Silk Road' that emphasizes not the transfer of wealth, but of knowledge between the Asian mainland and Japan.[1] This transfer of the written word, which Ōba

1 Ōba Osamu 大庭脩, *Kanseki yunyū no bunkashi: Shōtoku Taishi kara Yoshimune e* 漢籍輸入の文化史：聖徳太子から吉宗へ.

Akin, Alexander, *East Asian Cartographic Print Culture: The Late Ming Publishing Boom and its Trans-Regional Connections*. Taylor & Francis Group, 2021
DOI: 10.5117/9789463726122_CH04

shows was already in full swing by the Tang period (618-907 CE), suggests that when appraising intellectual events in one part of East Asia we should consider influences that affected the region as a whole. Developments in cartography are no exception.

This chapter examines the significance of Ming publishing to cartography in Chosŏn Korea. By this time, Chinese maps had been imported into Korea for centuries through both official and private channels.[2] Such maps even inspired poetry, as Korean readers contemplated their place in a Sinocentric worldview. The Koryo poet Yi Kyubo (1168-1241) responded to a Song map by comparing his peninsula's small size on the map with its cultural preeminence:

> Myriad kingdoms stretch across leaves of paper,
> while Samhan is tucked away like a clump of dirt.
> But do not belittle it, kind observer,
> for in my eyes it seems quite large.
> Then and now it pours forth talented men,
> and compares to China without shame.
> Such men a nation make – without them, nothing:
> the wide barbarian lands are but trifles.
> Have you not heard the Chinese call us Little China?
> Such a phrase is truly worth taking up.[3]

> 萬國森羅數幅牋，
> 三韓隈若一微塊。
> 觀者莫小之，
> 我眼謂差大。
> 今古才賢衮衮生，
> 較之中夏毋多愧。
> 有人曰國無則非，
> 胡戎雖大猶如芥。
> 君不見華人謂我小中華，
> 此語眞堪採。

2 The *Zhu fan zhi* 諸蕃志, by Zhao Rugua 趙汝适, an inspector of foreign trade during the Song period, lists books among the main exports to the Korean peninsula, where they were traded for such local goods as ginseng, cloth and pine seeds (Hirth and Rockhill, *Chau Ju-Kua*, p. 168).

3 Yi Kyubo, 李奎報, 題華夷圖長短句. Translation (quoted with permission) is by Graham Sanders, who originally posted it on November 27, 2012 on Facebook, bringing it to my attention.

Though the Korean interest in classical Chinese texts was longstanding, we shall see that a Ming geographical text served as the model for a vitally important government-sponsored Chosŏn work. The instructions of King Sŏngjong explicitly demonstrate the impact that the widely read *Da Ming yitong zhi,* discussed in Chapter One, had on Chosŏn notions of the gazetteer.

In contrast to the florescence of diverse geographical and cartographic materials in the late Ming, however, Chosŏn did not see a commercial publishing boom in such materials until much later, in the nineteenth century, and it continued adhering to models established early in the dynasty. Maps copied from the *Sŭngnam* were often reproduced by hand in miniature atlases, though they were occasionally reprinted by woodblock. The burgeoning market in texts in the late Ming discomfited some Neo-Confucian scholars even in Ming China itself, as has been colorfully discussed by Timothy Brook in *The Confusions of Pleasure*. In the early Chosŏn, we see the impact of precedent-setting Ming texts, but as the later Ming took a 'wayward' path with more works straying from the canon, divergences became evident in the nature of works that circulated in the Ming but were not copied in Chosŏn.

Complicating any analysis of such influence is the fact that the Chosŏn reading public was smaller than that of the vastly more populous Ming, so print runs were correspondingly small. Furthermore, the toll of warfare, including the Imjin wars of the late sixteenth century, the Manchu raids just a few years later, and the extensive toll of conflict in the twentieth century, has dramatically reduced the number of surviving copies of these already scarce texts. As a result many are known from only a single copy, with untold numbers lost completely. It is thus particularly dangerous in the Korean case to draw conclusions from the absence of certain types of evidence.

The circulation of texts in the seventeenth century appears to have relied to a greater extent on manuscript than did the contemporary intellectual market in late Ming and early Qing, and the spread of private commercial publishing along lines similar to the late Ming took place only in the nineteenth century. Thus Chosŏn offers a basis for comparison with the cases discussed in earlier chapters. The high rate of manuscript copying offers a test of the notion that mechanical reproduction is responsible for phenomena such as the conscious creation of a 'geo-body,' or the retention of stereotyped models. The standardization and repetitious perpetuation of certain maps, particularly the Ch'ŏnhado 天下圖 genre discussed below, or the dynastically legitimized maps in the *Sŭngnam*, took place *without* the ubiquitous presence of the woodblock press. This suggests that cultural orthodoxy could be as strict a guide of cartographic form as the successive copying of woodblocks.

Among the issues I examine in this chapter is the argument that crudely rendered maps in the *Sinjŭng Tongguk yŏji sŭngnam* 新增東國輿地勝覽 (Newly augmented geographical survey of the Land of the East [i.e., Korea]) were intentionally simplified for military security. After challenging this notion from the internal evidence of the text itself, including its listing of sensitive military installations, I suggest that an alternate explanation for the character of the *Sŭngnam's* maps is that their features are typical of cartography from early printed texts and do not require any such excuse for failing to fit into a teleological framework of progressive improvement. Features specific to woodblock-printed maps, as well as the relationship of such maps to other forms of cartography and geographical writing, cannot be overlooked when fitting the *Sŭngnam* into a narrative of Chosŏn cartographic development.

Although the discussion in this and the following chapter focuses mostly on works related to Ming precedents, I would like to draw attention to the existence of the inverse phenomenon: the importation of materials from abroad into Ming China. The book trade's growth during the late Ming publishing boom, combined with increased interest in military affairs and demand for texts on the broader regional situation, resulted in the domestic publication of more works on neighboring countries. The sources for such works included the writings of Chinese who traveled abroad, as well as occasional imported 'indigenous' materials from these countries. I examine in detail one such work, the *Chaoxian tushuo* 朝鮮圖說 (Annotated maps of Chosŏn), printed in 1600 by a Ming military officer. In sum, this chapter will show that Ming works set precedents for Chosŏn publications, but that the flow of cartographic texts was to some extent bidirectional.

A great deal of scholarship has focused on Korean primacy in certain areas such as the use of moveable type, while relatively less attention has been paid to issues like the development of commercial publishing, perhaps because it appears to have begun relatively late.[4] Boudewijn Walraven argues that an

4 In his dissertation (*The Apparition of the Rational Public: Reading Collective Subjectivity in the Korean Public Sphere*), Michael Kim addresses the role of lending libraries and bookstores in late Chosŏn society but admits that it remains unclear when *panggakpon* (坊刻本), private commercial publications, became common (p. 56). Kim recapitulates the prevailing emphasis on the early adoption of metal type printing in Korea, and again highlights the quality of state publications. He notes that during Sejong's reign only ten books had printing runs of more than one hundred copies (p. 55), but our knowledge of printing culture in Ming China suggests that many works were printed to order, rather than issued in large runs for distribution through market channels. Thus such figures reveal state imperatives more than the actual circulation of texts. Nor are the subjective impressions of contemporary writers ('there are plenty of books

evolutionary perspective in which manuscript succumbs to block printing, which is then superseded by moveable type, does not match Korea's experience, given that materials of all three types were produced simultaneously. He suggests that Chosŏn print culture achieved remarkable success on its own terms; to decry the comparatively recent development of commercial publishing as evidence of 'regrettable slowness in the advance to the promised land of full commercialization' is wrong, he argues, because Chosŏn Korea possessed its own sort of thriving book culture.[5] Whereas Kai-wing Chow's argument in *Publishing, Culture, and Power in Early Modern China* is that a late-Ming commercial publishing boom both reflected and stimulated cultural changes comparable to those in post-Gutenberg Europe, Walraven's essay does not specifically address which social, economic, and political factors shaped the direction of the Chosŏn publishing industry, or whether commercial publishing in turn transformed cultural discourse. The question of readership and circulation is only beginning to be explored in the Chosŏn context.

Scholarship on Korean cartography

Much of the basic work in reconstructing Korean cartographic history was done in the late nineteenth and early twentieth centuries by Japanese scholars and Western missionaries.[6] After the end of the Japanese colonial period, particularly after the Korean War, Korean scholars embarked on a search for positive historical precedents, examples to cite in promoting a proud and modern identity. As Yungsik Kim has discussed, this caused much work in the historiography of premodern Korean technology and science to emphasize independence from, and advancement over, Chinese, or Japanese influences. Kim describes this as a 'reaction to so-called colonialist historiography of the Japanese colonial period.'[7]

around' / 'there aren't enough books around') truly sufficient to demonstrate the availability of books in various regions or among different classes.

5 Walraven, 'Reader's Etiquette, and Other Aspects of Book Culture in Chosŏn Korea.'

6 For discussion and criticism of some late nineteenth and early twentieth century assessments of Korean cartography see Nakamura, 'Old Chinese World Maps Preserved by the Koreans.' Nakamura, who did research in Korea during the colonial period, often takes a tone that some readers find disturbing, exemplified by his comment that 'the sciences and arts of Korea were almost always slavishly modeled upon those of China' (p. 13). Thus his name is something of an anathema in Korean studies and his conclusions tend to be disputed, sometimes for good reason (see Ledyard, 'Cartography in Korea,' p. 263).

7 Yungsik Kim, 'Problems and Possibilities in the Study of the History of Korean Science,' p. 50.

Because of this reaction, there has been significant work published on men like the nineteenth-century cartographer Kim Chŏng-ho 金正浩, Korea's first private publisher of high-quality maps,[8] as well as on aspects of cartography thought to reveal essential facets of the Korean mind, particularly what Gari Ledyard calls 'a nationally conceived theory of geomantic analysis.'[9] A pathbreaking contribution was made by Yi Ch'an (known in English as Chan Lee) in 1977 with his comprehensive and fairly balanced *Han'guk ko chido*, which was expanded upon and to some degree superseded by his 1991 *Han'guk ui ko chido*, cited at various points in this chapter.

Greater prosperity and stability in South Korea have created more opportunities for this field of inquiry to grow. Civic pride is one of the factors underlying the publication of compendia of city maps; an impressive example is Hŏ Yŏng-hwan's collection of Seoul maps.[10] In recent years there has also been an explosion of interest in old Korean maps as both private and government-backed organizations attempt to use the cartographic record to support political movements, such as the campaign to use the term 'Eastern Sea' instead of 'Sea of Japan' or to support territorial claims against Japan.[11] Whatever the reasons for their compilation, this increase in the number of high-quality cartographic publications greatly facilitates research today. In the digital age there has also been a concerted effort to put important maps online, a project undertaken with particular alacrity by the Kyujanggak in Seoul.

Materials in English, on the other hand, have been comparatively limited in number. Gari Ledyard's 'Cartography in Korea,' a chapter in Harley and Woodward's *History of Cartography* series, marks an important effort to overcome this deficit and constitutes a major turning point in the field.

8 Reproductions of Kim's maps, the *Taedong yŏjido* 大東輿地圖 and the *Ch'ŏnggudo* 青邱圖, were issued during the 1970s, a high point of South Korea's promotion of historical glory. For important examples see the Han'guk yŏn'gu charyo ch'onggan edition of the *Taedong yŏjido* and the P'yŏnjip Minjok Munhwa Ch'ujinhoe facsimile of the *Ch'ŏnggudo*.

9 Ledyard, 'Cartography in Korea,' p. 235. The fundamental text for geomancy is the *T'aengniji*, available in Inshil Choe Yoon's translation (*Yi Chung-hwan's T'aengniji: The Korean Classic for Choosing Settlements*). This text was not illustrated with maps, but the influence of geomantic thinking on cartography has been a frequent theme in Korean scholarship. For an overview, with an example of a siting map from a genealogy, see Ledyard, 'Cartography in Korea,' pp. 276-279.

10 Hŏ Yŏng-hwan 許英桓, *Chŏngdo 600-yŏn Sŏul chido* 定都600年서울地圖 / *600 year's maps of Seoul*.

11 For a comprehensive overview of the 'Eastern Sea' claims see Lee Ki-suk, Kim Shin, and Soh Jung-chul. *East Sea in World Maps*. On Dokdo/Takeshima, see Choe Suh Myun, 'Tokdo in Old Maps.' While these are specialized works on these topics, almost any recent writing on Korean cartography or European cartography of Korea will touch on them at least briefly.

Ledyard reconstructs development and change within several genres of map production over the course of the Chosŏn. His main goals are to classify a broad array of map forms while articulating a history of cartography in the context of, but independent from, China and Japan, showing Korean maps as having unique or precocious features in comparison to other East Asian maps. Ledyard's impact in Korea seems to remain limited. *Uri yet chido wa kŭ arŭmdaum* 우리 옛지도와 그 아름다움 (published in English under the title *The Artistry of Early Korean Cartography*) contains numerous examples of nationalist claims that were undercut by Ledyard.[12] More recently, John Rennie Short has published an attractive illustrated history that is accessible to the lay reader.[13]

Given the proud claims that have been made about the steady advancement of cartography during the Chosŏn, it is easy to understand the dilemma posed by crudely printed maps such as the example illustrated in Figure 4:1, a provincial map in the *Sinjŭng tongguk yŏji sŭngnam* 新增東國輿地勝覽 (Newly augmented geographical survey of the Land of the East).[14] This work, hereafter referred to simply as the *Sŭngnam*, was a national gazetteer of the Chosŏn state first compiled in 1487, revised in 1499, and expanded in 1530, this latter edition being the earliest extant. Some scholars who have argued for the precocity of Korean cartography in this period have felt obliged to account for the simplicity of the maps in this work, asserting that these maps were intentionally abridged for purposes of national security. As I shall argue below, however, the internal evidence of the *Sŭngnam* itself rebuts this notion.

Comparison of these printed maps with large-format painted works from the court has led to a search for the 'explanation' of their crudeness. It is more appropriate to examine them in the context of woodblock-printed cartography, noting particularly the relationship between text and image. As we have seen in

12 This book was chosen by the Korean Literature Translation Institute in 2005 as one of one hundred influential books most deserving of translation into a foreign language. I was asked to edit the English translation, which was published by Tamal Vista Press. My main concern in so doing was to balance two competing academic needs. It is of great use to have a window into the ideas of scholars in South Korea who work with maps, but it is in the interests of novices to the field not to promote erroneous or highly nationalistic renderings of cartographic history. To what extent I succeeded in balancing these factors, by trimming material superseded or rejected by Ledyard and others, while leaving in some material that could be considered tendentious by some critics, must be judged by the reader. The articles it contains are certainly worthy of attention, citing numerous maps and texts previously unmentioned in English sources.

13 Short, *Korea: A Cartographic History*.

14 This title is commonly translated 'Augmented Survey of Korean Geography,' though this misses the reference to the *Fangyu shenglan* that would have been obvious to many contemporary readers.

Figure 4.1 Map of Ch'ungch'ŏng *Province*, located on the pages before Chapter 14 in the 1611 edition of the *Sinjŭng tongguk yŏji sŭngnam*

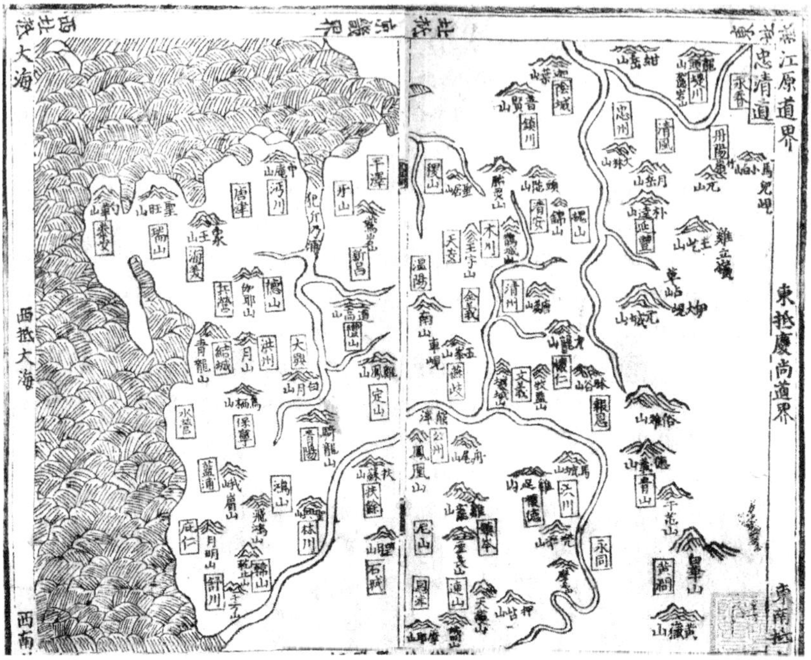

Newly Augmented Geographical Survey of the Land of the East
Courtesy of the Harvard-Yenching Library

our examination of Ming geographical texts and gazetteers, readers expected to find greater detail in the textual portions of such works than in the maps. In the case of the *Sŭngnam*, the maps served only to provide a basic framework that would have been of little use without the accompanying detailed written content. If we try to make any statement about Korean cartography based on the maps in this influential work, we must consider this broader context.

The *Sŭngnam*'s cartography

The *Sŭngnam* is a vast collection of geographical, socioeconomic, and cultural information from all corners of the realm.[15] Although the extensive

15 As we might expect of such a monumental text, much has been written about it in both Korean and English, the first substantial English translation of extracts appearing in 1897 (Hulbert, 'An Ancient Gazetteer of Korea').

data that comprise its contents derive from indigenous sources, its main precedents in terms of organization and structural philosophy are two Chinese works. One is the *Fangyu shenglan* 方輿勝覽 (Best views of the realm) by the scholar Zhu Mu 祝穆. This collection of literary samples from all over the Song empire, organized geographically, was intended for the use of poets and authors, who could use these passages to impart local color to their own work without actually traveling to faraway locales. Though first published in 1239, the *Fangyu shenglan* was reprinted numerous times and circulated widely in the Ming. The impact of this text on the *Sŭngnam* is recognized in the compilers' use of the term *shenglan* in the title, pronounced *sŭngnam* in Korean. A more direct influence on the *Sŭngnam* comes from the 1461 *Da Ming yitong zhi* 大明一統志 (Comprehensive gazetteer of the Great Ming's unification), which King Sŏngjong explicitly ordered the compilers to use as their model. This Ming gazetteer, discussed above in Chapter One, had a much more administrative focus than the *Fangyu shenglan*.

In 1485, Sŏngjong issued an edict calling for work to begin on revising an earlier draft of a Chosŏn gazetteer. The king wrote,

> Although it is said that this [manuscript] book, based on Zhu Mu's publication, cites important matters [of each locale] and supplements these with poetry and prose, seeking them broadly and recording them copiously, and truly brings some benefit to the country's literature, I fear that its [descriptions of] mountains and rivers and its historical notes are incomplete, and these men's work includes some rather muddled areas. You all are to compare and collate materials from all periods up to the present. The organization [of the new edition] is to follow that of the *Da Ming yitong zhi* in every respect.[16]

> 傳曰是書據祝穆之編提其事要兼采詩文博求而廣記之於國家文獻誠有所益然其間山川及古實恐脫略而諸人之作荒冗殺雜者頗有之卿等宜更讎校櫽括期至於精當其凡則一以大明一統志爲法。

Two years later in 1487 the revised edition was complete, having been delayed briefly by a drought. A postface indicates that work was able to proceed so quickly because there already existed local maps and gazetteers of even tiny, isolated parts of the realm; this suggests that the compilers

16 Kim Chongjik's postface, pp. 1a-1b; p. 1015 in the Tongguk Munhwasa facsimile edition.

were granted access to such materials held in the royal archives.[17] As this postface also indicates, the focus of the work shifted when it was reedited, moving from a more literary document to a text with a more administrative cast. This is reflected in the categories of bureaucratic information that have been adopted. Appendix 5 lists the content categories for both works; if we compare the content of the *Sŭngnam* with the *Da Ming yitong zhi*, the structural similarity is indeed striking. There is, of course, a different scale of administration due to the relative sizes of the two states. It is also important to note that the Chosŏn work has some unique features, such as a listing of prominent surnames in each prefecture.[18]

This brief survey helps to contextualize the characteristics of the *Sŭngnam*'s maps. Han Yŏng-u, former Dean of the College of the Humanities at Seoul National University, is among the many contemporary scholars who suggest that government secrecy played a role in ensuring the maps' crudeness. He writes that maps were kept under lock and key as classified government documents, and civilian possession of maps was prohibited. When the government issued maps and geographical works for the public, it deliberately made the details simple and crude.[19]

The *Sŭngnam* was intended for wide distribution; in a postface to its final volume, the scholar Im Sahong noted that King Sŏngjong had called for a work that would be found everywhere, 'from the Royal Archives at the top, all the way down to private libraries.'[20] Although there would indeed have been concern about detailed maps of sensitive locales falling into the wrong hands, the extremes encountered in periods such as the purges of Yŏnsan-gun at the turn of the sixteenth century, or the tense early years of relations with the Qing, should not be interpreted as representing a constant status quo. The Song polymath Shen Kua's eleventh-century record of an official in Yangzhou tricking Koryŏ envoys into surrendering their Chinese maps, and a parallel tale from the *Koryŏsa* recording the imprisonment of two local

17 Ibid., p. 3a; Tongguk Munhwasa edition, p. 1016. This reference to copious materials also highlights how much has been lost from the cartographic record.

18 Although Chinese national gazetteers from after the Song dynasty lack the category of local surnames, there are earlier Chinese precedents for this arrangement. The early Song *Taiping huanyu ji* 太平寰宇記 (Record of the world in the Taiping Era, edited by Yue Shi) notes elite names for certain locales.

19 Han Yŏng-u, *Uri yet chido wa kŭ arŭmdaum*, p. 41.

20 Im Sa-hong's postface, p. 1a; Tongguk Munhwasa edition, p. 1013. Note that Han argues elsewhere that the antique maps in the Kyujanggak were purchased from independent collections, and in the case of the *Hwadong kojido*, had been in private hands as early as the beginning of the seventeenth century – again undermining his argument about strict government supervision of maps (Han, *Uri yet chido wa kŭ arŭmdaum*, p. 257).

officials who had planned to export maps to Song China, have been taken as emblematic of the tight lid kept on cartographic materials by both sides.[21] It should be remembered, however, that relations between Koryŏ and Song were complicated by Koryŏ's diplomatic relations with the Song's northern enemy, the Liao. There was little trust between these states, as compared with the close ties of much of the Ming/Chosŏn period. Attitudes about the transfer of specific texts between states varied with political trends.

Gari Ledyard concedes that the medium of woodblock printing affected the nature of the maps in the *Sŭngnam*, but he still adheres to the explanation that military secrecy was the reason for what he terms 'cartographic reticence:'[22]

> Because Sŏngjong intended to give the *Sŭngnam* a broad distribution, militarily useful information had to be strictly absent. This is probably the reason for the underdeveloped treatment of the northern borders on both the general map and the maps of the relevant provinces.[23]

The question of security is, however, more complex than it might at first appear. It is worth noting that security is not always bound up with secrecy: The earliest extant printed map of Seoul was circulated in a 1751 decree issued by King Yŏngjo instructing local citizens in civic defense.[24] Furthermore, maps were only one of many forms in which geographical information was presented. Aside from maps, textual materials could relate to security. Since the maps comprise a mere nine plates in this work, which has a total of some two thousand pages of dense text, we cannot fully interpret the maps without examining this accompanying text. The cartographic poverty of the *Sŭngnam* would be difficult to attribute to security concerns if other types of sensitive information are included in the text, and this is precisely what we find.

Communications were part of the Chosŏn defensive network; there was a network of beacon stations, called *pongsu*, situated in prominent locations

21 See, for example, Han, *Uri yet chido wa kŭ arŭmdaum*, pp. 17-18. Hendrik Hamel, in the *Beschryvinge van't Koninghrijck Coeree* (A description of the Kingdom of Korea), notes that 'Korean cartographers represent the country as an oblong rectangle, like a playing card, although there are several points reaching out into the sea' (Hamel, *Description*, p. 52), suggesting that even these unwelcome European shipwreck victims, the ultimate outsiders, may have been able to see Korean maps during their captivity from 1653 to 1666.

22 Ledyard, 'Cartography in Korea,' p. 294.

23 Ibid., p. 295.

24 Han, *Uri yet chido wa kŭ arŭmdaum*, p. 72.

of high elevation, each within view of the next station, for transmitting vital signals from the northern frontier or the seaboard to the capital. If we estimate that it took three minutes to ignite a fire bright enough to be seen from the next station, then it would have taken about four hours for news of an attack to reach Seoul from Kyŏnghŭng, the most remote station on the far northeastern frontier. News was communicated by code.[25] These beacons were links in a chain that had to remain unbroken for the system to serve properly. The capture of several of these beacons, or even one strategic one, could slow communications dramatically, giving troublemakers precious time to act before the capital was even aware, let alone ready to respond. Naturally, a publication obsessed with maintaining secrecy would not be expected to mention these beacons, let alone specify their locations. Yet, this is precisely what the *Sŭngnam* does.

Take the region of P'yŏngyang for example, situated in P'yŏngan province, bordered on the north by the Yalu River. The *Sŭngnam*'s map of the province notes the presence of the Yeren Jurchens crouching just across the border, while to the west lies a vulnerable coastline. Yet we see that beacons are listed in surprising detail:

> *Pusan* beacon: to the north it communicates with Tokcha Mountain in Sunan county; to the south it communicates with Chagyak Mountain. *Chagyak Mountain* beacon: 14 *ri* west of the prefectural seat. To the south it communicates with Sojildang-jŏm, in the north it communicates with Pusan. *Sojildang-jŏm* beacon: situated within the walled prefectural seat. To the south it communicates with Hwasa Mountain, to the north it communicates with Chagyak Mountain. *Chosa Mountain* beacon: 26 *ri* south of the prefectural seat. To the south it communicates with Unbong Mountain in Chunghwa-gun; to the north it communicates with Sojildang-jŏm. *Pulgok* beacon: 100 *ri* west of the prefectural seat [...] (etc.)[26]
>
> 釜山烽檖：北應順安縣獨子山南應作藥山。 作藥山烽檖： 在府西十四里，南應所叱堂帖北應釜山。所叱堂帖烽檖： 在府城内，南應晝寺山北應作藥山。晝寺山烽檖：在府南二十六里，南應中和郡雲峰山北應所叱堂帖。佛谷烽檖： 在府西一百里...

25 Pratt, *Korea: A Historical and Cultural Dictionary*, p. 23.

26 Chapter 51, p. 13a; Tongguk Munhwasa edition, p. 925. The bold names are in larger script in the original. As Sixiang Wang has drawn to my attention, the field is far from unanimity in its assessment of the degree to which the idealized beacon system matched practical application.

By going through the entries for each prefecture in the gazetteer, linking up the outermost beacons listed in one prefecture with those of the next, one can trace the entire network of these beacon routes, which are supposed to have extended from the northern frontier or the southern seaboard all the way to the capital. These data are complete enough to enable modern scholars to reconstruct maps of the beacon system.[27]

This level of detail about a vital security network, published in a work intended for broad distribution, appears difficult to reconcile with the 'security' thesis. Furthermore, this is only the tip of the iceberg. If a conniving enemy had obtained a copy of this text, he could also have determined the locations of post stations with their horse relays, not to mention the sites of important bridges to be captured in an invasion, the locations of fortified garrisons – even the heights of garrison walls, the number of wells contained within them, and so on.

Why provide these details in the first place? In this period, before the Imjin wars and after the troubled northern border had quieted significantly from the earlier days of the Chosŏn regime, it is possible that the inclusion of such material was not considered reckless at all; in fact, it might just as well serve as a boast of the dynasty's secure entrenchment. This possibility suggests that we must reconsider the entire orientation of a work like the *Sŭngnam*. With its vast selections about local culture, famous personages, prominent families in each province, valuable regional products, and so on, the *Sŭngnam* as a whole reads more as a celebration of the flowering Chosŏn dynasty, a way of showing it has come into its own, than as a document preoccupied with guarding state secrets. This interpretation is similar to that of Kim Tai-jin, who argues that 'The unified and synthetic compilation of Korean geography [in this work] signifies the stability and active will of the nation and its ruler.'[28]

What does the *Sŭngnam* itself tell us about the purpose of its maps? Kim Chongjik's postface includes a brief note: 'The general map in the first volume records the sacrificial peaks and streams, famous mountains and major rivers. Each of the eight provincial maps includes only the garrisoned mountains of the prefectures and counties. The maps are also supplemented with information on what lies in the four directions [beyond the edges of

27 For an example, see Pratt, *Korea: A Historical and Cultural Dictionary*, p. 22. Granted, even with these locations it would still require much effort to reach the actual stations.

28 Kim Tai-Jin, *A Bibliographical Guide to Traditional Korean Sources*, p. 179. Han Yŏng-u's argument that maps printed in books were simplified for reasons of security would be considerably stronger if there were also examples of xylographic maps showing fine detail and, unlike the *Sŭngnam*, explicitly restricted from public circulation.

the map]' (卷首總圖則錄祀典所載嶽瀆沮名山大川。八道各圖則只錄州縣之鎮山。其四至四到並附著于圖面).[29] In purpose and style, the maps are similar to those found in the *Da Ming yitong zhi* (Gazetteer of the Great Ming's unification, see Figure 1:1).[30]

All of this shows that the nature of the maps in the *Sŭngnam* should not be evaluated in isolation. Since we have seen that its production was partly a response to the *Da Ming yitong zhi,* we should compare its cartographic quality and the relationship of words to text not by the standards of lovely large-format painted maps lavished with color, which we know to have been available to the king at his court, but rather by the standards of woodblock printed texts in the Ming and elsewhere. In short, the appearance of these maps is simply typical of woodblock-printed books of the time.

Post-*Sŭngnam* works

Ming publications imported to Chosŏn Korea included reprints of older texts, which were just as influential upon Korean geographical works as were more recent books. Many of these geographical texts had no maps at all. For example, Han Paek-kyŏm (1552-1615), one of the 'fathers' of *Sirhak* or Practical Learning, searched the Chinese dynastic histories of the Han to the Song – none of which is illustrated – for notes about the various states that had existed on the Korean peninsula, and compiled his findings in the *Tongguk chiriji* 東國地理誌 (Geographical treatise on the Eastern Kingdom). The text consists mainly of quotations from these dynastic histories. Han's purpose seems to have been to make these materials available to readers who could not access or afford imported copies of the entire histories, or who were mainly interested in their discussions of Korea. The circulation of this work was facilitated by its publication in a woodblock edition in 1640.[31]

More contemporary to the *Sŭngnam* is the *Haedong chegukki* 海東諸國記 (Record of various eastern countries), compiled by Sin Sukju 申叔舟 in 1471 but surviving in its earliest extant copy in a version from 1506.[32] It was

29 Third postface, p. 2b; Tongguk Munhwasa edition, p. 1015.

30 The Ming work also portrays sacrificial mountains, though for reasons of scale the main map limits these to the most important marchmounts, those peaks where the emperor should ideally perform sacrifices (though in practice this happened very rarely).

31 Han Paek-kyŏm 韓百謙, *Tongguk chiriji* 東國地理誌.

32 On this work see Robinson, 'Centering the King of Chosŏn: Aspects of Korean Maritime Diplomacy, 1392-1592.' Some of its maps are discussed in Ledyard, 'Cartography in Korea,' pp. 269-273. This and the *Kanyang nok* discussed below are paired in a heavily annotated volume

considered vital for the state to know enough about neighboring lands to fulfill the duties of propriety in dealing with them. The preface notes the importance of understanding the customs and history of these lands, which have substantially different cultural traditions. At this time even the local rulers of each *kuni* in Japan retained significant independence.

Aside from depictions of Korean ports such as that at Pusan, the maps printed in the *Haedong chegukki* include images of the places that had diplomatic or trade relations with Chosŏn, particularly Japan (in various views, with emphasis on its different regions) and the Ryukyus.[33] Its representations of the Japanese archipelago are the earliest known to survive in printed form.[34] The maps show sailing routes in the form of lines threading through the wave patterns of the sea.

Like the *Sŭngnam*, this book reflects a relatively confident mindset during the early Chosŏn, when relations with Japan in particular were more complex and reciprocal than in the aftermath of Hideyoshi's destructive campaigns. A decidedly less neighborly view of Japan is presented in the *Kanyang nok* 看羊錄 (Record of a Shepherd), a record by Kang Hang 姜沆 of eight years spent in Japan after his capture by Hideyoshi's forces in 1597. This work includes a simple woodblock printed map of Japan.[35] Several editions of this work appeared during the Chosŏn dynasty, and it is contained within Kang Hang's collected works, the *Suŭnjip* 睡隱集.[36] As with late Ming works on northeastern frontier affairs, books on neighboring lands continued to reproduce old maps even as their textual content was updated. Much of the change that took place in Chosŏn views of Japan between the late fifteenth century and the turn of the seventeenth can be seen in the contrast between the *Haedong chegukki* and the *Kanyang nok*, yet the maps alone reveal nothing of this transformation.

To argue for the broad regional influence of a Ming work like the *Da Ming yitong zhi* is not to argue that the international flow of information was always free. The sixteenth century writer Ŏ Sukkwŏn relates the story of an interpreter, Kim Isŏk, who was spotted buying the *Da Ming yitong zhi* in Beijing in 1522. Though this book had already reached Chosŏn about half a century earlier, the director of the Ming court's Bureau of Receptions saw Kim buying it and arbitrarily decided that such a book should not fall into

(unfortunately not a facsimile edition), *Haedong chegukki / Kanyang nok* 海東諸國記 / 看羊錄, edited by Yi Ŭr-ho 李乙浩.

33 Reproduced in Yi, *Haedong chegukki / Kanyang nok*, p. 35.

34 In the following chapter we will see a later Japanese copy.

35 Reproduced in Yi, *Haedong chegukki / Kanyang nok*, p. 168.

36 The Harvard-Yenching Library possesses editions from 1658 and 1868.

the hands of foreigners. He then ordered the Chosŏn diplomatic compound to be closed off, preventing its officers from coming and going freely.[37]

Despite such occasional capricious exercises of censorial power, however, the flow of Ming books into Chosŏn Korea and the intensity of their influence there continued apace. Indeed, elsewhere in Ŏ's work we see mention of proud adherence to the models set by Ming publications. He recalls that while on a diplomatic visit to Beijing in 1546 he heard two Ming diplomats comparing Chosŏn unfavorably to the Ryukyu Islands and Annam because the official dress of those two states more closely resembled that of the Ming. Ŏ countered by stating that the Chosŏn was closer to the Ming model where it really mattered: 'In general, Korean books on astrology, *geography*, medicine, fortune-telling, and rules of calculation are exactly like those of China' (emphasis mine).[38] From the Ming perspective, display-worthy scrolls or screens showing the extent of its reach would have served as excellent propaganda. This may explain why maps such as the *Yuditu* with Wang Pan's preface from the turn of the seventeenth century were apparently openly sent to Chosŏn.[39] The colophon states that this map 'shows the magnificence of the Heavenly Court's unification of the world in our times,' words which, when spoken abroad, would have been most welcome to Ming ears.[40]

Within Chosŏn Korea there were occasional crackdowns on geographic texts, including the banning of private ownership of the *Sŭngnam* itself during one of the purges of Yŏnsan-gun.[41] These instances may be cited as evidence of security concerns about such texts, but in fact they seem to stand out as exceptions during periods of security panic. A number of years after Yŏnsan-gun was deposed, a new edition of the *Sŭngnam* appeared and the work became available to at least some sectors of the public again. The vicissitudes of this book alone show that while restrictions on geographical information did arise in some circumstances, such restriction cannot be assumed to be the default in Chosŏn Korea.

37 Lee, *A Korean Storyteller's Miscellany: the P'aegwon Chapki of Ŏ Sukkwŏn*, p. 74.

38 Ibid. p. 140. Note the ironic similarity to Nakamura's comment about 'slavish' imitation of China, mentioned above, which has earned him such criticism today.

39 On this map see Han, *Uri yet chido wa kŭ arŭmdaum*, pp. 223-261. Han refers on p. 224 to the first publication of this map by Marcel Destombes, in 'Une carte Chinoise du XVIe siècle découverte a la Bibliothèque Nationale,' an article in which it was overlooked that this was in fact a Korean copy of the map.

40 Han, *Uri yet chido wa kŭ arŭmdaum*, p. 226.

41 Kim Tai-jin, *A Bibliographical Guide to Traditional Korean Sources*, p. 177.

Limits of Ming and Jesuit influence

Around the turn of the seventeenth century, the diversity of maps in print within Ming China had reached unprecedented heights when Jesuit worldviews were added to the mixture of administrative and educational, historical, and religious perspectives that had already been in ferment. Many of these maps reached Chosŏn Korea, where their merits were debated. In Chosŏn the response to some of these works was heavily shaded by their unorthodox taint. Jesuit and Buddhist world maps were not popular in Chosŏn, albeit for different reasons.

While it is sometimes implied that Jesuit ideas had a dramatic impact in Chosŏn almost immediately after their incorporation into Ming books,[42] evidence of significant attention to them does not appear until the eighteenth century. There are records of early importations of such materials, including two versions of Ricci's world map brought back by diplomats in 1603 and 1604, just one year after each was printed.[43] In 1631 the Jesuit missionary João Rodrigues, temporarily stationed in the north in support of a Ming military campaign, met the Chosŏn diplomat Chŏng Tuwŏn and gave him numerous books including the *Zhifang waiji* and five copies of Ricci's world map.[44] There does not seem to be surviving evidence of reproduction or extensive discussion of these works at the time, however.

There is a brief mention of Ricci in Kim Suhong's *Ch'ŏnha Kogŭm Taech'ong P'yŏllam-do* 天下古今大總便覽圖 (Comprehensive map of the ancient and modern world) of 1666.[45] A caption placed in the general vicinity of the Gobi Desert uses his Chinese name to refer to him as 'Li Madou from Europe,' and Ricci's division of one degree into sixty minutes is mentioned in the preface. Nevertheless, the author has not adopted Jesuit mensurate ideas, for in his comparison of different geodetic measurements he sighs,

42 For example, Hŏ Yŏng-hwan writes, 'After the seventeenth century, with the advance of western missionaries into the Orient, European maps of the globe made their way into Korea via China. This stimulus further activated interest in geography and cartography, leading to production of more accurate and technically advanced maps' (Hŏ, *Sŏul ŭi ko chido*, p. 87).

43 Ledyard, 'Cartography in Korea,' p. 249. A 1603 edition of Ricci's map entitled *Liangyi xuanlantu* (Mysterious map of the two forms) is known from a copy that entered Korea; its importance and the content of its prefaces are discussed extensively in D'Elia, 'Recent Discoveries,' pp. 120-158. Since D'Elia's article, another example in Korea and an example in China have become known (Day, 'The Search for the Origins...,' p. 112).

44 Cooper, *Rodrigues the Interpreter*, pp. 348-349; Ledyard, 'Cartography in Korea,' p. 249.

45 Yi, *Han'guk ŭi ko chido*, pp. 26-27.

'How should we evaluate them and find the true one?'[46] The map itself is of the prevalent Sinocentric type with historical places depicted alongside contemporary features. A.L. Mackay argues that 'Kim's critical questioning of the quantitative measurements of the dimensions of the universe and of distances and units on the earth indicates that some escape from the religious format was in prospect.'[47] To be clear, however, there is no evidence here that the Jesuits were taken seriously, only that a confusing variety of claims was accessible to Kim Suhong.

The earliest surviving evidence of precise reproduction of Jesuit cartography is from 1708, when an elaborately decorated version of Ricci's map produced by Adam Schall von Bell was copied at the behest of King Sukchong.[48] The colophon attached to this copy reveals a cautious attitude, however: Ch'oe Sŏkchŏng states that 'The Western theories are far reaching but devious and boastful; they stray into the unattested and uncanonical.'[49] A 1770 reprint of Aleni's 1623 map is another clear example of borrowing from the Jesuits, a trend that apparently became more common only around this time, in limited circles.[50]

In the case of a terrestrial globe held at Koryŏ University, argued by Joseph Needham to demonstrate Korean use of Jesuit sources as early as 1669, Ledyard convincingly argues that the globe was created about a century later.[51] Ledyard touches on the question of why such things never caught on more broadly, noting that reluctance of Confucian scholars to accept Western cartography derived in part from pressure towards orthodoxy in Korea and also from the religious baggage that Jesuit materials carried in their eyes. He also suggests that Western maps had little for Koreans to relate to, while East Asian maps were full of allusions to familiar places.

46 Mackay, 'Kim Su-Hong and the Korean Cartographic Tradition,' p. 36. Mackay's general argument in this article is that Kim's map represents a stage between the mandalic traditional map and modern cartography.

47 Ibid., p. 31.

48 See Yi, *Han'guk ŭi ko chido*, pp. 32-33. Minako Debergh discusses an almost identical copy, also dated 1708, now held in Osaka, in 'La Carte du Monde du P. Matteo Ricci (1602) et sa version Coréenne (1708) conservée à Osaka.'

49 Ledyard, 'Cartography in Korea,' p. 254.

50 Illustrated in Yi, *Han'guk ŭi ko chido*, pp. 40-41.

51 Ledyard, 'Cartography in Korea,' p. 253. The institution called Koryŏ University in Needham's time is now better known as Korea University.

The *Ch'ŏnhado* (Map of All Under Heaven)

Buddhist worldviews are not entirely absent from the Chosŏn record; Nakamura notes the existence of a Korean copy of a map of Jambudvīpa resembling that in the *Tushu bian*.[52] There appears to be a link between the Buddhist worldview and the most distinctive genre of world maps that circulated in Chosŏn during this period, the *Ch'ŏnhado* genre. This type of map portrays a central landmass surrounded by an inner sea, beyond which lies a ring of land with countries from the *Shanhai jing* (such as the homes of One-Eyed Men and Amazons), followed by an outer sea. It was the most common worldview produced for common consumption in the latter half of the Chosŏn dynasty, reproduced in both block-printed and hand-painted atlases. The proposition that this genre derived from maps of Jambudvīpa has been most explicitly outlined by Nakamura.[53]

Ledyard dismisses this genealogy of the *Ch'ŏnhado* as derivative of Nakamura's disdain for Korea, and argues instead that *Ch'ŏnhado* maps actually drive from the *Kangnido*. He illustrates a series of missing links by which the latter could have transmuted into the former, though his case would be stronger if any examples of these missing links still existed. Ledyard concludes that the 'seemingly reversed typological development' of the crude *Ch'ŏnhado* genre, coming as it did later than much more detailed and accurate earlier maps, can be explained as the transformation into a more 'balanced' image from the *Kangnido*, which 'did not and could not relate to the traditional geographic culture of Korea or of East Asia as it existed at the beginning of the fifteenth century.'[54] Bae Woo Sung points in another direction, suggesting that the popularization of the *Ch'ŏnhado* arose from the shock of exposure to the Jesuit worldview, which led Korean scholars to turn to the *Shanhai jing* and other classic geographical texts for intellectual reassurance.[55]

It is important to look at the *Ch'ŏnhado* not as a separate map but as an integral part of the atlases in which it was usually incorporated.[56] These atlases, which became common around the seventeenth century, often used

52 Nakamura, 'Old Chinese World Maps Preserved by the Koreans,' p. 13.

53 Ibid., p. 18.

54 Ledyard, 'Cartography in Korea,' p. 266.

55 Bae, 'Yet chido wa segye,' pp. 171-182.

56 For an example of one such complete atlas see Yi, *Han'guk ŭi ko chido*, beginning p. 156. This genre is also discussed in Richard Pegg's *Cartographic Traditions in East Asian Maps*, pp. 68-82, which includes small color reproductions of three complete woodblock-printed versions from the nineteenth century.

Figure 4.2 Map of Ch'ungch'ŏng *Province* in a Chosŏn-era manuscript atlas

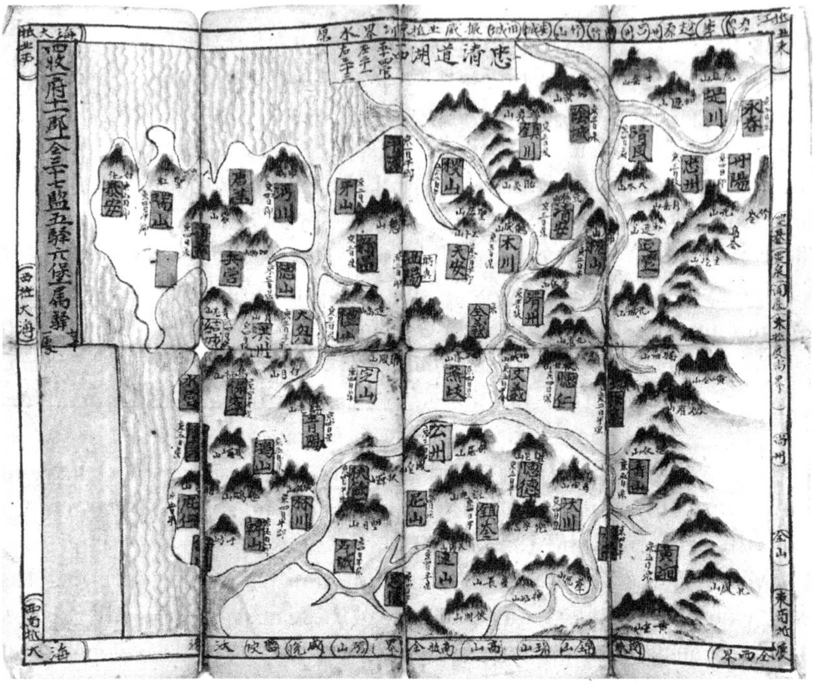

Undated, but late Chosŏn
Collection of the author

a *Ch'ŏnhado* as the frontispiece. For the rest of their depictions of Chosŏn and its provinces, however, they followed the *Sŭngnam* precedent. For example, if we compare the map in Figure 4:2, Ch'ungch'ŏng Province as depicted in a late Chosŏn-era manuscript atlas, to the map from the *Sŭngnam* in Figure 4:1, it is clear that the map not only copies the general style, but the specific arrangement of features as they had been standardized in the national gazetteer, though centuries had passed since the *Sŭngnam*'s maps were composed. This loyalty to the official gazetteer is a standard feature of such privately made atlases, in both manuscript and printed form. Some versions of these atlases are supplemented at the end with images of Japan and the Ryukyus copied from other printed texts.[57]

In the absence of a highly developed publishing industry until a relatively late period, standardized maps based on the *Sŭngnam* continued to

57 For example, note the use of a Ryukyu map from the *Chouhai tubian* as shown in Bae, 'Yet chido wa segye,' p. 168.

circulate for at least three hundred years, usually in manuscript copy. The fact that these maps became familiar, stereotypical images despite the lack of widespread commercial printing poses a unique challenge to the linkage of a recognizable national 'footprint' to print capitalism. We have already seen in China that the standardization of an instantly recognizable 'national' map through cartographic mass-production arose long before European colonialism touched its shores. In Korea, however, such a standardized outline was promulgated even in the absence of a substantial commercial publishing industry, usually by brush on paper or silk.

Reverse influence: The case of the *Chaoxian tushuo*

As I have noted, the flow of cartographic information was by no means limited to a unidirectional export of Ming works and their absorption by neighboring states. Chosŏn maps became a subject of increasing interest in the late Ming not only for the military specialist but also for the educated layman who wished to keep up with current affairs. This was especially true in the shadow of the Imjin wars, when Ming military engagements with Japan took place on Korean soil, and northern Chosŏn territory also became significant for its proximity to the rising Jurchens. We will now turn to examine an instance of Ming appropriation of a Chosŏn atlas. The circumstances under which this took place also shed some light on the mechanisms by which maps could cross borders.

The National Central Library in Taipei holds a short text called *Chaoxian tushuo* 朝鮮圖說 (Annotated maps of Chosŏn).[58] Zheng Ruozeng published a book by this title, but its contents are completely different.[59] This *Chaoxian tushuo* is in fact a copy of a Korean atlas that was brought to China and reprinted with slight modification by Li Chengxun 李承勛, a Ming military official who had participated in the mopping-up operations that took place at the conclusion of the Imjin war (1592-1598). It has no apparent connection with Zheng Ruozeng.

58 The Harvard-Yenching Library's microfilm of this work was produced in 1978 by the National Central Library in Taipei from the original edition in its holdings. The Harvard-Yenching copy is catalogued under Zheng Ruozeng's name alone, while the National Central Library in Taipei has listed it under Zheng with Li Chengxun as a secondary author. In fact, there does not appear to be any connection whatsoever with Zheng.

59 For comparison, a facsimile of a Kangxi-era reprint of Zheng Ruozeng's work by this name may be found in Volume 19 of the *Chaoxian shiliao huibian* 朝鲜史料汇编, pp. 337-434.

This atlas reminds us that the Korean peninsula's role in the Chinese strategic universe made it a site of special geographical interest, especially after the mid-Ming. In contrast to the view that maps were always closely guarded tools of the Chosŏn state, as exemplified in well-known instances of cartographic secrecy in the Song/Koryŏ and Qing/Chosŏn relationship,[60] the closer diplomatic and military ties of the late Ming appear to have fostered a more open relationship, including the sharing of maps. It is difficult to imagine Ming officers moving troops through Korea without fairly detailed information from their Chosŏn hosts about geography and local conditions. The *Chaoxian tushuo* suggests that in times of tension with common enemies, maps could cross boundaries between allies – unequal participants in a tributary relationship, but allies nonetheless.

This *Chaoxian tushuo* comprises eighteen sheets printed from woodblocks. The first sheet is purely textual, offering a historical overview of the Korean peninsula's dynastic succession and its relations with Chinese states. This is followed by a rather crowded map depicting the entire Korean peninsula and some adjacent islands. The remaining pages bear maps of each *to* 道, or province, followed in each case by a listing of the province's prefectures and counties (州, 縣) with their relative positions and distances apart. The numbers given in these listings are based on the distances provided in the *Sŭngnam*. The similarity of the maps in this work to those in the *Sŭngnam* further confirms the source of this information.

Although the maps closely resemble those in the *Sŭngnam*, they possess additional notation along the top margin and sometimes the sides, including details of post stations, local commanderies, and other institutions, depending on the region covered. The practice of copying maps derived from the *Sŭngnam*, either in manuscript or in woodblock print, with additional annotations in the margins appears to have been widespread in the sixteenth century. A small woodblock-printed domestic atlas held by Yŏngnam University, for example, has marginal notes added by hand.[61] Comparing our *Chaoxian tushuo* to other atlases of this type, its maps appear strikingly similar to – indeed, nearly identical to – those contained in a manuscript atlas in the Yoon Hyung-doo collection.[62] As Yi Ch'an notes, because this manuscript atlas shows Chinwŏn (珍原) and Changsŏng (長

60 On such tensions see for example Han, *Uri yet chido wa kŭ arŭmdaum*, pp. 17-18, as well as Ledyard, 'Cartography in Korea,' p. 241, and pp. 301-305 for details regarding the Mukedeng incident during the Qing.

61 *Yŏngnamdae pangmulgwan sojang Han'guk ŭi yet chido*, pp. 21-25.

62 Partially illustrated in Yi, *Han'guk ŭi ko chido*, p. 88 and 388.

Figure 4.3 The first provincial map in Li Chengxun's 1600 *Chaoxian tushuo*

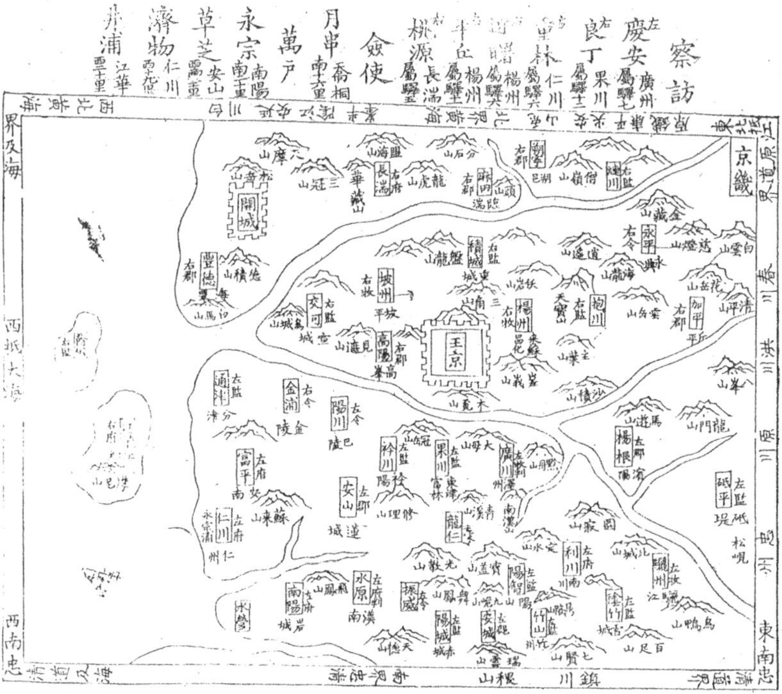

Annotated maps of Chosŏn, with the capital depicted as a square with crenellated walls (pp. 3a-3b)
Courtesy of the Harvard-Yenching Library

城) as separate administrative units, it has been dated to before 1600, when Chinwŏn was absorbed into Changsŏng.[63] Li Chengxun's version of the atlas preserves this feature, which is not surprising given that the changes were instituted only very recently at the time of its publication.

Li's version differs, however, in that it contains a note at the bottom of the eighth page explaining the special use of *sik* 息 in Chosŏn as a unit of measure signifying 30 *ri*. This is a usage that would have puzzled a reader in China, so this explanatory note, together with the historical summary on the opening page, shows that Li considered the needs of a Chinese audience in reproducing these Korean maps.

In keeping with this consideration of his readers, Li's historical introduction to the *Chaoxian tushuo* was intended to provide a Chinese audience with an outline of the Korean peninsula's past, thereby helping to frame

63 Yi, *Han'guk ŭi ko chido*, p. 387.

the accompanying vision of its present. The preface presents the following narrative:

> Chosŏn began with Tangun. Long ago, the Heavenly God Hwanin's *sŏjo* (secondary son) Ung descended to Taebaek Mountain. He joined with a bear, and they produced Tangun. Tangun was a contemporary of Yao [one of the legendary Chinese sage kings]. At the time of Yu's Meeting at Tushan [where representatives of all lands sent tribute to Yu], he sent his son Puru to attend.
>
> Later, King Wu (1046-1043 BCE) of the Zhou dynasty enfeoffed Jizi (*K.* Kija) in Chosŏn, and Tangun's clan came to an end. At the time of Empress Gao of the Han dynasty (d. 180 BCE), Wei Man (*K.* Wiman) of Yan staged an attack on Chosŏn, and Kija's clan, which had been enfeoffed by the king (Wu), came to an end. At the time of the Han emperor Wu (140 BC-87 BC), [the Chosŏn king][64] Ugŏ killed a Han envoy. Emperor Wu sent General of Large Ships Yang Pu from Qi across the Bohai Sea, and General of the Right Xun Zhi from Beiping through Liaodong, to chastise Chosŏn. People in Chosŏn then killed Ugŏ in order to submit (i.e., to placate the Han dynasty). Emperor Wu divided the territory into four commanderies called Nangnang, Imdun, Hyŏnt'o and Chinbŏn. Then came the (Three) Hans, Mahan, Chinhan, and Pyŏnhan. The Three Hans fell and three kingdoms arose called Silla, Koguryo and Paekche. Of the three kingdoms Koguryŏ was the strongest, time and again causing trouble along its borders. Although Silla was small, over generations it cultivated its customs and was exceedingly proper in its reception of envoys, and it eventually united the Three Hans (i.e., the Korean peninsula). This succession of states lasted a millennium. When Silla fell the Koryŏ king Wang Kŏn arose. The Koryŏ lasted for 500 years. The Koryŏ fell and the Chosŏn king Yi Sŏnggye arose in the twenty-fifth year of Hongwu (1392).
>
> Respectfully recorded by Li Chengxun of Kuocang, 括蒼, Zhedong, in the last month of autumn of the *gengzi* year of Wanli (1600).
>
> 朝鮮始自檀君。昔天神桓因庶子雄降于太白山與熊假合是生檀君。檀君與堯同時。禹會塗山遣子夫婁朝焉。後周武王封箕子於朝鮮桓氏亦衰絕。漢高后時燕人衛蒲襲朝鮮王準箕氏遂亡。漢武帝時朝鮮王右渠襲殺漢使。武帝遣樓船將軍楊僕自齊渡渤海，右將軍荀彘自北平出遼東以伐朝鮮。朝鮮人殺右渠以降。武帝分其地為四┌：曰樂浪，曰臨屯，

64 The words in brackets here have been partially inked out on the original, perhaps indicating that the one who inked them out thought it incorrect to refer to Ugŏ as a 'Chosŏn king.'

曰玄菟，曰真蕃；或稱韓：曰馬韓，曰辰韓，曰卞韓，三韓衰而三國興，曰新羅，曰高句麗，曰百濟。三國獨高句麗最强大，世為邊患。新羅雖小世修風教奉藩節甚謹，卒混一三韓，歷國千載。新羅衰而高麗王王建興，高麗王歷五百年而朝鮮王李成桂興是歲洪武二十五年也。萬曆庚子季秋，浙東栝蒼李承勛謹識。

The sources for this narrative are somewhat obscure. The portion covering the period between Kija's enfeoffment and the Three Kingdoms contains information and wording reminiscent of that found in Yue Shi's early Song geography, the *Taiping huanyu ji*.[65] The subsequent developments would have been available in more recent chronicles, though the precise wording of most of the narrative as it appears here may well be Li's own.

The first few lines of the narrative summarizing the Tangun legend are distinctive, especially for their emphasis on Puru's attendance as a delegate to Yu's court at the Meeting of Tushan. The sentence noting this incident is almost identical to a sentence in Chapter 154 of the *Sejong sillok* 世宗實錄 (Veritable record of the Sejong reign), which in turn quotes a passage from the *Tangun kogi* 檀君古記 (Ancient record of Tangun).[66] Given that Li's narrative emphasizes the peninsula's historical connections with China, it is perhaps not surprising to see this claim that Tangun's own son had expressed fealty to Yu.

This brings us to a central question: Who was Li Chengxun and how did he come to publish a Chosŏn cartographic text with a historical narrative for a Ming audience? The indication of his native place, Kuocang, at the end of the preface translated above enables us to identify him with certainty. He was evidently considered too minor a figure to cite in the dynastic history of the Ming, but he was sufficiently respected at the local level to warrant brief notices in provincial and prefectural gazetteers. He was also involved in the publication of at least two works of military significance prior to the *Chaoxian tushuo*.

Perhaps the more famous of these works is his edition of the *Jixiao xinshu* 紀效新書 (New book of effective disciplines), a compilation of military strategies he reprinted in 1588. This is one of the most famous military texts of the late imperial period, originally authored by Qi Jiguang 戚繼光, who died one year before Li's fourteen-volume edition appeared. Because

65 Yue Shi, *Taiping huanyu ji*, Chapter 172b.

66 The *Sejong sillok* refers to Puru as a *t'aeja* 太子 (Chapter 154: 4a. *Sejong sillok*, vol. 5, p. 683 in the *Chosŏn wangjo sillok* series, Yŏngin ch'ukswaep'an edition). Li's preface calls him simply 子; aside from this, the two sentences are identical. It is not clear which *Tangun kogi* the sillok means; a facsimile of a work by this title included in a recent collection contains similar information written in a different form on p. 14b (*Tangun ko kirok sagong*, p. 29).

the *Jixiao xinshu* was reprinted numerous times and Li was involved in only one of these editions, his name tends to be forgotten in discussions of the text, but for our purposes it serves to highlight the direction of his interests.[67] Li also authored the brief *Mingjian ji* 名劍記 (Record of famous swords).[68] As its title suggests, this is a collection of short descriptions of almost sixty blades either belonging to famous individuals throughout history, or allegedly possessing amazing qualities. The details provided for each sword, usually just a few sentences, are culled from a wide range of historical records and classical texts. The author is identified on the first page as '括蒼李承勛' (Li Chengxun of Kuocang). Together with Li's reprint of the *Jixiao xinshu*, this text clearly reflects a combined passion for military arts and history. Li also wrote a supplement to the *Mingma ji* 名馬記 (Record of famous horses) by Guo Zizhang 郭子章, further adding to his roster of military-historical contributions.[69]

The Yongzheng edition of the *Zhejiang Tongzhi* 浙江通志, a provincial gazetteer of Zhejiang, provides a brief biography of Li. It traces his military career beginning in the 1560s. Among other deeds, his biography credits him with having been instrumental in the 1582 capture of Ma Wenying, a 'leading villain' in Zhejiang. Li rose through the ranks to take on assignments in naval defense that brought him north along the coast to Shandong, and, as part of the Ming's military intervention against Toyotomi Hideyoshi's invasion of the Korean peninsula, to Chosŏn. He arrived in 1599, after the Japanese forces had withdrawn but in the midst of rumors that they might strike again. In an undated vignette, his biography describes a standoff at Pusan, facing the Japanese island of Tsushima where he was assisting in defensive preparations. The biography says that when the forces at Pusan were firing guns in practice drills they could be heard on the other side, and the Japanese on Tsushima laid low, offering some respite to the Korean side. During this time Li received a notice of his mother's death. He begged leave to hurry back for her funeral, but after an examination of legal precedents, he was denied. Unable to return, he instead cut his thigh and sent an offering.[70]

67 For a brief discussion of Li's involvement and his reasons for reprinting the book while working in Zhejiang, see Fan Zhongyi, *Qi Jiguang bingfa*, p. 20.

68 A Shunzhi year 3 (1646) edition is collected with other volumes in the *Shuofu* 說郛, *xu juan* 36 (vol. 158).

69 A facsimile of a Ming edition of this work is available in the *Xuxiu Siku quanshu* 續修四庫全書, vol. 1119.

70 Ji Zengyun . *Zhejiang Tongzhi*, *juan* 174, pp. 17a-17b. Reproduced in *Yingyin Wenyuange Siku quanshu*, vol. 523, p. 564. This biography is cited in the *Zhejiang Tongzhi* as being based on that in an earlier edition of the *Chuzhou fuzhi* 處州府志.

Li is celebrated in this hagiographic account not only for his military service but also for his filial piety. A later gazetteer of 1877 from Chuzhou, the Zhejiang prefecture in which Kuocang is situated, summarizes Li's career in briefer terms, omitting this particular episode, but gives his courtesy name as Xiyong 錫庸.[71]

Korean sources relevant to Li Chengxun, particularly the *Sŏnjo sillok* 宣祖實錄, actually provide a greater wealth of information about him, and from a more critical perspective, than can be gleaned from these brief notices.[72] After having set sail for Chosŏn in the fifth lunar month of year 32 of Sŏnjo's reign (the summer of 1599), Li met the king at court on the twenty-second day of the seventh month (September 11). He is described in the *sillok* as approximately sixty years of age at this time.[73] References to Li in the veritable record total more than two dozen, many of which are minor, but they reveal that the Korean court, while showing every courtesy when he was present, actually held him in relatively low regard as a shallow and vulgar man.[74] He comes under criticism repeatedly for his failure, in the eyes of the king's advisors, to comply with the Confucian rites of mourning.

In contrast to the moving account of his cutting his own thigh cited above in the *Zhejiang Tongzhi*, the *Sŏnjo sillok* records that he dressed inappropriately after hearing of his mother's death and tasted meat and wine when it was still forbidden to a son in mourning.[75] Li is said to have argued that the king should be willing to share wine with him (rather than the more 'proper' tea the king offers) because it is permissible in China if fifty days have passed since the parent's death; besides, he has already tasted

71 Pan Shaoyi 潘紹詒, *Guangxu Chuzhou fuzhi* 光緒處州府志, *juan* 18, p. 33b.

72 Li is also noted in other Chosŏn works, including the *Ponjo kiryak* (本朝紀略, Outline history of our dynasty), an anonymous early eighteenth century text which has been made available online by the Kyujanggak, and he is included among the Ming military figures listed in a manuscript called the *Imjin p'illok* (壬辰筆錄, Record of the Imjin wars) also held by the Kyujanggak in Seoul. All of these texts serve to remind the scholar of Chinese history that it can be perilous to overlook Korean sources. It should be noted that there was another Chinese official named Li Chengxun in Korea at the same time, who died while there. Our Li Chengxun is referred to in the *sillok* by his title, *tidu* 提督, and can thus be differentiated from his ill-fated contemporary.

73 *Sŏnjo sillok* 115: 17a (7th month, 22nd day.) For ease of reference I will also include page numbers in the *Chosŏn wangjo sillok* series (Seoul: T'amgudanng, 1984) in the following format: vol. 23: 649.

74 *Sŏnjo sillok* 118: 17b (vol. 23: 693).

75 *Sŏnjo sillok* 118: 17a (vol. 23: 693). His actions and those of his subordinates were cited as examples of the decay of ritual propriety among the Chinese; for example, see the criticism of his assistant's participation in a mourning ritual for the king's mother in *Sŏnjo sillok* 127: 6b (vol. 24: 90).

wine during another ritual.[76] The criticisms of Li's conduct recorded in the *sillok* contrast significantly with the tone of the Chinese gazetteer entries, which were, after all, written to celebrate a local hero and were probably based on his self-reported exploits.

The fact that Li already knew of his mother's death on November 29, 1599, when he stirred controversy at court with his drinking of wine, helps us date the Pusan vignette cited above to some time before that date. Despite the court's criticisms of Li behind his back, the king expressed unlimited gratitude to his face at every audience, continuing to do so until the time of Li's farewell on November 2, 1600.[77] Li's sojourn in Korea lasted, in total, approximately a year and a half.

Even in the absence of further details of Li's literary or editorial activities while in Korea, it is clear that his presence there afforded him the opportunity, perhaps even the official responsibility, to obtain contemporary maps of Chosŏn. While its intended audience and arena of circulation are unknown, the existence of this work is clear evidence of the direct appropriation of Chosŏn cartography during the Imjin war.

The unhappy late sixteenth-century role of Chosŏn as a contested arena between the spheres of Chinese and Japanese influence was reprised in some ways at the end of the nineteenth century. Therefore it is not surprising to note that another Chosŏn geographical text was published in China at this time by an officer who had brought it back from Korea. The *Chaoxian dili xiaozhi* 朝鮮地理小志 (Concise geographical survey of Chosŏn), a translated version of the classic Korean *T'aengniji* 擇里志, was printed in 1885 by the Qing general Wu Changqing 吳長慶, who had led a Chinese expedition into Chosŏn at the time of the 1882 Military Mutiny. While the details of this work have been published elsewhere by Sun Joo Kim,[78] it is worth citing as another example of the means by which the changing relationship between regimes in Korea and China affected the movement of geographic and cartographic materials between these states.

Conclusion

Printed maps from the Ming played a significant part in the development of Chosŏn's cartographic culture. In the field of geography, the most profound

76 *Sŏnjo sillok* 118: 18a (vol. 23: 693).

77 *Sŏnjo sillok* 129: 21b (vol. 24: 130).

78 Sun Joo Kim, 'A Glance at the *Chosŏn chiri soji*.'

influence can be seen in the use of the Ming national gazetteer as the model for the *Sŭngnam*. In turn, the *Sŭngnam's* series of national and provincial maps set the precedent for the miniature atlases that were reprinted and, more frequently, copied by hand, until almost the end of the dynasty. Moreover, many of the visual and physical qualities of Korean woodblock-printed cartography that have been interpreted as revealing uniquely Korean characteristics are, when viewed in broader perspective, actually quite common elsewhere.

After the formative influences of the early decades of the dynasty, the scope of popular Chosŏn cartography does not seem to have reflected the diversity of later Ming works. Although we must always be reticent about drawing too firm a conclusion based on the absence of evidence, as the sixteenth century progresses, we see a growing divergence as Chosŏn maps show few of the changes fostered by the Ming commercial publishing industry. Neither do we see pronounced Buddhist cartographic influences or imitation of European models. In Chosŏn this might have been seen, from the perspective of many a *yangban* scholar, as evidence of greater respect for orthodoxy and the shunning of fads.

Yungsik Kim laments that work in the field of the history of science in Korea 'can best be characterized by the word isolation.' He writes, 'Most Korean scholars work in isolation from historians of science in the outside world. As a result, Korean science is frequently studied in isolation from the science of the rest of East Asia, let alone that of the world.'[79] As this chapter shows, familiarity with Ming geographical and cartographic works can help elucidate the nature of a Chosŏn text like the *Sŭngnam*, while attention to Chosŏn publications can likewise help resolve questions surrounding works published elsewhere, like the *Chaoxian tushuo*, that made use of these sources for their information about the Korean peninsula. Historians of China, Korea, Japan, Vietnam, and other East Asian nations can better understand the trends that affected their own regions of interest by working together to see how these developments moved about the region as a whole. Familiarity with Ming geographical and cartographic works can help elucidate the nature of a Chosŏn text like the *Sŭngnam*, while attention to Chosŏn publications can likewise help resolve questions surrounding Chinese and other works that made use of these sources for their information about the Korean peninsula. Despite the natural and political barriers separating Chosŏn Korea from its neighbors, it was never truly isolated, so why should we study it in isolation now?

79 Yungsik Kim, 'Problems and Possibilities in the Study of the History of Korean Science,' p. 49.

Works cited

Primary sources

Chosŏn wangjo sillok 朝鮮王朝實錄. Seoul: T'amgudang, 1984.

Fan Zhongyi 范中義. *Qi Jiguang bingfa* 戚繼光兵法. Beijing: Shishi Chubanshe, 1998.

Han Paek-kyŏm 韓百謙. *Tongguk chiriji* 東國地理誌. Facsimile edition. Seoul: Ilchogak, 1982.

Ji Zengyun 嵇曾筠, editor. *Zhejiang tongzhi* 浙江通志. Facsimile reprint in *Yingyin Wenyuange Siku quanshu*, vols 519-526. Taibei: Taiwan shangwu yinshuguan, 1983.

Kim, Chŏng-ho 金正浩. *Ch'ŏnggudo* 青邱圖. Seoul: Minjok Munhwa Ch'ujinhoe, 민족문화추진회, 1971.

—. *Taedong yŏjido wŏndo* 大東輿地圖原圖. Seoul: Kyŏnghŭi Taehakkyo Pusŏl Han'guk Chŏnt'ong Munhwa Yŏn'guso 慶熙大學校附設韓國傳統文化研究所, 1977.

Li Chengxun 李承勛. *Chaoxian tushuo* 朝鮮圖說. S.n., Autumn of 1600 (萬曆庚子季秋).

—. *Jixiao xinshu* 紀效新書. Republication of Qi Jiguang's edition, Wanli 16 [1588].

—. *Mingjian ji* 名劍記. In the *Shuo fu* 說郛, vol. 158. Shunzhi year 3 [1646], Xu juan 36.

—. *Xu Mingma ji* 續名馬記. Facsimile of Ming edition, in *Xuxiu Siku quanshu* 續修四庫全書, Zi bu (370 volumes), vol. 1119: 319-338. Shanghai: Shanghai guji chubanshe: [1995-1999].

No Sasin 盧思慎, et al. *Sinjŭng Tongguk yŏji sŭngnam* 新增東國輿地勝覧. Hansŏng: Mallyŏk 39 [1611].

—. *Sinjŭng Tongguk yŏji sŭngnam*. Seoul: Tongguk Munhwasa facsimile edition, Tan'gi 4291 [1958].

—. *Sinjŭng Tongguk yŏji sŭngnam*. P'yŏngyang: Chosŏn Minjujuŭi Inmin Konghwaguk Kwahagwŏn Ch'ulp'ansa, 1959.

Pan Shaoyi 潘紹詒 and Zhou Rongchun 周榮椿, eds. *Guangxu Chuzhou fuzhi* 光緒處州府志. Guangxu year 3 [1877].

Sim Paek-kang, ed. *Tangun ko kirok sagong*. Seoul: Minjok Munhwa Yŏn'guwŏn, 2001.

Yi Ŭr-ho 李乙浩, ed. *Haedong chegukki / Kanyang nok* 海東諸國記 / 看羊錄. Seoul: Taeyang Sŏchŏk, 1972.

Yue Shi 樂史. *Taiping huanyu ji* 太平寰宇記. Hangzhou: Gujiu shudian, 1982.

Zheng Ruozeng 鄭若曾. *Chaoxian tushuo* 朝鮮圖說. In *Chaoxian shiliao huibian* 朝鲜史料汇编 (20 volumes), vol. 19: 337-434. Beijing: Quanguo tushuguan wenxian suowei fuzhi zhongxin, 2004.

Secondary sources

Bae Woo Sung 배우성. 'Yet chido wa segye' 옛지도와 세계. In *Uri yet chido wa kŭ arŭmdaum* 우리 옛지도와 그 아름다움, edited by Han Yŏng-u, An Hwi-jun, Pae, U-sŏng, 171-182. Seoul: Hyohyŏng Ch'ulp'an, 1999.

—. 'Worldviews & Cartography.' In *The Artistry of Early Korean Cartography*, edited by Alexander Akin, 91-129. Larkspur, CA: Tamal Vista Press, 2008.

Choe Suh Myun. 'Tokdo in Old Maps.' *Korea Observer* 29, no. 1 (Spring 1998): 187-203.

Cooper, Michael. *Rodrigues the Interpreter: An Early Jesuit in Japan and China*. New York and Tokyo: Weatherhill, 1974.

Day, John D. 'The Search for the Origins of the Chinese Manuscript Copies of Matteo Ricci's Maps.' *Imago Mundi* 47 (1995): 94-117.

D'Elia, Pasquale M. 'Recent Discoveries and New Studies (1938-60) on the World Map in Chinese of Father Matteo Ricci S.I.' *Monumenta Serica* 20 (1961): 82-164.

Debergh, Minako. 'La Carte du Monde du P. Matteo Ricci (1602) et sa version Coréenne (1708) conservée à Osaka.' *Journal Asiatique*, 274, No. 3-4 (1986): 417-454.

Destombes, Marcel. 'Une Carte Chinoise du XVIe Siècle Découverte a la Bibliothèque Nationale.' *Journal Asiatique* 262 (1974): 193-212.

Hamel, Hendrik. *Hamel's Journal and a Description of the Kingdom of Korea, 1653-1666* (revised edition). Translated by Br. Jean-Paul Buys. Seoul: Royal Asiatic Society, Korea Branch, 1998.

Han Yŏng-u 한영우, An Hwi-jun 안휘준, Pae U-sŏng 배우성. *Uri yet chido wa kŭ arŭmdaum* 우리 옛지도와 그 아름다움. Seoul: Hyohyŏng Ch'ulp'an, 1999.

Hirth, Friedrich, and W.W. Rockhill. *Chau Ju-kua: His Work on the Chinese and Arab Trade in the Twelfth and Thirteenth Centuries, Entitled Chu-fan-chï*. St. Petersburg: Printing Office of the Imperial Academy of Sciences, 1911. Reprint, Taibei: Literature House, 1965.

Hŏ Yŏng-hwan 許英桓. *Chŏngdo 600-yŏn Sŏul chido* 定都600年서울地圖 / *600 year's maps of Seoul*. Seoul: Pŏmusa, 1994.

—. *Sŏul ŭi ko chido* 서울의古地圖. Seoul: Samsŏng Ch'ulp'ansa, 1989.

Hulbert, Homer B. 'An Ancient Gazetteer of Korea.' *The Korean Repository* 4 (1897): 407-416.

Kim, Michael. *The Apparition of the Rational Public: Reading Collective Subjectivity in the Korean Public Sphere*. PhD Dissertation, Harvard University, 2004.

Kim, Sun Joo. 'A Glance at the *Chosŏn chiri soji*.' In *Studies on the Korean Materials in the Harvard-Yenching Library*, edited by Choong Nam Yoon, 173-185. Seoul: Kyŏngin Munhwasa, 2004.

Kim Tai-Jin. *A Bibliographical Guide to Traditional Korean Sources*. Seoul: Asiatic Research Center, 1976.

Kim, Yungsik. 'Problems and Possibilities in the Study of the History of Korean Science.' *Osiris*, 2nd Series, Vol. 13, 'Beyond Joseph Needham: Science, Technology, and Medicine in East and Southeast Asia' (1998): 48-79.

Ledyard, Gari. 'Cartography in Korea.' In *The History of Cartography, Volume 2, Book 2: Cartography in the Traditional East and Southeast Asian Societies*, edited by J.B. Harley and David Woodward, 235-345. Chicago and London: University of Chicago Press, 1994.

Lee, Ki-suk, Kim Shin, and Soh Jung-chul. *East Sea in World Maps*. Seoul: The Society for East Sea, 2002.

Lee, Peter H. *A Korean Storyteller's Miscellany: The P'aegwon Chapki of Ŏ Sukkwŏn*. Princeton, NJ: Princeton University Press, 1989.

Mackay, A.L. 'Kim Su-Hong and the Korean Cartographic Tradition.' *Imago Mundi* 27 (1975): 27-38.

McCune, Shannon. 'World Maps by Korean Cartographers.' *Journal of Social Sciences and Humanities* 45 (June 1977): 1-8.

Nakamura Hiroshi. 'Old Chinese World Maps Preserved by the Koreans.' *Imago Mundi* 4 (1947): 3-22.

Ōba Osamu 大庭脩. *Kanseki yunyū no bunkashi: Shōtoku Taishi kara Yoshimune e* 漢籍輸入の文化史: 聖徳太子から吉宗へ. Tōkyō: Kenbun Shuppan, 1997.

Pegg, Richard A. *Cartographic traditions in East Asian maps*. Honolulu: University of Hawaii Press, 2014.

Pratt, Keith. *Korea: A Historical and Cultural Dictionary*. London: Curzon Press, 1999.

Robinson, Kenneth. 'Centering the King of Chosŏn: Aspects of Korean Maritime Diplomacy, 1392-1592.' *The Journal of Asian Studies* 59, no. 1 (February 2000): 109-125.

Sanders, Graham. Personal communication granting permission to use his translation of an Yi Kyubo poem posted on Facebook. March 13, 2013.

Short, John Rennie. *Korea: A Cartographic History*. Chicago and London: University of Chicago Press, 2012.

Walraven, Boudewijn. 'Reader's Etiquette, and Other Aspects of Book Culture in Chosŏn Korea.' In *Books in Numbers: Some Papers Read at the Conference in Celebration of the Seventy-Fifth Anniversary of the Harvard-Yenching Library*. Cambridge, MA: Harvard-Yenching Library, 2003.

Yi Ch'an (Lee Chan) 李燦. *Han'guk ko chido* 韓國古地圖. Seoul: Han'guk Tosŏgwanhak Yŏn'guhoe 韓國圖書館學研究會, 1977.

—. *Han'guk ŭi ko chido* 韓國의古地圖 (*Old Maps of Korea*). Seoul: Pŏmusa, 1991.

[Yŏngnam University Museum]. *Yŏngnamdae pangmulgwan sojang Han'guk ŭi yet chido* 영남대박물관소장韓國의옛地圖. Kyŏngsan: Yŏngnam Taehakkyo Pangmulgwan, 1998.

Yoon, Inshil Choe, trans. *Yi Chung-hwan's T'aengniji: The Korean Classic for Choosing Settlements*. Sydney: Wild Peony Press, 1998.

5 Japanese Cartography between East and West

Abstract
This chapter examines the role of influential Ming texts such as the *Sancai tuhui* (Illustrated Compendium of the Three Fields of Knowledge) and the *Huang Ming zhifang ditu* (Administrative Atlas of the Imperial Ming) in Japan, even in the context of Japan's comparatively direct contact with the cartographic traditions of the Dutch, the Portuguese, and other European lands. The role of Jesuit cartography and the long-lasting influence of Buddhist cosmology are considered within this trans-regional context.

Keywords: Ming, Japan, Jesuit cartography, *Ishō Nihon den*, *Nansenbushū bankoku shōka no zu*, *Wa-Kan sansai zue*

There was significant cartographic influence in Chosŏn Korea from the earlier Ming period, but very little reflection of the explosion of diversity that took place later in the Ming dynasty or the addition of European materials to the existing repertoire. Japan, also linked to the networks of East Asian textual circulation, offers a host of contrasts in its intersections with late Ming publications as well as Western cartography.

While Japan had much looser political relations with the Ming than did Chosŏn, it engaged in a significant amount of both officially authorized and private trade. Products of the late Ming publishing boom reached Japanese hands in processes that have been chronicled by authors such as Ōba Osamu and Peter Kornicki.[1] In the early decades of the Tokugawa, Japanese readers were fully aware that their islands had been the subject of writings on the continent, and there were readers who were keen to peruse

1 Kornicki addresses book imports primarily from the Qing (p. 296), for it was after the imposition of censorial regulations that lists of imported books began to be compiled. Books imported during the late Ming generally left no trace of their entry and can only be identified from their influences or citations in Japanese works.

Akin, Alexander, *East Asian Cartographic Print Culture: The Late Ming Publishing Boom and its Trans-Regional Connections*. Taylor & Francis Group, 2021
DOI: 10.5117/9789463726122_CH05

these accounts. Scholars like Matsushita Kenrin gathered together texts and maps produced under various Chinese and Korean dynasties, reproducing and critically evaluating them for a local audience. Imported Ming works also played a significant role in shaping the direction of cartography in Japan, serving as direct inspiration for the geographical content and organization of certain encyclopedias and other popular works. Even after a series of bans on Christianity and European works came into force during the 1630s, discussion of the earth's sphericity and the 'Five continents theory' introduced from Europe continued in Japan, not only because Japanese authorities treated maps differently from other categories of Western texts, but also because syncretic Ming works that had assimilated Jesuit writings introduced such ideas completely shorn of Christian content. I will illustrate this phenomenon in the *Wa-Kan sansai zue* 和漢三才圖會 (Japanese and Chinese illustrated compendium of the three types of knowledge), which was assembled beginning in the 1680s and found its way into print in 1713, seven years before shogun Yoshimune began relaxing prohibitions on European technical works.

Within the field of cartographic history the Ming contribution to Japanese cartography is often overlooked in favor of attention to European influences. Coincidentally, it was at almost the same time that the Ming instituted a maritime trade ban to crack down on piracy, hampering the outflow of books from Ming territory, that this new cast of characters arrived on the scene in Japan. The alighting of three Portuguese men in a ship blown off course in 1543 quickly escalated into trade relations that took advantage of the vacuum left by the Ming ban. In 1549, God followed gold as Francis Xavier inaugurated the Jesuit mission on this distant soil. Portuguese traders and Jesuit missionaries brought practical maps, in the form of portolan-style charts used in navigation, as well as instructive materials including world maps and geographical texts. Consequently, unlike Chosŏn, Japan was not reliant upon Chinese sources for transmissions of the European 'five-continent' worldview in translation; it had its own European visitors and residents with both secular and missionary intent.

Furthermore, for reasons touched upon below, some of Japan's rulers had a rather intimate understanding of the practical uses of maps and the uses to which surveying had been put in other countries. Particularly different from either the Ming or the Chosŏn situation is Japan's interest in developing trans-regional overseas trade networks in the early seventeenth century. This had cartographic ramifications because there is significant evidence of the adoption of Portuguese-style portolan charts by the pilots of Japanese licensed ships. Matsuki Satoru 松木哲 notes that the earliest extant maritime maps produced in Japan are copies of European imports.[2]

The contours of Tokugawa Japan's own publishing boom differ dramatically from what took place in Chosŏn Korea.[3] Japan saw a virtual explosion of commercial publishing in the seventeenth century, somewhat later than in Ming China but earlier than in Chosŏn Korea. The role of the state in printing didactic texts for the public began relatively late in Japan, so that Buddhist institutions were instead a major force in the early production of both printed and manuscript texts, sometimes with government sponsorship. Furthermore, whereas Chosŏn government publications self-consciously followed early Ming models throughout much of the dynasty, large-scale commercial publishing in Japan emerged at a point where there was already great diversity in available models. Japan had extensive commercial and cultural ties with countries other than Ming China; even after the expulsion of the Portuguese and the Spanish it was far more internationally connected than Chosŏn. Japan was able to maintain trade and even certain cultural ties with Europe through the Dutch East India Company.

James R. Bartholomew has described the Tokugawa shogunate as having an 'almost institutionalized suspicion and paranoia'[4] about novel ideas, though he writes elsewhere that 'Japan's dominant Confucian intellectual tradition was loosely structured and relatively tolerant.'[5] The apparent contradiction lessens when we compare the relatively open manner in which cartography *in general* was treated, to the government's response to certain *particular* cartographic questions, such as Matsudaira Sadanobu's suppression of the results of a coastal survey of Sakhalin Island.[6] There were areas of sensitivity, but for reasons I will explore in this chapter, cartography showed less impact from the state's exclusionist policies than one might expect. European worldviews, while hotly contested in some circles, were not as fundamentally incompatible with Japan's view of its place in the world as they had been to observers in the

2 See Matsuki Satoru, 'Kōro zu ni tsuite' 航路図について (A note on maritime maps), pp. 105-106. Examples of Japanese portolans (maritime maps based on European prototypes, with unmistakable compass roses) survive in several examples, and later copies simplified for publication in books as curiosities also remain extant. See Nakamura, 'The Japanese Portolanos of Portuguese Origin of the XVIth and XVIIth Centuries,' as well as Peter Shapinsky's 'Polyvocal Portolans: Nautical Charts and Hybrid Maritime Cultures in Early Modern East Asia.'

3 For an overview of recent scholarship in many subfields of Japanese cartographic history, including the publishing industry during the Edo period, see Wigen et al., *Cartographic Japan: A History in Maps*.

4 Bartholomew, *The Formation of Science in Japan*, p. 38.

5 Ibid., p. 3.

6 Ibid., p. 37. The survey had demonstrated that it was an island rather than a peninsula.

Ming. The key fact that Japan had always been on the periphery, whether in the ancient Buddhist perspective or in the Sinocentric world order, actually made it possible to perceive Japan as its own sort of center.[7]

Buddhist cosmology and cartography played a much more significant role in Japanese printed cartography than in either Chosŏn Korea or Ming and Qing China. Precisely because this cosmology endured for so long, Buddhist cartographic publications were forced to either come to terms with or justify their resistance to newer knowledge as understanding of the broader world deepened. In this chapter I will examine some of the European influences that distinguished Japan's experience from those of the Ming and Chosŏn, followed by a discussion of direct Japanese responses to late Ming works. I will also examine works that combined Chinese and European materials and even worked to synthesize them with Buddhist cartography.

Historiographical approaches

Some have argued that Japan was more highly developed than Ming and early Qing China in terms of both understanding and reproducing Western cartography. For example, Helen Wallis asserted of the Jesuit influence that 'The many world maps published in Japan in the hundred years after 1645 testify to the great interest of the newly awakened middle classes in the new knowledge of the world, and to their curiosity about foreign countries, despite the government's policy of seclusion [...] Ricci's geographical teachings brought western knowledge to Japan more effectively than to China, because they reached more receptive minds, in a wider group of the community. Divorced from their Christian context, they were assimilated into a native Japanese form of cartography, although much simplified in the process.'[8]

This passage illustrates a longstanding preoccupation with the extent and quality of Japanese openness to foreign ideas. Japanese scholars have undertaken this project as assiduously as have others. Nakamura Hiroshi represents a common view when he characterizes the Japanese as astute students of European cartography, but bemoans the 1636 ban on Christian imports that, he argues, dramatically curbed further development.[9] Unno

7 On the repudiation of the Ming diplomatic system and later Japan-centric conceptions see Toby, *State and Diplomacy in Early Modern Japan*, pp. 34, 211-230.

8 Wallis, 'The Influence of Father Ricci on Far Eastern Cartography,' p. 45.

9 Nakamura, 'The Japanese Portolanos of Portuguese Origin of the XVIth and XVIIth Centuries,' p. 44.

Kazutaka, one of the most prolific scholars of historical cartography in Japanese, sees a somewhat less intense assimilation of European cartography during the 'Christian century' (the period of significant missionary influence prior to the 1636 ban). By the same token, however, he emphasizes Japan's continuing contact, direct or indirect, with new developments in cartography even after the 1630s. Unno's deep expertise in Chinese and Korean cartography, as well as indigenous Japanese trends since the eighth century, helps him to untangle the unruly hubbub of these years when manifold influences suddenly found their way into print.

The question of European influence in Japanese mapmaking is intertwined with analysis of the social and state forces imbricated in cartography. Philip C. Brown's work on surveying in the Tokugawa period poses a challenge to state-centric narratives of Japanese surveying and cartography.[10] Brown argues that there was little demand in early modern Japan for the most advanced surveying techniques, for they did not suit the needs of local lords who preferred to resolve territorial disputes through negotiation or occasionally by force. He suggests that cadastral surveys were rather sloppily conducted, the precise contours of the land being beside the point for the true purposes of such surveys.[11] We read in the work of Hugh Cortazzi that there was no cumulative improvement in effectiveness over the course of the first several cadastral surveys of the early Tokugawa period.[12]

Aside from these factors, the potential military applications of surveys were not lost on the suspicious shogunate. Furthermore, relying on precise measurement would have impinged upon the flexibility of the complex feudal relations between various *daimyo*, and between the local level and the center, that were vital to keeping such an administrative structure running smoothly. In other words, technology that could have settled territorial

10 Brown, 'Never the Twain Shall Meet: Land Survey Techniques in Tokugawa Japan.'

11 Brown enumerates several ways in which social and market forces were unsupportive of the development of surveying, including samurai indifference to the application of science, censorship because of concerns about the use of the technology against the government, lack of educational institutions for imparting skills, and a social system that limited even the small number of trained surveyors to work almost exclusively for their local lords. The latter situation is explicitly contrasted with that in England, where surveyors could travel in search of work and hence honed their techniques in the hope of beating rivals to contracts (Brown, 'Never the Twain Shall Meet,' p. 77). Brown holds that the nature of the market for surveying is the essential 'key to understanding Japan's failure to develop further the native methods of triangulation or to adopt and exploit new, more sophisticated measuring techniques from Europe' (ibid., p. 79).

12 Cortazzi notes that the effort of the shogun Tsunayoshi to compile new fief maps in 1697 actually produced a less accurate cartographic series than had the surveys of 1603 and 1644 (Cortazzi, *Isles of Gold: Antique Maps of Japan*, p. 33).

disputes was available, but went unused in favor of using a system that relied on interpersonal and interfamilial relations.[13]

There is another reason why it can be misleading to interpret the early cadastral surveys as the appearance of an early modern phenomenon in seventeenth century Japan. Although these surveys were indeed responses to calls from the state, the method by which they were undertaken remained similar to that recorded for the efforts of previous governments.[14] Though we misleadingly use the word 'survey' in both cases, the compilation of maps reflected the feudatory system of governance in that they were made by combining individual maps submitted from the local level into a larger patchwork, rather than a comprehensive survey by a single authority, as would have been the case with a traveling team of surveyors appointed by the central authorities. The new regime was forming a unified vision, but doing it by piecing together preformed fragments submitted from afar.

How do we reconcile this fact with the appearance of detailed survey-based urban maps that proliferated even as rural and coastal surveys were frowned upon? My suspicion is that the answer to this question further challenges the emphasis on the state as the primary motivating force behind early modern cartography. While the state's apparatus for maintaining internal boundaries could maintain a certain fluidity, as it was the relationship between the local lord and the center that meant most to the shogunate, the needs of property owners in the city required a different kind of spatial record. As A.S. Sadler anecdotally notes in his discussion of Edo's early settlement, rapid inflation in land prices prompted a concern to establish clear boundaries.[15]

13 A collection of village administrative maps (*muraezu* 村絵図) described at length by Kimura Tōichirō supports Brown's arguments to a large degree, with most of the hand-painted maps lacking defined scale. However, we do see certain differences in a few cases where border disputes or other contentious issues were being addressed. Not only are the maps more carefully drawn, but the cartographic specialists (*eshi* 絵師) responsible for their compilation record their names as a sort of guarantee. (Kimura Tōichirō 木村 東一郎, *Kinsei muraezu kenkyū* 近世村絵図研究). Nevertheless, the detail on even these maps is not methodically or precisely plotted.

14 This custom goes back as far as a 646 edict for provinces to submit maps to the central government, based on Tang precedent (Unno, 'Cartography in Japan,' p. 354).

15 Sadler, *The Maker of Modern Japan: the Life of Tokugawa Ieyasu*, p. 232. Berry notes in her biography of Hideyoshi that in the sixteenth century, 'Of the twenty-two documented appeals for justice [to Imagawa judges] that provoked land surveys, as many as seventeen were initiated by cultivators' (Berry, *Hideyoshi*, p. 29). This is another example of bottom-up forces, as opposed to state motivation for greater control, leading to land surveys. Berry's argument that these surveys helped make 'arbiters into lords' is a convincing example of the linkage between power and mapping, but the agency here looks like the opposite of what we might expect based on a Foucauldian argument.

The financial interests and potentially competing claims of *chōnin* (townsmen) as well as the estates set up in the capital by various feudal lords must have been an important factor in the intensive mapping of the new capital. We may also consider the fact that much of the settlement was literally laid out upon wasteland, a blank canvas so to speak, and one that occasionally saw large areas wiped out by fire.[16] The government's desire for a 'panopticon' to survey the population could potentially have been accomplished in other ways than survey mapping, such as through a mutual responsibility system like the *baojia* sometimes used in China, but such methods could not have satisfied the needs of non-state actors. The impetus to create survey-based maps thus emerged at its most developed stage in the newly and densely settled urban setting.[17]

For a country with such an early history of printing, commercially published maps seem to have appeared rather late in Japan; separately printed sheet maps only survive from 1624 on.[18] These earliest printed maps of Japan, depicted with a dragon encircling the islands, were used as talismans against earthquakes. They appeared in the beginning of almanacs such as the *Ise Koyomi* 伊勢暦, and their use continued with little modification until the late nineteenth century in various calendrical publications. The depiction of Japan was often supplemented with an image of Mt. Sumeru, the giant mountain that stood at the center of the earth in Buddhist cosmology.[19] The first commercially published atlas of Japan is the 1666 *Nihon bunkei zu* 日本分形圖, with a series of small, subdivided maps. In spite of these rather late beginnings of printed cartography, Japan soon saw a flood of such works, and

16 A detailed survey of Edo plotted against a network of compass roses, commissioned after the disastrous fire that destroyed much of the city in 1657, is illustrated in Unno, 'Cartography in Japan,' p. 402. Yonemoto notes that this survey by Hōjō Ujinaga became the standard template for commercial maps in following years (Yonemoto, *Mapping Early Modern Japan*, p. 17).

17 Kornicki notes that the city maps that were printed commercially 'had little to contribute to the development of cartographic knowledge in Japan,' being not primary sources but rather mass-produced works bringing established knowledge to a broader market, but by the late seventeenth century their increasing popularity indicated not only 'the spread of spatial literacy' but 'the increasing functional significance of maps in Tokugawa society' (Kornicki, *The Book in Japan*, pp. 60-61).

18 On the 1624 map see Unno, 'Maps of Japan used in Prayer Rites or as Charms,' p. 68. Such examples are preceded of course by maps printed abroad in the Chosŏn *Haedong chegukki*, noted in Chapter Four, and in Xue Jun's *Riben kaolue* 日本考略 of 1523 (see Okamoto Yoshitomo 岡本良知, *Jūrokuseiki ni okeru Nihon chizu no hattatsu* 十六世紀における日本地図の発達 (The development of Japanese maps in the sixteenth century), pp. 10-12), highlighting the possibility that now-lost pre-1624 printed maps circulated abroad, but quite likely just exemplifying the transfer of manuscript maps to woodblock print in those countries.

19 Unno, 'Maps of Japan Used in Prayer Rites or as Charms.'

at a time when domestic peace also offered greater opportunities to build domestic readership for Ming and European maps and geographic texts.

European cartography in Japan

In Japan, commercial contact with Europe predated Jesuit influence by a brief interval, and even during the high tide of Christian missionary work in Japan, practical affairs drove local leaders' interest in Western cartography. A significant proportion of European-influenced maps from Japan reflect this secular background, as exemplified by surviving portolan charts. The secular and sacred often moved hand in hand, as has been noted by George Elison; the Ming closure of trade with Japan in the mid-sixteenth century opened the door for Portugal to supply Japanese markets with Ming goods, and the Jesuits were in turn able to open doors for activity in Japan by promising such ships to ports that welcomed the Catholic message.[20] As for cartography, although a changing political situation eventually led to the discouragement of certain practices, the Azuchi-Momoyama period in the last quarter of the sixteenth century and the early Tokugawa period witnessed a dramatic appropriation of European ideas.

As had been the case in Ming China, but not in Chosŏn Korea, some Japanese were able to develop close personal or working relationships with Europeans and thus to better understand the meaning and uses of their maps. Direct contact with merchants and political representatives was much more common in Japan than in Ming, where the Jesuit use of maps as instructional symbols was the main route by which the new cartography entered broader society. This may be why practical applications for maps, including navigation, seem to have been appreciated by the Japanese rulership from as early as the time of Oda Nobunaga. Oda held conversations with Jesuits in 1569 and 1579 in which geographical questions played a major role; in 1579 he showed the Portuguese Jesuit Gnecchi-Soldo Organtino a globe, asking whence he had come and sounding out his motives for traveling from such a distant land.[21]

20 Elison, *Deus Destroyed: The Image of Christianity in Early Modern Japan*, pp. 86-87. Note that the discussion of European cartography here is only a sketch; for serious discussion and illustration of early influence see Unno Kazutaka 海野一隆, *Tōzai chizu bunka kōshōshi kenkyū* 東西地図文化交渉史研究 (Research on the history of cartographical exchange between the East and the West). There is a wide-ranging and extensive literature on European maps of Japan, which falls beyond the subject of this chapter.

21 Ayusawa Shintarō, 'Geography and Japanese Knowledge of World Geography,' p. 276.

Awareness of European cartography took a major leap under Toyotomi Hideyoshi. In 1591, the sons of *daimyō* who had gone on Japan's first delegation to Rome delivered upon their return a copy of the 1570 edition of Abraham Ortelius's world atlas as well as three volumes of the *Civitatis Orbis Terrarum*. Folding screens and paintings based on maps from this atlas still survive, for Hideyoshi asked the painter Kanō Eitoku 狩野永徳 to make reproductions.[22] Hideyoshi was also preoccupied with the activities of Europeans in Asia; he spoke with João Rodrigues at length about Europe and its interests in India.[23]

Nakamura Hiroshi shows that Japanese maps from Hideyoshi's reign are essentially faithful copies of Portuguese maps, but he argues that within half a century Japanese surveys had begun to refine the shape of Hokkaido and other regions using the newly acquired technologies. Nakamura thus characterizes Japan as the only Asian country to have learned cartographic techniques fully from the Europeans, even while blaming the exclusion policies for stifling further development.

As Tokugawa Ieyasu approached his goal of uniting Japan under his rule, he took as one of his many advisors the Englishman William Adams (1564-1620) who had arrived in Japan in 1600. Adams's advice on foreign policy, in particular the decoding of European politics and the motives of both religious and merchant visitors, supplemented Ieyasu's already active interest in geographical and cartographic knowledge. Under Ieyasu significant cartographic steps were taken to prepare for the conquests of the Ryukyus and Taiwan. Shimazu Iehisa's 島津家久 envoys to the Ryukyus in 1608 reminded their king that tribute was expected, and the envoys drafted maps of the islands before their return.[24] It is not clear whether these maps were used in the invasion of the Ryukyus the following year, but the islands' defeat was terribly lopsided. In the same year as the Ryukyu

22 Ibid., p. 277. See also Unno, 'Cartography in Japan,' p. 377. An example of a decorative screen on this theme, held at the Namban library, that reproduces Ortelius' Latin title 'Typus Orbis Terrarum' across the top of a painted world map is illustrated in Cortazzi, *Isles of Gold*, Plate 33. On this example, roughly dated to the early seventeenth century, the shift in perspective to put East Asia in the center is reminiscent of Ricci's Sinicized maps. Thus Ortelius' title has been paired with a map taken from elsewhere. Cortazzi notes that a map of Japan that has been found among the Medici papers may have been presented to the Grand Duke of Tuscany by the Japanese envoys (Cortazzi, *Isles of Gold*, p. 23. An illustration of the map may be found in George Kish, 'Some Aspects of the Missionary Cartography of Japan during the Sixteenth Century,' p. 42). If true, this demonstration that the envoys were familiar with and interested in maps may have prompted the reciprocal gift of the European atlases when they were in Padua.

23 Cooper, *Rodrigues the Interpreter*, p. 83.

24 Sadler, *The Maker of Modern Japan*, p. 246.

invasion, Ieyasu commanded that the island of Taiwan be surveyed from tip to tip to decide on a favorable port. In this case the islanders defended themselves rather more successfully, allowing some prisoners to be taken but rebuffing conquest.[25]

Ieyasu agreed to a Spanish request to survey Japan's shores in 1611, apparently thinking of the benefits that would accrue to himself from having detailed coastal maps. It was not until William Adams informed him that in a European context this would constitute an aggressive act that Ieyasu seems to have drawn the link between this coastal survey and the rumors then circulating about Spanish designs on Japanese territory.[26] Even so, in 1613 when Sebastian Vizcaino asked for permission to survey the coasts in search of the mythical Islands of Gold and Silver thought to lie nearby, Ieyasu granted permission in hopes that such a systematic search would bring the fabled wealth of these lands under his rule.[27]

The knowledge that European sailors relied on such maps themselves, and the intimate realization that understanding these maps could play a role in the security of their own realm, seems to have led much of Japanese officialdom to accept this sort of cartography in a way not appreciated in Ming China. To at least some degree, cartography was viewed as independent of the other cultural freight of the European Christians. Even as bans on Christian works were instituted, cartographic materials continued to be treated as a separate category.[28]

So it is that we see Ieyasu, who had already shown a curiosity about more esoteric geographical questions before coming to ultimate power, seeking practical data about the world beyond Japan from maps.[29] The records of the Tokugawa house note Ieyasu's studies of world maps in 1611, the first such mention in an official chronicle.[30] It was Ieyasu's familiarity with cartography and the complex motivations behind European activity in Asia that led to his series of proscriptions on Christian activity in Japan.

25 Ibid., pp. 247-248.

26 Ibid., pp. 242-244.

27 Sadler, *The Maker of Modern Japan*, pp. 265-266; Cortazzi notes than any maps produced have been lost, along with the detailed record of the journey (*Isles of Gold*, p. 32).

28 The obliteration of the Jesuit seals on the copy of Ricci's 1602 world map held at Kyoto University graphically demonstrates one way of at least superficially separating such maps from their religious associations.

29 João Rodrigues records that his discussion with Ieyasu in 1593, during which Rodrigues defended the notion that there was but one world in contrast to the many worlds proposed by the Buddhists, led to talk of other topics related to geography and navigation (Cooper, *Rodrigues the Interpreter*, p. 101).

30 Ayusawa Shintarō, 'Geography and Japanese Knowledge of World Geography,' p. 277.

Before addressing this series of bans, however, let us briefly examine the cartographic relevance of Jesuit activities in Japan.

Jesuits and their maps[31]

Jesuit works found in Japan included publications originally created for use in the Ming, as well as new works specifically developed for a Japanese audience. Ricci's world map was among the Jesuit publications from China that were requested by Francesco Pasio and other missionaries.[32] Aside from using maps as gifts for the prominent and influential, geography was part of the indoctrination given to converts, as well as to Europeans (often traders or soldiers) who decided to enter the religious order while they were abroad. In Xavier's instructions for the mission, explanation of natural phenomena was the third of ten categories of instruction to be taught before baptism.[33] Pedro Gomez taught formal classes in natural science in the 1580s at the Jesuit school at Funai.[34] Geography may have played some role in education at other Jesuit institutions as well, given that Alessandro Valignano, Visitor of the Jesuit missions in the Indies, had included the sciences as part of the curriculum at the schools established for educating the sons of Japanese gentry and exposing them to Christianity.[35] Non-Jesuit missionaries in Japan mocked this practice. Alphonsus Navarette, a Dominican friar who died in Japan in 1617, is quoted by Helen Wallis as stating that 'I, for my part, would prefer to see our Friars in China with Crosses around their Necks, rather than with Maps and Clocks in their Hands.'[36]

The Jesuits began operating a press in Japan in 1588, and Japanese printers were even trained in Europe to operate it, though the press was sent via Manila to Macau after Ieyasu's 1614 expulsion order.[37] It seems to have played a relatively minor role in Macau, as the Jesuits continued to use woodblock printing.

31 For works on Jesuit cartography and its influence in Japan which go into far greater detail than appropriate here, see Ayusawa Shintarō, *Mateo Ritchi no sekai zu ni kansuru shiteki kenkyū* マテオ「リッチの世界図に関する史的研究, and other works cited herein.

32 Laures, *Kirishitan Bunko: A Manual of Books and Documents on the Early Christian Mission in Japan*, p. 101.

33 Elison, *Deus Destroyed*, p. 36.

34 Cooper, *Rodrigues the Interpreter*, p. 60.

35 Ibid., pp. 63-64. Marcia Yonemoto indicates that the Jesuits used maps in their academy at Kyoto after 1605 (Yonemoto, *Mapping Early Modern Japan*, p. 16).

36 Wallis, 'The Influence of Father Ricci on Far Eastern Cartography,' p. 43.

37 Boxer, *The Christian century in Japan: 1549-1650*, p. 198.

In Japan, as in the Ming, Jesuit arguments about the topology of the earth were part of their strategy for demonstrating the rationality of their stance. Some Japanese critics of the Jesuits rejected their arguments along lines reminiscent of Ming polemics, albeit with slight differences (Japan as 'Land of the Gods' rather than a Middle Kingdom), but the reaction to the theories of the round earth and the five continents was generally more favorable in Japan than in Ming. In Japan the absence of a single overriding cartographic paradigm for the world's image resulted in a more open hearing for European ideas. The fact that Europeans themselves traversed enormous distances, sometimes with Japanese aboard their ships, while trusting such charts was known to many Japanese (though some had their doubts, as we shall see). The introduction of portolan charts, the maps used by European seafarers, and the short-lived flourishing of Japanese *shuinsen* 朱印船, 'vermillion seal ships' licensed to trade at ports as distant as Southeast Asia until the 1630s, demonstrated that geodesy and cartography were sciences with practical applications. Even after the various prohibitions against Christianity began to descend in stages (beginning with Hideyoshi's 1587 decree, continuing after the establishment of the Tokugawa shogunate with a harsher wave in 1612 and 1614, climaxing in the maritime prohibitions of 1633-1639), geographic works continued to show the influence of such utilitarian maps.[38]

The writings of Fabian Fucan (1565?-1621), a Japanese convert who served as one of the Jesuits' primary defenders before reversing course and becoming a vocal apostate, help us to understand the process of the Jesuit encounter with Japan. Fucan's story can also tell us something about the influence of Western geographical theory during this period and its perceived separation from Christian religion.

Men like Fucan were of special value to the Jesuits in Japan, who needed not only to attract converts but also to convey their side effectively in oral debate with local critics. Oratorically gifted Japanese converts were indispensable. A record of a debate in the summer of 1606 between Fucan and the Neo-Confucian-inspired thinker Hayashi Razan 林羅山 reveals the role of geography and the natural sciences in the interplay of arguments, although our only record of the debate is from Hayashi's scornfully dismissive perspective.[39]

38 As is widely acknowledged today, the formerly common term *sakoku* 鎖國 or 'locked country' is an inadequate and anachronistic term for the post-1639 policies of isolation. See Toby, *State and Diplomacy in Early Modern Japan*, pp. 12-22.

39 Despite Hayashi's curt impression of Fucan, we know that the Jesuits thought him to be a valuable and convincing speaker. This is communicated in contemporary records, such as an October 18, 1606 report to the General of the Society of Jesuits which describes Fucan as 'very well

As the debate began, after inquiring about a 'portrait of Deus,' Hayashi leapt right to the area of our concern, mocking the notion of a spherical earth. His objections recall those of Ming critics. To Fucan's citation of overseas voyages by Japanese as evidence of the earth's form (perhaps a reference to the youthful Japanese ambassadors to Rome who visited several European cities between 1584 and 1586), Hayashi retorted 'On the ocean there are winds and waves, and ships heading west may actually go north or south or even east, the men on board not recognizing the direction, thinking the while that they are going west.'[40] Hayashi cites Zhu Xi as the ultimate authority for the shape of the universe, describing it as a round heavenly vault rotating like a bowl over a flat earth. As for the technological implements the Jesuits brought with them, Hayashi laughs them off as spurious imitations of Chinese mechanisms, or in the case of a prism, a clever novelty 'for the dazzlement of the common people,' a trick meriting the execution of the would-be dazzlers.[41]

In 1620, having been stung by the Jesuits' refusal to grant him entrance to the priesthood and bearing the scars of anti-Japanese prejudice which had suffused the acts and attitudes of some European priests, Fucan published a recantation of his Catholic faith. Entitled *Ha Daiusu*, 'Deus Destroyed,' this pamphlet renounced the views for which Fucan had become known, and condemned what he described as ominous future plans for undermining the authority of the shogunate and colonizing Japan.[42] In the midst of his blunt rejection of their metaphysics and political aims, however, Fucan never takes issue with the claims of the Jesuits about geography. As with the authorities of the shogunate who later promulgated the expulsion edicts, Fucan separated explanations of naturalistic phenomena from religious ideas.

Another prominent anti-Jesuit thinker, Mukai Genshō 向井元升 (1609-1677), also admitted the roundness of the earth in spite of the general perfidy

versed in the doctrines of Japan, so that there is no bonze who could surpass him, nor any that would dare dispute with him' (quoted in Elison, *Deus Destroyed*, p. 149). On the 1606 debate see also Ayusawa, 'Geography and Japanese Knowledge of World Geography,' p. 278. Kiri Paramore has argued that Hayashi's record of the debate is a piece of propagandistic fiction written after Fucan's death (Paramore, 'Hayashi Razan's Redeployment of Anti-Christian Discourse: The Fabrication of Haiyaso'). Even if Paramore is correct, and Hayashi puts words in Fucan's mouth that he never uttered, Hayashi's arguments against his 'pseudo-Fucan' straw man would reflect his own real positions.

40 Elison, *Deus Destroyed*, p. 150. A translation of the entire record of the debate may be found on pp. 149-153.

41 Ibid., p. 151.

42 For a full translation of *Ha Daiasu*, see Elison, *Deus Destroyed*, pp. 259-291.

of Christianity; he was the editor and commentator for the *Kenkon bensetsu* 乾坤辨說 (On the Universe) of Christovão Ferreira, an apostate European who converted to Buddhism and remained in Japan after the suppression of Christianity.[43] Indeed, the cartographic importations of the European Christians retained such respect that they could even be deployed against their own religion. Long after the Jesuits had been banned from Japan, Arai Hakuseki 新井白石 (1657-1725) cited the proximity of Judea to western India in his argument that Christianity was merely an offshoot of Buddhism.[44]

Ironically, perhaps, European maps played a role in the downfall of the Jesuit mission in Japan. The problems began during Hideyoshi's rule. As a man of great ambition capable of dreaming of conquest on a grand scale, Hideyoshi may have been especially attuned to the possible schemes of others along these same lines. Advisors who had other reasons to dislike the presence of Christians in Japan (a presence which had been tolerated largely for economic reasons despite a 1579 ban) could point to the expansion of Spanish and Portuguese colonies around the world as a warning to their lord.

In 1596 the Spanish trading ship *San Felipe* was wrecked at Urado. Its contents were seized, and rumors spread about soldiers and weapons that had supposedly been discovered on board. The captain and other officers were debriefed by Masuda Nagamori 增田長盛, one of Hideyoshi's *bugyō* 奉行, or commissioners, who had obtained and examined the ship's nautical charts.[45] How, he inquired, feigning surprise at the depictions on the charts, had Spain expanded its imperial domains to such a vast extent, engulfing much of the Americas as well as the Philippines? The implications for Japan were clear. Hideyoshi was informed of this situation and may have studied the maps himself, for when he spoke with João Rodrigues in April of the following year he showed familiarity with the great span between Spain's colonies in Mexico and the Philippines, voicing his suspicion that the ulterior

43 Elison, *Deus Destroyed*, p. 449 (footnote 73). On the *Kenkon bensetsu* see Yajima Suketoshi, 'The European Influence on Physical Sciences in Japan,' pp. 341-342, or at greater length, José Miguel Pinto dos Santos, 'The "Kuroda plot" and the Legacy of Jesuit Scientific Influence in Seventeenth Century Japan.' On Ferreira and the controversies surrounding his apostasy and alleged revocation before martyrdom, see Cieslik, 'The Case of Christovão Ferreira.'

44 Elison, *Deus Destroyed*, p. 240.

45 Cooper, *Rodrigues the Interpreter*, pp. 157-158. Though these charts are no longer extant, it was common on navigational materials of the time to use flags to denote the colonial powers in charge of various ports on the coastline, as shorthand to indicate which places could offer a friendly harbor. Thus it would have been apparent at a glance, to someone who understood this practice, how many ports were Spanish. This suggests that Masuda was not encountering such materials for the first time but rather may have been only pretending to be surprised by the extent of empire as a way of bringing this topic into his interrogation.

purpose of the missionaries in his own land was to lay the groundwork for yet broader future conquest. Furthermore, although the Jesuit mission in Japan was under Portuguese sponsorship, Hideyoshi observed that the Spanish king, the ruler of this vast colonial empire, now also ruled Portugal, did he not? Rodrigues admitted this was so.[46] The recent advent of Franciscan missionaries from the Spanish Philippines to Japan could only have added to this concern about Spanish movements.[47]

After the bans

Though the prohibition of Christian activities occurred in several stages, the so-called Kan'ei Edict of 1630 was the first significant one for matters related to cartography. The *Tianxue chuhan* 天學初函 (Primer of heavenly studies), a multivolume collection of Jesuit texts published in Ming China, on topics ranging from the nature of the soul to Western hydraulic engineering, was specifically proscribed.[48] This collection included the *Zhifang waiji* 職方外紀 (Account of lands not officially recorded), a Jesuit geographical treatise by Giulio Aleni. For the purposes of the censors, Aleni's work was more than a simple geography, for it contained descriptions of the culture of the lands of Christendom. In 1720, when the prohibition on technical texts was relaxed, the *Zhifang waiji* was included among the newly permitted titles. It was banned again in 1775, though enforcement varied in rigor and effectiveness, even in cases of more explicitly Christian works.[49] Some banned works continued to circulate secretly, as in the case of the *Zhifang waiji*.[50]

46 Cooper, *Rodrigues the Interpreter*, pp. 143-144.

47 Among the Jesuits, it was common to blame the Dutch for Japanese suspicions. Alexander de Rhodes, for example, gives the following narrative: 'These two [Dutch] scoundrels, finding themselves at the court of the king of Japan, showed him a map of the world with the Philippines on one side and Macao on the other, which the king of Spain then held in China as the king of Portugal, and said to the king, "Do you rightly see, Sire, how far Spanish control has extended? [...] You are the only one left to be taken"' (Hertz, *Rhodes of Viet Nam*, p. 96).

48 See Ōba Osamu, 'Sino-Japanese Relations in the Edo period, Part Three: The Discovery of Banned Books,' especially pp. 60-62. On the Kan'ei Edict see also Shio Sakanishi, 'Prohibition of Import of Certain Chinese Books and the Policy of the Edo Government.'

49 Ōba Osamu, 'Sino-Japanese Relations in the Edo Period, Part Three: The Discovery of Banned Books,' pp. 71-73.

50 Ayusawa, 'Geography and Japanese Knowledge of World Geography,' p. 282. The dating of such copies is problematic. An example of a manuscript copy of a map taken from the *Zhifang waiji*, tentatively dated to the 'early Edo period,' is illustrated on p. 272 of Okamoto Yoshitomo, *Jūrokuseiki ni okeru Nihon chizu no hattatsu*.

Despite the prohibition of Christian learning, there is extensive evidence of continued open circulation of European-influenced cartography as a field of knowledge in itself, separate from any religious dimension, at the same time that locally produced maps of Japanese subjects were proliferating. Ayusawa Shintarō divides the cartographic output of the post-ban period into five categories: objective, Buddhist, Chinese, Japanese, and legendary.[51] He uses 'objective' to refer to European-introduced cartography, and notes its persistence despite the curtailment of direct communication.

The post-ban persistence of Western style world maps is easily illustrated. The earliest extant printed map of the world published in Japan was an imitation of Ricci's, a Nagasaki imprint from 1645 titled *Bankoku Sōzu* 萬國総圖 (Comprehensive map of the Myriad countries).[52] The period from 1644 to 1788 in Japan has been called the 'Age of Matteo Ricci's Map' by Kurita Mototsugu because so many world maps during this time reflect his influence.[53] Unno views the genre as somewhat more diverse, noting that after 1645 other maps were brought into consideration to compare with Ricci's version and make alterations. An unknown European work with illustrations of people from around the world was used as a model for the manuscript *Bankoku sōzu*, which differs from Ricci's map in orientation and numerous other details.[54] Unno notes the popularity of this printed map, which was put out by other publishers after 1646.[55] In the meantime Ricci's version, especially as reproduced in Ming books, continued to circulate.[56] Copies also continued to appear in altered form or to illustrate texts with no connection to Ricci.[57]

51 Ayusawa Shintarō, 'The Types of World Map Made in Japan's Age of National Isolation,' p. 123.

52 Unno, 'Cartography in Japan,' pp. 405-407. Unno notes here that copies of Riccian maps appear to have been given as certificates to students of surveying before printing for the market began. In Ayusawa's 'The Types of World Map Made in Japan's Age of National Isolation,' an earlier date of 1643 is given for the appearance of these printed maps (p. 123), but the primacy of the *Bankoku sōzu* is supported in a later article by the same author (Ayusawa, 'Geography and Japanese Knowledge of World Geography,' p. 280) and in Beans, *A List of Japanese Maps of the Tokugawa Era*, p. 11.

53 Ayusawa, 'The Types of World Map Made in Japan's Age of National Isolation,' p. 125.

54 Illustration in Unno, 'Cartography in Japan,' p. 406.

55 Unno, 'Cartography in Japan,' p. 407. For a much more extensive review of Japanese maps derived from Jesuit sources, see Unno's *Tōzai chizu bunka kōshōshi kenkyū*, especially pp. 117-175 on the classification of European-inspired Japanese maps.

56 Direct copies of Ricci's map are listed in Unno, 'Cartography in Japan,' pp. 404-405, followed by discussion of various copies and derivatives up to p. 410.

57 For several examples see Ayusawa, 'Geography and Japanese Knowledge of World Geography,' p. 281, and Ayusawa, 'The Types of World Map Made in Japan's Age of National Isolation,' pp. 123-127.

The *Bankoku sōkaizu* 萬國總界圖 (Map of the world's myriad countries), a dramatic example of this genre, was produced in Edo in 1708 by Ishikawa Ryūsen 石川流宣.[58] This map, which depicts the world in an elongated oval tilted so that east is at the top, is comprised of several woodblock prints arranged to form a strip just over one meter tall and half as wide. Printed originally in black, the map was often hand-tinted. On some parts of the map the comments about foreign countries are similar to those on Ricci's, such as a note about the lack of knowledge of Magellanica. Following the tradition established by Chinese reworking of Jesuit materials, Japan has been placed near the center of this map. The distance to the land of 'Redhairs,' the Dutch, is given as 12,500 *ri*.

On the example held at the UC Berkeley's East Asian Library, faint red lines have been drawn in by a user's hand to indicate sea routes, both to the island of Ezo (Hokkaido) in the northeast and from Japan to Chosŏn. A number of place names have been highlighted in red as well, including places of political or cultural significance (such as Nanjing) as well as places with which Japan had formerly had trade relations (Luzon, parts of Southeast Asia, and the Netherlands). The reason for these annotations is unknown, but perhaps this example was used for teaching, or for self-education, about the locations of places important to anyone studying Japan's foreign contacts at the time.

The colorful folding screens (*byōbu*) decorated with representations of the world map that had been so popular in the late sixteenth century continued to be produced even as the ban developed. The functions of such screens could have been various; a *byōbu* showing the lands beyond Japan could show that its owner was a veritable man of the world. A decorative map of Japan, often placed on the other side of the same screen, could hint at scholarly erudition, such as familiarity with literary traditions associated with the *Tōkaidō* highway connecting Edo and Kyoto.

One such screen held at UC Berkeley's East Asian Library is a typical example in that it bears an image of the world on one side and a decorative map of the stations of the *Tōkaidō* on the other.[59] Though this folding screen

58 Ishikawa Ryūsen 石川流宣, *Bankoku sōkaizu* 萬國總界圖. I have examined the example held at the UC Berkeley East Asian Library, catalogued as Map AO.

59 In the Berkeley East Asian Library this is map Byōbu 1, catalogued under the title *Sekaizu narabini Nihonzu byōbu*. A detailed scan is posted in the library's online collection of Japanese maps. I am indebted to Yuki Ishimatsu for facilitating a viewing of the screen in person and for bringing to my attention an article by Yamori Kazuhiko that discusses its relationship to similar pieces in other collections (Yamori Kazuhiko 矢守一彦, 'Bākurē hon Nihon zu byōbu' バークレー本日本図屏風'). Yamori determines, based on an analysis of the place names used on the map, that its date of production cannot be pinned down any more specifically than the first

is smaller than Ricci's world map and lacks many of its details and its text, it gives us an inkling of the function of transferring maps to such a format. By presenting it in an aesthetically appealing form lending itself to display, a mapmaker could promote his view of the world to new audiences, making this an ideal form for Jesuit propaganda – not least because later artists' copies, made simply as decorative objects, could further perpetuate its visual message. Such folding screens were markers of status, and an example like this added the appeal of the exotic. Later copies like this one would remove Ricci's textual arguments, in this case leaving just a few dozen place names, and turn his carefully considered explanatory diagrams into mere colorful ornaments.

The top left corner of the screen shows a diagram of the concentric crystalline spheres of the heavens, originally used to explain the motion of heavenly bodies before the acceptance of heliocentric theory, though in this case the explanatory text from the Ming Jesuit originals has been effaced. At top right is a similarly decontextualized depiction of the earth with ships on four sides, originally meant to illustrate how the sphericity of the earth allows a ship to travel in a single direction and return to its point of departure, all the while allowing observers on the ship to sense that the earth is below rather than above them. These illustrations, as found on the original maps, addressed two of the notions the Jesuits found most difficult to convince people of in China. (In the case of the crystalline spheres, we may say in retrospect that such skepticism was well justified.) Deprived of their explanatory text, however, were these diagrams still intelligible? Did the copyist understand what they meant? It is difficult to know. At bottom right and left are polar views of the northern and southern hemispheres.

On the other side of this same screen is half of a two-part map showing the stations of the *Tōkaidō*. The pairing of the world and Japan on two sides of a screen demonstrates an awareness of Japan's place in a broader universe, one in which it was determined (beginning with Hideyoshi) to play a role of increasing importance. One similar screen from the 1590s adds a depiction of the route of Hideyoshi's invading ships from Tsushima to Chosŏn Korea.[60]

The persistence of European maps in the Japanese context after 1639 raises the question, to what degree were these maps still associated with

half of the seventeenth century. When Yuki Ishimatsu showed me the map, I saw that there was a place where the paper was torn, and at least one European-style numeral was printed on the other side. This suggests that old scraps of paper were used in its construction; it is possible that a comparison of that fragment of text with known printed works could help illuminate the circumstances of the map's production.

60 Cortazzi, *Isles of Gold*, p. 27.

Europeans? Unno suggests that Japanese scholars thought Ricci was Chinese because he used a Chinese name on his maps.[61] Anyone who had read the preface on his original map or a faithful copy thereof would have known that he had come from Europe, of course, but many copies did not reproduce this information. A nineteenth century Japanese print of Ferdinand Verbiest depicting him as a bearded Chinese man hailing from the village of Dongxi 東溪村 shows that such confusion may indeed have existed.[62]

Another book discussed below, the *Wa-Kan sansai zue*, exemplifies the nature of the Jesuit-introduced information still available via Ming publications several decades after the final expulsion of the Christian missionaries. By this point in the early eighteenth century, it is clear that much cartographic information had been separated from its religious context and preserved despite the bans.[63] In the meantime, however, Ming geographical and cartographic publications unrelated to the Jesuits or Europe were attracting the notice of Japanese thinkers.

Ming works and the Buddhist/European synthesis

By the late seventeenth century Japanese scholars had access to a growing spectrum of late Ming geographical and cartographic publications, including

61 Unno, 'Cartography in Japan,' p. 410.

62 Stephenson, 'Chinese and Korean Star Maps and Catalogs,' p. 574.

63 We may also ask to what extent Christian converts in Japan were influenced by the geographic notions their teachers imported along with their faith. The Christian communities that went 'underground' naturally leave rather few sources, but passages from the *Tenchi hajimari no koto* 天地始之事 (*The Beginning of Heaven and Earth*) suggest that, long after the Jesuit teaching institutions had been closed, belief in Christianity did not necessarily mean greater awareness of geography beyond the seas, or even awareness that the lands depicted in the Christian texts were indeed separated from Japan by seas. This text, a nineteenth-century written version of what is believed to have been an earlier oral tale, came to the notice of Bernard Petitjean in 1865 soon after the French priest established contact with the Kakure Kirishitan or 'Hidden Christians,' a sect that had escaped eradication during the suppression of Christianity. (For a summary of the debates surrounding the origin and transmission of the *Tenchi hajimari no koto*, as well as an overview of the work's significance in the context of underground Christian practice, see the Introduction in Whelan's *The Beginning of Heaven and Earth: The Sacred Book of Japan's Hidden Christians*). In the *Tenchi hajimari no koto*'s version of the nativity, the three kings of Mexico, Turkey, and France voyage on foot to Bethlehem (ibid., p. 51). This version relays an awareness of the existence of countries such as Mexico and France, far beyond the frontiers of any land mentioned in the biblical text, but the belief that a king from Mexico could walk to Bethlehem indicates that even the most general notion of geographical location was absent. Further illustrating this point is the stunning revelation in the text that Mary, prior to giving birth to Jesus, had herself walked from Japan to Bethlehem (ibid., pp. 49-50).

Ming reprints of older texts as well as newly compiled works. These ranged from books in which maps were but a small portion of the whole to large, separate sheet maps. The imported works included the full range of ideological positions, from acceptance of Jesuit arguments to Chen Zushou's hostile reaction and championing of the Zhifang tradition.

Matsushita Kenrin's 1693 *Ishō Nihon den* 異稱日本傳 (Records of Japan from foreign sources) is a compilation of selections about Japan from numerous imported books, both Chinese and Korean. In this respect it resembles Han Paek-kyŏm's 1640 *Tongguk chiriji*, with several significant differences: It is much larger and more comprehensive, makes use of more recent texts which include maps, and is peppered throughout with critical commentary. It allows us to glimpse not only what was being read in Japan, but also the reaction of at least one scholar.

The *Ishō Nihon den* consists of three main sections. The first covers Chinese sources from the Han (beginning with the *Shanhai jing* and the *Shiji*) to the end of the Yuan dynasty, and comprises three fascicles.[64] None of these sources is illustrated with maps. The second section focuses on Ming works and is more than double the length at eight fascicles, some of them illustrated. This reflects not only the increase of publications in the Ming but their expanding concern with Japan. The third section, with four fascicles, contains selections from the books of Shiro (斯盧), an antiquarian term for Silla, used here to mean books from Korea in general.

Sources incorporated in the Ming section include some of those I have addressed in Chapter One. The *Sancai tuhui*'s map of Japan is omitted because, a note says, it is no different from the map in the *Tushu bian*,[65] but the latter work is represented extensively. Not only is there a reproduction of the *Tushu bian*'s main map of Japan with a paragraph of commentary,[66] but there is a copy of its map illustrating sixteenth century *wokou* raids on the Ming coast (Figure 5:1).[67]

64 Most of these works would have been available in Ming reprints rather than much scarcer earlier copies, though Matsushita does not provide details of provenance.

65 *Ishō Nihon den*, 中之三, p. 21a.

66 The map is reproduced on 中之五, pp. 6a-7b. The commentary on p. 8a notes that the map is more detailed than that in the *Sancai tuhui* but surprisingly contains errors in the names of domains.

67 The map, copied from *juan* 57 of the *Tushu bian*, is reproduced on 中之五, pp. 28a-28b. It is reproduced here from a Genroku 6 [1693] edition by permission of the East Asian Library at UC Berkeley. A version of this map had also previously appeared in the second chapter of the *Chouhai tubian*.

Figure 5.1 A portion of a map showing the sites of *wokou* pirate raids on the Ming coast, from the 1693 edition of the *Ishō Nihon den*

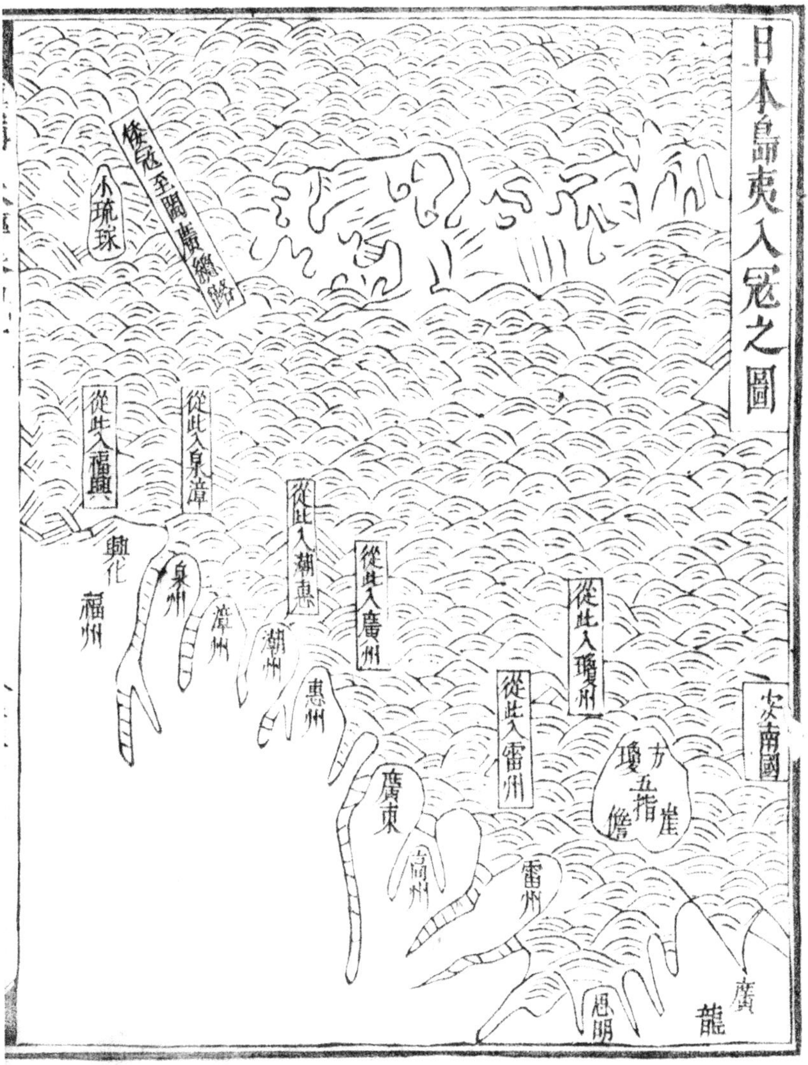

Records of Japan from Foreign Sources, *juan* 中之五, p. 28a
Courtesy of the East Asian Library, University of California, Berkeley

The insulting terms used for the Japanese (夷, 倭) are retained intact on this reprinted map; the goal of a work like this was to let the reader know how Japan was viewed abroad, so even aspects of this depiction that might arouse indignation are preserved. Matsushita does exact a certain degree of revenge, if we may call it that, when he describes these foreign maps as error-strewn, as in the case of a cartographic crudity reprinted from the *Wubei zhi* 武備志 (Treatise on military preparations).[68]

As in Ming China, one feature of Japan's own seventeenth century publishing boom was to collect dispersed and difficult-to-find references and combine them into single works to increase their accessibility. The *Ishō Nihon den* exemplifies the continuing significance of late Ming woodblock-printed cartographic works in Japan, several decades after the fall of the dynasty, as subjects of critical synthesis.

Buddhist cartography

A remarkable feature of cartography in Japan is the persistence of the Buddhist tradition, as reflected in maps of Jambudvīpa, the Buddhist 'Southern Continent' (called in Japanese *Nansenbushū* 南贍部洲). Buddhists had long been important conduits of geographical information from the continent, particularly during the height of cultural contact with the Tang dynasty. Despite the external origin of these concepts, a vastly greater number of Buddhist world maps survive from Japan than from elsewhere in East Asia. This is probably due both to their disfavor in non-Buddhist intellectual circles in Ming and Chosŏn, and the fact that their use by temples in Japan led to their reproduction, careful storage, and hence greater preservation today. Peter Kornicki also notes the importance of Buddhist institutions and individuals in the publishing industry in Japan, and this, along with much greater respect for Buddhism in educated circles, is reflected in cartography as well.[69]

Muroga Nobuo and Unno Kazutaka outline three major stages in the Japanese cartographic depiction of Buddhist worldviews, beginning with Jambudvīpa maps on which the names of places are based on records of the Tang monk Xuanzang's journey to the West. The oldest extant example of

68 The map from the *Wubei zhi* is reproduced on 中之六, pp. 10b-13b. The commentary begins on p. 14a. Note that the *Wubei zhi* is the work now best known for containing the earliest extant map of Zheng He's western voyages. As for the last maps included in the *Ishō Nihon den*, those taken from the early Chosŏn *Haedong chegukki*, they have been discussed in Chapter Four. The outward flow of Japanese maps to Ming and Chosŏn began fairly early, as exemplified by these images in books published abroad, but these are very general Gyōki-type maps. It is difficult to know whether more detailed maps of local regions, cities, and so on reached either courts or the public overseas.

69 For example, one of the early and important Kyoto firms established by Murakami Kanbē (a firm that survives today) was affiliated first with the Jōdo and later with the Nichiren sect, publishing numerous Buddhist texts alongside Sinological and historical works (Kornicki, *The Book in Japan*, pp. 208-209).

this map type is the 1364 manuscript *Gotenjiku zu* by the Buddhist priest Jūkai, held by the Hōryūji temple in Nara,[70] but Muroga and Unno discuss two possible earlier examples, an 806 copy reputedly of Xuanzang's own map and a later map compiled by a certain Yin-pu for presentation to the Koryŏ king before 1154.[71] The next general wave of maps shows an expanded world-continent adapted to include China, in the manner of the Ren Chao map discussed in Chapter Two of this book. Lastly the authors posit a third stage of Buddhist map development, one that includes even Europe on the periphery, but which marks the final stage of Buddhist cartography.[72]

It was a Buddhist priest, Gyōki (668-749), whose representation of Japan as a bulbous archipelago with delineated provinces and roads indicated by lines became the standard image of the archipelago for several centuries.[73] The Gyōki-type map, extant in a number of versions from the fourteenth century onward, was held in some temples and used, among other purposes, in rituals intended to protect Japan from calamities, and later printed for wider distribution. As such, it was intended not only as a basically informative map about the relative positions of provinces and major roads to the capital, as shown by its inclusion in Buddhist didactic works like the *Shūgaishō* 拾芥抄, but as a cartographic logo with numinous powers. In no event was it a practical map for administrative use.[74] Over time it became a standardized antiquarian symbol to represent Japan even when better maps were available; its very quaintness made it a common theme on porcelain, netsuke, and other decorative objects even after the end of the Tokugawa and into the present day.

With greater exposure to information about Europe and the rest of the world, some Buddhist maps after the seventeenth century adapted to a

70 Muroga and Unno, 'The Buddhist World Map in Japan and its Contact with European Maps,' p. 49.

71 Ibid., p. 50.

72 Ibid., pp. 68-69. They here describe the first stage as showing a 'yearning for the Holy Land of Buddhism,' and the next stage as a form of resistance to a Sinocentric worldview.

73 Although Gyōki, a priest who had come to Japan from the Korean peninsula and traveled widely both to preach and provide instruction in basic civil engineering, is credited with this type of map, there is no firm evidence that he personally compiled a prototypical example from which the others are descended (Cortazzi, *Isles of Gold*, pp. 4-5).

74 On ritual uses see Unno, 'Cartography in Japan,' pp. 367-368. See also Unno, 'Maps of Japan Used in Prayer Rites or as Charms,' pp. 72-75. The depiction of a dragon encircling the Japanese archipelago, which is common on earthquake charms of the seventeenth century, appears as early as c. 1305 on a partial map held by the Kanazawa Bunko. (Illustrated in Cortazzi, *Isles of Gold*, Plate 4.) Thus the printing boom in this case helped to proliferate a long-established manuscript motif.

changing understanding of Jambudvīpa, as in the case of a Buddhist world map printed in 1809, which supplements the heavily altered continent of Jambudvīpa with the Americas depicted as a separate landmass.[75] We can imagine the discussions that might have taken place between scholars viewing such maps alongside works like an 1808 manuscript map of the northern and southern hemispheres, a perspective first introduced in the corners of Matteo Ricci's imported maps at the turn of the seventeenth century.[76]

Unno has argued that the success and influence of Buddhist maps such as the 1710 *Nansenbushū bankoku shōka no zu* 南瞻部洲萬國掌菓之圖 (Visualized map of Jambudvīpa's myriad lands),[77] published in Kyoto in 1710 by the priest Hōtan 鳳潭 (1654-1738),[78] lay not in 'religious conviction or in a belief that the Buddhist worldview was correct, but in their traditional image of Asia and in the place-names, especially in the interior of China and India, that were missing from the maps derived from European models.'[79] Elsewhere he writes,

> The history of cartography shows us that there are certain types of maps where the geographical content remains unaltered through a succession of copyings. For instance, nineteenth-century copies of the maps of the Five Indies [...] which are still part of the tradition of some Buddhist temples in Japan, show the same features as the oldest extant example, made in 1364. This does not imply, of course, that the Japanese people's knowledge of India remained unchanged during those six centuries; it is just that

75 Kōbe Shiritsu Hakubutsukan 神戸市立博物館, *Kochizu no sekai: Nanba Matsutarō-shi shūshū: Nihon zu, dōchūzu, sekaizu* 古地図の世界: 南波松太郎氏 收集: 日本図┌道中図┌世界図, Plate 78.

76 Ibid., Plate 89. One remarkable illustration of Kornicki's point on the parallel existence of manuscript and print cultures is the 1602 version of Matteo Ricci's map held by the Miyagi Provincial Library in Sendai. A printed original, which belonged to the collection of Date Masamune 伊達政宗 (1567-1636), is accompanied by a hand-drawn copy of the same map. The manuscript version has been made more hospitable to the Japanese reader by using *kana* alongside the Chinese characters for place names (Day, 'The Search for the Origins of the Chinese Manuscript of Matteo Ricci's Maps,' p. 112).

77 I have used Unno's translation for the title of this map, although he somewhat obscurely translates the compound 掌菓 ('fruit in the palm') in the title. The text at the bottom right corner of the map indicates that with its aid, observing the world becomes as easy as viewing a mango in the palm of one's hand (如掌中菴羅菓), clarifying the significance of these characters in the title.

78 For details of the dating of Hōtan's map and its relationship to predecessors, see Muroga and Unno, 'The Buddhist World Map in Japan and Its Contact with European Maps,' pp. 60-63.

79 Unno, 'Cartography in Japan,' p. 429.

Figure 5.2 The 1710 *Nansenbushū bankoku shōka no zu* (Visualized map of Jambudvīpa's myriad lands)

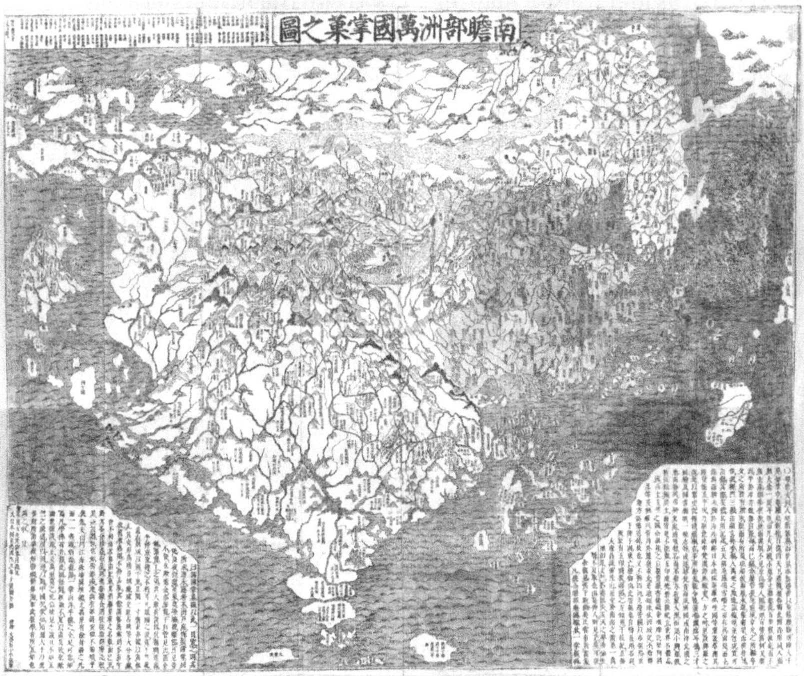

Published in Kyoto in 1710 by the priest Hōtan
Courtesy of David Rumsey Map Collection, David Rumsey Map Center, Stanford University Library

> these manuscript temple maps reflect no concern for new knowledge. It would seem that they were made to provide a realistic atmosphere of pilgrimage to the sacred places of India, birthplace of Buddhism, not as a source of geographical information.[80]

The message conveyed by the list of references provided at the top of Hōtan's map, and the essay inscribed along the bottom of the map, clearly demonstrate both familiarity with and *rejection* of the worldviews, both Confucian/Sinocentric and European, that had been introduced to Japan through Ming publications. This testifies to the self-confidence of the Buddhist establishment in Japan.

Hōtan's large folding map (Figure 5:2) reproduces a view of the world that combines the Jambudvīpa vision with some awareness of European

80 Unno, 'Maps of Japan Used in Prayer Rites or as Charms,' p. 65.

countries.[81] Muroga and Unno note that many of the toponyms used on this manuscript map are taken from late Ming encyclopedias.[82] Ultimately, however, this manuscript map and its kin can trace their heritage to Tang-imported Buddhist maps.[83]

In this depiction the view of Jambudvīpa obviously dominates the scene, with the swastika-like spiral of the world's four great rivers unfurling from a central point. The map cites over a hundred sources, which are listed in a long horizontal box at the top. The majority of the works cited do not contain maps at all. The first three categories of texts are religious, including twelve sutras (經), six treatises (論), and four vinaya (律).[84] These are followed by several treatises by individual monks about their travels to the Western Regions, as well as a handful of other works, before we come to a list of standard historical sources including the *Shiji* 史記, twelve dynastic histories reaching up to the Yuan, and the Ming *Yitong zhi* – only the last of which is illustrated by maps.

The list of works consulted goes on to name a number of gazetteers of Chinese places, such as Guangzhou, as well as of other foreign states, including Champa (the *Zhancheng zhi* 占城志) and the Ryukyus (the *Liuqiu zhi* 琉球志). Numerous late Ming publications, including the *Sancai tuhui* and the *Tushu bian*, both of which contain illustrations and explanations of the five-continent worldview that this map rejects, are then listed alongside more supportive works like the *Fozu tongji*, whose Buddhist cosmological maps have been discussed in Chapter Two. The list includes one late Ming miscellany, the *Wanbao quanshu* 萬寶全書 (Compendium of a myriad treasures), as well as specialized geographical works like Pan Angxiao's 潘昂霄 fourteenth century *Heyuan ji* 河源記 (Treatise on the source of the [Yellow] River). Various local gazetteers and other texts, some with seemingly marginal geographical significance, wrap up the list, including even the collected works of Tang writers like Liu Zongyuan 柳宗元 and Han Yu 韓愈 as well as Song writers like Su Shi 蘇軾 and Yang Wanli 楊萬里.[85] The latter

81 Unno notes that the Hōtan map was reprinted with the unchanged date of 1710 until 1815 (Unno, 'Cartography in Japan,' p. 429), which helps to explain the relative plentitude of this map in institutional collections and on the market.

82 Muroga and Unno, 'The Buddhist World Map in Japan and Its Contact with European Maps,' p. 60.

83 See discussion in Unno, 'Cartography in Japan,' pp. 371-376, and his Appendix 11.3 (p. 460) for a list of extant copies.

84 The sutras, treatises, and vinaya are of course the 'three baskets' for which the Tripitaka is named.

85 Yonemoto cites Jansen, *China in the Tokugawa World* (p. 37) as evidence of the paucity of Chinese gazetteers in Japan during the shogunate of Tokugawa Yoshimune. The citation of some

may have been included as references for descriptions or arguments their authors made in passing about geography, but there is no indication of what precisely was taken from them. This list reminds us that maps were only one of many sources for geographical information, particularly in a period when the character of maps was such that their textual accompaniment was vital to their interpretation.

The preface by Hōtan that begins at bottom right posits a direct challenge to non-Buddhist geographical works, emphasizing the limitless expanse of the universe and the limited perspective of those who, due to their only fragmentary awareness of this vast creation, are 'like frogs in a well.' How can those Chinese works, whose coverage does not even extend to Chosŏn, Japan, the Ryukyus or Southeast Asia, possibly encompass the countless continents spread across the vastness of space? Hōtan lists errors of placement for various foreign states in the *Sancai tuhui*, the *Tushu bian* and the *Da Ming yitong zhi*, all of which derive from the failure of the compilers of those Ming works to consult Buddhist sources. Now that this new map has been published, however, Hōtan tells the reader that it is possible to have the world in one's hand without stepping out of the door. Students of Buddhism who fail to consult it, he warns, will find themselves unable to fully digest the gist of the Buddhist sutras or treatises they read. But the audience is not limited to Buddhists alone, for Hōtan chastises the Confucian reader, too. No one will be convinced by their arguments if they prattle about the extent of the world while leaving out the Five Indies!

Particularly striking in Hōtan's approach is the emphasis on the limitless expanse of the universe, the known world comprising only a speck of the whole. We are reminded of the shogun Ashikaga Yoshimitsu's chastisement of the Hongwu emperor, founder of the Ming dynasty, reminding him that it was delusional for the Ming to consider itself the center of the world when it was actually only a peripheral corner.[86] The tenacity of Buddhist cartography, and challenge to a Sinocentric perspective, was one of several factors responsible for Japan's relative openness to the introduction of European maps depicting heretofore-unknown lands and continents.[87]

Jambudvīpa maps showed further development up through the nineteenth century. One colorful example even depicts a hot air balloon swooping across

gazetteers by Hōtan shows that at least a few were accessible to him (*Mapping Early Modern Japan*, p. 197 n. 8).

86 Boxer, *The Christian Century in Japan*, p. 250.

87 D. Max Moerman's forthcoming book, *The Japanese Buddhist World Map: Religious Vision and the Cartographic Imagination*, includes detailed discussion of this map in Chapter Five.

the sky north of Europe, shown as lying to the northwest of Jambudvīpa.[88] In addition to the creation of maps, efforts to resist Western cosmology extended even to the creation of complex mechanical devices to represent the movement of the heavenly bodies across the vault of heaven above Jambudvīpa, which was depicted as lying upon the surface of a flat earth. The monk Entsū 円通 (1754-1834), desiring to refute the theories of cosmic motion presented by European orreries, devised a superficially similar system in accord with Buddhist cosmology. His disciples hired Tanaka Hisashige 田中久重, the founder of a company that later became Toshiba Electronics, to construct several mechanical models based on Entsū's theory.[89]

Both Buddhist and Western worldviews evolved after their contact on Japanese soil, but in different directions. Buddhist maps showed increasing realization of a world beyond the sutras, as places from Western sources were added, at first as islands in the south, then over time shifting to something approximating Europe's actual spatial relationship to India. Nevertheless, Buddhist maps remained within the framework of the Jambudvīpa worldview, even well into the nineteenth century. Western-style maps were replicated and continued to be respected, but they became more out of date as time passed without much fresh input. Thus it is not entirely clear if a 1709 manuscript copy of a 1637 European-style map was meant to represent current knowledge, or merely to serve as a facsimile of an antique.[90] This uncertain status for Western cartography only changes, albeit rather dramatically, in the mid-nineteenth century, around the same time that Buddhist cartography fades from significance.

Ming antecedents, Ricci, and the *Wa-Kan sansai zue* 和漢三才圖會 (Japanese and Chinese illustrated compendium of the three fields of knowledge)

Though in its title and its general purpose the *Wa-Kan sansai zue* hearkens to a Ming encyclopedia of a century earlier, the *Sancai tuhui* 三才圖會,

88 An illustration of this map is in Kōbe Shiritsu Hakubutsukan, *Kochizu no sekai*, Plate 80. See also Muroga and Unno, 'The Buddhist World Map in Japan and Its Contact with European Maps,' pp. 64-65.

89 For a discussion of these models and Entsu's theory, see Yamada Keiji 山田慶児, 'Ryūkoku Daigaku Ōmiya Toshokan shozō shukushōgi zu, setsu oyobi mokei ni tsuite' 龍谷大学大宮図書館所蔵 縮象儀-図「説および模型-について.

90 This map is illustrated in Kōbe Shiritsu Hakubutsukan, *Kochizu no sekai*, Plate 96. The notes to the plate mention that another copy of this map is known from 1837.

the relationship between these works is too complex for the former to be called a Japanese version of the Ming text.[91] It is as much a full-fledged combination of disparate threads as any of the late Ming books discussed in Chapter One, harvesting extensively from predecessors to present a more-or-less new configuration of 'essential' knowledge. The diversity of sources and the method of compiling entries by borrowing from different texts are mentioned in the introductory explanatory comments.[92] Works like this, especially their descriptions of exotic countries and their people, shaped the way foreigners and faraway countries were portrayed, paving the way for fiction like that of Hiraga Gennai 平賀源内.[93]

The *Wa-Kan sansai zue* was the product of some three decades of work, mostly by the Osaka-based Terajima Ryōan 寺島良安, to assemble an encyclopedia comparable to its highly influential Ming namesake but addressed to a Japanese audience. The text of the *Wa-Kan sansai zue* reached its current form in 1712, though woodblock appears to have first been put to paper in 1713. Based on the claim that it had been in process for thirty years, Inden Emi calculates that its compilation must have begun in about 1682.[94] It was thus an artifact of the cultural fluorescence of the Genroku period, but it was also a product of the 'restricted' era before the 1720 loosening of restrictions on foreign information. Inden shows through analysis of passages on Southeast Asia that information on foreign territories continued to reach Japan during this period, either directly or indirectly.[95] I will, in turn, show in the passage below that the *Wa-Kan sansai zue* reproduced words written by Matteo Ricci long after the Japanese ban on Jesuit publications. The book attributes these quotations from the introduction to Ricci's map as coming from the *Sancai tuhui*, rather than from Ricci's original.

The text of the chapter on *chi* 地 (Earth) retains the map and many pieces of text from the *Sancai tuhui* that I have discussed in Chapter Three, but is

91 Inden Emi elaborates on the relationship between the *Sancai tuhui* and the *Wa-Kan sansai zue* as reflected in their somewhat different treatment of foreign subjects, though maps are not included in this discussion. Inden Emi 位田絵美, '*Wa-Kan* sansai zue ni miru taigai ninshiki-Chūgoku no *Sansai zue* kara Nihon no *Wa-Kan sansai zue* e' 和漢三才図会にみる対外認識 ―中国の『三才図絵』から日本の『和漢三才図絵』へ―.

92 Terajima Ryōan 寺島良安, *Wa-Kan sansai zue* 和漢三才圖會, *hanrei* p. 1.

93 See Yonemoto, *Mapping Early Modern Japan*, pp. 105-106.

94 Inden, '*Wa-Kan* sansai zue ni miru taigai ninshiki,' p. 60. Inden also notes that, based on the wide variety of both printed and manuscript copies now known, not to mention Meiji-period moveable type editions, the *Wa-Kan sansai zue* was one of the more widely available encyclopedic texts (p. 60).

95 Inden, '*Wa-Kan* sansai zue ni miru taigai ninshiki,' pp. 63-64.

quite different in its combination of various threads of discussion.[96] Like the *Tushu bian*, the *Wa-Kan sansai zue* includes text and images reflecting contradictory viewpoints. Terajima begins with a definition of Earth taken from the *Shuo Wen* 說文, and lists alternate forms for the character. He summarizes ideas about the movement of the earth and heavens, and the roles played by opposing cosmological forces (the substance of the earth tends toward *yang* in the north and *yin* in the south, thus mountains are numerous in the north; however, the substance of the heavens tends toward *yang* in the south and *yin* in the north, thus the sun moves in the south). After giving a diameter of 90,000 *ri* for the globe, comparing it to a walnut shell, and quoting from the *Bowu zhi* 博物志 (Treatise on curiosities)[97] and the *Guang bowu zhi* 廣博物志 (Expanded treatise on curiosities)[98] on its physical characteristics and structure, Terajima's next passage will sound familiar:

> The *Sancai tuhui* says the land and the sea are basically round in shape, and together they form a globe, which is situated in the celestial sphere like the yolk of a chicken's egg is suspended in the white. It is said that the earth is *hō* 方, but this refers to its being fixed and immobile, not to a square shape.
>
> 三才圖會云地與海本是圓形而仝為一球居天球之中如鷄子黄在青内。有謂地為方者乃語其定而不移之性非語其形體也.

Terajima goes on to summarize the listing of the continents that appeared in the *Sancai tuhui*, accompanied by a simplified Riccian map (compare Terajima's map in Figure 5:3 to the *Sancai tuhui* version, Figure 3:3). In his mention of the Red Sea, the character 紅 has mistakenly been replaced with 江, just as in the *Sancai tuhui*. He then describes the climate zones and time intervals discussed in the *Sancai tuhui*, all derived from Ricci's map preface.

No Qing works are cited here, though by the time of publication almost seventy years had passed since the establishment of that dynasty.[99] The repetition of abbreviations and the replacement of 紅 with 江 from the

96 Terajima, *Wa-Kan sansai zue*, pp. 602-603.

97 Attributed to Zhang Hua 張華 (232-300).

98 A late Ming text attributed to Dong Sizhang 董斯張 (1586-1628).

99 Perhaps this can be attributed in part to the decline in imports that coincided with the establishment of the Qing; Kornicki notes that by this time 'imports were insignificant in quantity in the context of a mature publishing industry in Japan that was producing books for a mass home market' (Kornicki, *The Book in Japan*, p. 299).

Figure 5.3 The world map as depicted in Chapter 55 of the *Wa-Kan sansai zue*

Japanese-Chinese illustrated compendium of the three fields of knowledge
Image taken from the Yoshikawa Kōbunkan 1906 facsimile of the 1713 edition, p. 602

Sancai tuhui redaction of Ricci's preface strongly suggests that Terajima worked from this text alone without reference to the original. It appears that he had no idea that these words were borrowed from the pen of a Jesuit missionary. This underscores the importance of Ming repetition of these ideas in their transmission to later days and other lands.

A section dedicated to China begins a long geographical survey illustrated with maps. The title for this section uses *Chūka* 中華 rather than *Shin* 清 or *Chūgoku* 中國, but a prefatory paragraph reviews the multiplicity of names by which it has been known and the tendency for Japanese to refer to it as either *Kan* 漢 or *Tō* 唐, the great Han and Tang dynasties having lent their names to 'China' long after their fall.[100] Following the lead of the *Sancai tuhui*, the series of maps here begins with a 'Unified map of the Chinese and Barbarians' 華夷一統圖. A note at the left edge of the map simply states that this is a copy of a Ming Wanli period map. Of the maps in the

100 Ibid., p. 673.

'Chūka' section, this is the only one taken directly from the *Sancai tuhui*; the provincial maps are distinctively simpler and may indeed derive from other unidentified maps.

The text accompanying the introductory map, a brief outline history of the administrative organization since the time of the Sage Kings, is cited as coming from the *Sancai tuhui*.[101] It shows only minor changes, such as the elision of words to shorten sentences.[102] A reader familiar with the *Tushu bian* might notice that this text appears there as well, in the same form as the *Sancai tuhui*, and in fact it is widely shared between various compendia.[103]

Terajima goes on to provide a simple gazetteer of China, province by province. When we compare the text accompanying these maps, we find that the style of the writing in the *Sancai tuhui* – basic administrative and natural features discussed in a manner reminiscent of the *Guang yutu* or a gazetteer – has been thoroughly rewritten and supplemented with descriptions of important people and cultural notes of interest to a Japanese audience. For example, the listings of coastal provinces begin with their compass direction and distance by sea from Kyushu, and Terajima has taken care to note products that would be familiar in the Japanese market.

Following the pattern set by the *Sancai tuhui*, after the discussion of China and its provinces Terajima then inserts maps and brief descriptions of Japan,[104] Chosŏn,[105] and the Ryukyus.[106] These are followed by a new addition that was naturally not included in Ming cartographic sources, the island now known as Hokkaido, referred to here as Ezo 蝦夷[107] though also listing several other names used through history.

It is only after all of these maps and textual descriptions, some eighty pages, that the Jambudvīpa world view is presented, but presented it is, cut

101 Ibid., p. 674.

102 Wang Qi and Wang Siyi, *Sancai tuhui*, *dili* 地理 *juan* 1, p. 5b.

103 Zhang Huang, *Tushu bian*, *juan* 34, p. 1a. Indeed, a search of the *Siku quanshu* electronic database for the phrase '*chuangzhi jiuzhou tongling wanguo* 創制九州統領萬國' (the establishment of the Nine Regions and the bringing together of the myriad states, attributed to Di Ku), reveals that they appear in similar form in other works as well, ranging from Chapter 171 of the *Tongdian* 通典 to Chapter 40 of the *Tongzhi* 通志.

104 Terajima, *Wa-Kan sansai zue*, p. 748.

105 Ibid., p. 750, citing the fifteenth-century *Tongguk t'onggam* 東國通鑑 of Sŏ Kŏjŏng 徐居正 as a source in the text, but taking its map from the *Sancai tuhui* (which in turn borrowed it from the *Guang yutu*).

106 Ibid., p. 752. This map also comes from the *Sancai tuhui* but derives from the one first made widely available in the *Guang yutu*, recognizable for its depiction of the royal buildings in outlandishly exaggerated scale.

107 Ibid., p. 753.

into several pieces to fit onto the page.[108] This map is not taken from any Ming work, but resembles instead the 1710 Hōtan map, which would have been a fairly recent publication at the time of printing. The text associated with these maps blends secular and sacred; the brief discussion of the Western Regions mentions the Han military mission of Zhang Qian, but shifts to Buddhist sources when *Tenjiku* 天竺 (India) comes under discussion. This section is followed by pictures of most of the thirty-three Buddhist patriarchs.[109]

Terajima's juxtaposition of all of these materials reflects uncertainty about the degree to which the different strands of available thought should be blended, or in cases of direct contradiction, which should be believed. Elsewhere he has tried to find ways to understand Buddhist ideas as versions of the same truth (as in the case of trying to identify Mount Sumeru with the Kunlun mountains), and presented challenges to Buddhist cartography, but he finds it impossible to illustrate his discussion of the Indies without using a Buddhist map. It seems that there was no tradition other than the Buddhist for representing the Western Regions and the Five Indies, as the European maps did not emphasize these places that were culturally important to a Japanese audience. We are reminded of the way the *Sancai tuhui* preserved the squared-off image of Ming territory from long-familiar maps while 'nesting' it within a new Riccian worldview. Perhaps a subtle point is made by the placement of the Ricci-style map from the *Sancai tuhui* and related text at the beginning of the section on Earth, while relegating the Jambudvīpa maps to the end of the section, without any direct textual explanation of the maps. The close-ups showing the adaptation of the Jambudvīpa image to more recent Western knowledge slightly complicate the picture, however, reminding us of the persistence of an evolving Buddhist worldview in Japan at this late date, a persistence which Terajima did not wish to completely deny.

After all of these other maps have been presented, Terajima provides a region-by-region breakdown of Japan, with occasional maps breaking up a dense textual description of locally important places, similar to what we might expect in a Chinese local gazetteer but with great emphasis on religious institutions such as temples. Japan has come last among the lands discussed, but it is discussed in far greater detail and at greater length.[110]

Works such as this *Wa-Kan sansai zue* exhibit some of the features I have addressed elsewhere – the direct appropriation of images and text from Ming

108 Ibid., pp. 754, 756-757.

109 Ibid., pp. 758-763.

110 The facsimile edition has 388 pages, as compared with 74 for the entire section on China.

works, the 'naturalization' of Western knowledge without citation, and the encyclopedic collection of information from a broad range of sources. The Buddhist component of the *Wa-Kan sansai zue* is a reflection of this latter trend, recalling the manner in which a Sinocentric worldview continued to be important in Ming publications and was even presented side by side with Jesuit views in some Ming works. The inclusion of material such as Ricci's world map in the *Sancai tuhui* helped this material to circulate in Japan, and it features prominently in the *Wa-Kan sansai zue*.

Conclusion

As I have suggested of Chosŏn in the previous chapter, a survey of Japanese cartography in the Tokugawa period looks very different if we focus on maps printed in books rather than on picturesque, and much scarcer, large-format maps. In books we see the republication and discussion of a broad range of cartographic sources, including images from encyclopedias and more specialized works. The characteristics of such woodblock-printed maps in books must be considered in the context of the prose surrounding them.

Published cartography burst onto the scene comparatively suddenly but forcefully in Japan during the seventeenth century, at a time when there had already been substantial contact with cartographic perspectives that challenged Sinocentric and Buddhist views of the world. This may partly explain the greater impact of European cartography in early Edo Japan as compared with Ming and Chosŏn. In the notoriously conservative realm of cartography, the Japanese had fewer fixed preconceptions preventing them from considering new paradigms. Knowledge that had been gained quite early, especially in the realm of surveying techniques, was not necessarily applied later in ways that would have been expected at the same time in parts of Europe, but the period of its fluorescence in the seventeenth century inspired survey-based urban maps which publishers promoted as part of the print culture of the Genroku period and later. I agree with Marcia Yonemoto that the popularization of maps was less a manipulative move by the state to make the population accept the terms of its new panopticon, than the result of the map's role in popular culture for pleasure or for navigation around the expanding urban zones.

I have argued that some of the singularities in the way cartography was appreciated in Japan resulted from a more intimate understanding of cartographic application by certain rulers. In comparing a figure like Tokugawa Ieyasu to his Ming contemporaries, we are struck by his understanding of both conquest and commerce. Ieyasu's alleged claim that 'From my youth I have not valued silver or gold or treasures'[111] is belied by his numerous enterprises – such as the mines that he did not neglect even after becoming shogun – not to mention his alacrity in proffering offers of trade to various European parties even as he clamped down on their missionary endeavors. He understood with some intimacy the purposes to which maps were put in Europe, and this also alerted him and his successors to the dangers inherent in allowing unauthorized surveys.

Two important trends that have attracted our attention are the persistence of Buddhist worldviews and their simultaneous circulation with the 'five-continent' worldview that came to Japan through Chinese-language works by Jesuits, and through Ming publications that incorporated such maps, but also, very importantly, through contact with Europeans themselves. Even after foreign intercourse was limited, the presence of the Dutch at Deshima was a constant reminder of the broader world, a world that Buddhist maps in Japan gradually accommodated.

Much of the focus of scholarship on cartography in the seventeenth and eighteenth century has been on its transformations, but it is also important to observe what did *not* change during Japan's publishing boom. Just as reprints of older books were part of the Ming boom, Japanese printers made a number of stable and familiar tropes, such as the Gyōki-style map, far more accessible without updating them. In the case of maps used as earthquake charms, the point was precisely *not* to change them. Works like the *Wa-Kan sansai zue* helped to maintain the currency of one version of a Jesuit map borrowed directly from a Ming encyclopedia of the previous century, with all traces of its Christian connection erased. Such is the power of publishing not only to fluoresce, but also to freeze.

Works cited

Primary sources

Ishikawa Ryūsen 石川流宣. *Bankoku sōkaizu* 萬國總界圖. Edo: Suhara Mohē, Hōei 5 [1708].

Matsushita Kenrin 松下見林. *Ishō Nihon den* 異稱日本傳. Sesshū 攝州: Morita Shōtarō 毛利田庄太郎, Genroku 6 [1693].

111 Sadler, *The Maker of Modern Japan*, p. 389.

Terajima Ryōan 寺島良安. *Wa-Kan sansai zue* 和漢三才圖會. Facsimile of 1713 edition. Tokyo: Yoshikawa Kōbunkan 吉川弘文館, Meiji 39 [1906].

Wang Qi 王圻, and Wang Siyi 王思義. *Sancai tuhui* 三才圖繪. Facsimile of c. 1609 edition. Shanghai: Shanghai guji chubanshe 上海古籍出版社, 1988.

Zhang Huang 章潢. *Tushu bian* 圖書編. Facsimile of Wanli 41 [1613] edition. Taibei: Chengwen chubanshe 成文出版社, 1971.

Secondary sources

Ayusawa Shintarō. 'Geography and Japanese Knowledge of World Geography.' *Monumenta Nipponica* 19, no. 3/4 (1964): 275-294.

—. 'The Types of World Map Made in Japan's Age of National Isolation.' *Imago Mundi* 10 (1953): 123-127.

—. *Mateo Ritchi no sekai zu ni kansuru shiteki kenkyū* マテオ┌リッチの世界図に関する史的研究 (A Historical Study of Matteo Ricci's World Map). Yokohama: Yokohama Shiritsu Daigaku, 1953.

Bartholomew, James R. *The Formation of Science in Japan: Building a Research Tradition*. New Haven, CT: Yale University Press, 1989

Beans, George Harry. *A List of Japanese Maps of the Tokugawa Era*. Jenkintown, PA: Tall Tree Library, 1951.

Berry, Mary Elizabeth. *Hideyoshi*. Cambridge, MA: Council on East Asian Studies; Harvard University Press, 1989.

—. *Japan in Print: Information and Nation in the Early Modern Period*. Berkeley: University of California Press, 2006.

Boxer, Charles Ralph. *The Christian Century in Japan: 1549-1650*. Berkeley: University of California Press, 1951.

Brown, Philip C. 'Never the Twain Shall Meet: Land Survey Techniques in Tokugawa Japan.' *Chinese Science* 9 (1989): 53-79.

Cieslik, Hubert. 'The Case of Christovão Ferreira.' *Monumenta Nipponica* 29, no. 1 (Spring 1974): 1-54.

Cooper, Michael. *Rodrigues the Interpreter: An Early Jesuit in Japan and China*. New York and Tokyo: Weatherhill, 1974.

Cortazzi, Hugh. *Isles of Gold: Antique Maps of Japan*. New York: Weatherhill, 1983.

Day, John D. 'The Search for the Origins of the Chinese Manuscript Copies of Matteo Ricci's Maps.' *Imago Mundi* 47 (1995): 94-117.

Elison, George. *Deus Destroyed: The Image of Christianity in Early Modern Japan*. Cambridge, MA: Council on East Asian Studies, Harvard University, 1988.

Hertz, Solange. *Rhodes of Viet Nam: The Travels and Missions of Father Alexander de Rhodes in China and Other Kingdoms of the Orient*. Westminster, MD: The Newman Press, 1966.

Inden Emi 位田絵美. '*Wa-Kan* sansai zue ni miru taigai ninshiki – Chūgoku no *Sansai zue* kara Nihon no *Wa-Kan sansai zue* e' 和漢三才図会にみる対外認識 — 中国の『三才図絵』から日本の『和漢三才図絵』へ—. *Rekishi hyōron* 歴史評論 592 (August 1999): 58-75.

Jansen, Marius B. *China in the Tokugawa World.* Cambridge, MA, and London: Harvard University Press, 1992.

Kimura Tōichirō 木村 東一郎. *Kinsei muraezu kenkyū* 近世村絵図研究. Tokyo: Komiyama Shoten, 1962.

Kish, George. 'Some Aspects of the Missionary Cartography of Japan during the Sixteenth Century.' *Imago Mundi* 6 (1949): 39-47.

Kōbe Shiritsu Hakubutsukan 神戸市立博物館. *Kochizu no sekai: Nanba Matsutarō-shi shūshū: Nihon zu, dōchūzu, sekaizu* 古地図の世界: 南波松太郎氏 収集: 日本図┌道中図┌世界図. Kōbe: Kōbe Kenkō Kyōiku Kōsha 神戸健康教育公社, 1983.

Kochizu ni miru sekai to Nihon: chizu wa kataru yume to roman: kaikan isshūnen kinen tokubetsuten 古地図にみる世界と日本: 地図は語る夢とロマン: 開館一周年記念特別展. Kōbe: Kōbe-shi Kenkō Kyōiku Kōsha, Shōwa 58 [1983].

Kornicki, Peter. *The Book in Japan: A Cultural History from the Beginnings to the Nineteenth Century.* Leiden and Boston: Brill, 1998.

Laures, John. *Kirishitan Bunko; A Manual of Books and Documents on the Early Christian Mission in Japan.* 3rd edition. Tokyo: Sophia University, 1957.

Matsuki Satoru 松木哲. 'Kōro zu ni tsuite' 航路図について. In *Kochizu ni miru sekai to Nihon: chizu wa kataru yume to roman, kaikan isshūnen kinen tokubetsuten* 古地図にみる世界と日本: 地図は語る夢とロマン, 開館一周年記念特別展. Kōbe: Kōbe-shi Kenkō Kyōiku Kōsha, Shōwa 58 [1983].

Moerman, D. Max. *The Japanese Buddhist World Map: Religious Vision and the Cartographic Imagination.* Honolulu: University of Hawaii Press, forthcoming.

Muroga, Nobuo, and Unno Kazutaka, 'The Buddhist World Map in Japan and Its Contact with European Maps.' *Imago Mundi* 16 (1962): 49-69.

Nakamura Hiroshi. 'The Japanese Portolanos of Portuguese Origin of the XVIth and XVIIth Centuries.' *Imago Mundi* 18 (1964): 24-44.

Needham, Joseph. 'Astronomy.' In *Science and Civilisation in China, Volume 3: Mathematics and the Sciences of the Heavens and the Earth*, 171-461. Cambridge: Cambridge University Press, 1959.

Ōba Osamu 大庭脩; translated by Joshua A. Fogel. 'Sino-Japanese Relations in the Edo Period (Part Three: The Discovery of Banned Books).' *Sino-Japanese Studies* 9, no. 1 (October 1996): 56-74.

—. *Kanseki yunyū no bunkashi: Shōtoku Taishi kara Yoshimune e* 漢籍輸入の文化史: 聖徳太子から吉宗へ. Tōkyō: Kenbun Shuppan, 1997.

Okamoto Yoshitomo 岡本良知. *Jūrokuseiki ni okeru Nihon chizu no hattatsu* 十六世紀における日本地図の発達. Tokyo: Yagi Shoten 八木書店, Shōwa 48 [1973].

Pinto dos Santos, José Miguel. 'The "Kuroda Plot" and the Legacy of Jesuit Scientific Influence in Seventeenth Century Japan.' *Bulletin of Portuguese-Japanese Studies* 10/11 (June/December 2005): 97-191.

Paramore, Kiri. 'Hayashi Razan's Redeployment of Anti-Christian Discourse: The Fabrication of Haiyaso.' *Japan Forum* 18, no. 2 (July 2006): 185-206.

Sadler, Arthur Lindsay. *The Maker of Modern Japan: The Life of Tokugawa Ieyasu.* Rutland, VT: C. E. Tuttle, 1978.

Sakanishi, Shio. 'Prohibition of Import of Certain Chinese Books and the Policy of the Edo Government.' *Journal of the American Oriental Society* 57, no. 3 (September 1937): 290-303.

Shapinsky, Peter D. 'Polyvocal Portolans: Nautical Charts and Hybrid Maritime Cultures in Early Modern East Asia'. *Early Modern Japan* 14 (2006): 4-26.

Stephenson, F. Richard. 'Chinese and Korean Star Maps and Catalogs.' In *The History of Cartography, Volume 2, Book 2: Cartography in the Traditional East and Southeast Asian Societies,* edited by J.B. Harley and David Woodward, 511-578. Chicago and London: University of Chicago Press, 1994.

Takahashi Tadashi 高橋正. 'Tōzen seru chūsei Isurāmu sekaizu.' *Ryūkoku Daigaku ronshū* 374 (1963): 77-95.

—. 'Chūsei Isurāmu sekaizu' 中世イスラーム世界図. In *Chizu no shisō* 地図の思想, edited by Hasegawa Kōji 長谷川孝治, 53-55. Tōkyō: Asakura Shoten, 2005.

Toby, Ronald. *State and Diplomacy in Early Modern Japan: Asia in the Development of the Tokugawa Bakufu.* Princeton, NJ: Princeton University Press, 1984.

Unno Kazutaka. 'Cartography in Japan.' In *The History of Cartography, Volume 2, Book 2: Cartography in the Traditional East and Southeast Asian Societies,* edited by J.B. Harley and David Woodward, 346-477. Chicago and London: University of Chicago Press, 1994.

—. 'Maps of Japan Used in Prayer Rites or as Charms.' *Imago Mundi* 46 (1994): 72-75.

—. *Tōzai chizu bunka kōshōshi kenkyū* 東西地図文化交渉史研究 [Monographs on the history of cartographical exchange between the East and the West]. Osaka: Seibundō, 2003.

Wallis, Helen. 'The Influence of Father Ricci on Far Eastern Cartography.' *Imago Mundi* 19 (1965): 38-45.

Whelan, Christal. *The Beginning of Heaven and Earth: The Sacred Book of Japan's Hidden Christians.* Honolulu: University of Hawaii Press, 1996.

Wigen, Kären, Fumiko Sugimoto, and Cary Karacas, eds. *Cartographic Japan: A History in Maps.* Chicago and London: University of Chicago Press, 2016.

Yajima Suketoshi. 'The European Influence on Physical Sciences in Japan.' *Monumenta Nipponica* 19, no. 3/4 (1964): 340-351.

Yamada Keiji 山田慶兒. 'Ryūkoku Daigaku Ōmiya Toshokan shozō shukushōgi zusetsu oyobi mokei ni tsuite' 龍谷大学大宮図書館所蔵 縮象儀-図「説および

模型-について. *Bulletin of the International Research Center for Japanese Studies* 16 (September 1997): 59-71.

Yamori Kazuhiko 矢守一彦. 'Bākurē hon Nihon zu byōbu' バークレー 本日本図屏風.' *Fukui kenshi shiori* 福井県史しおり, Shiryō hen 16 ue 資料編16上 (February 28, 1990): 1-5.

Yonemoto, Marcia. *Mapping Early Modern Japan: Space, Place, and Culture in the Tokugawa Period, 1603-1868*. Berkeley: University of California Press, 2003.

—. 'The "Spatial Vernacular" in Tokugawa Maps.' *The Journal of Asian Studies* 53, no. 3 (August 2000): 647-666.

Conclusion

This book has examined the role of the late Ming publishing boom in the evolution of cartography in East Asia, investigating the ways in which the expansion of publishing that began in the mid-sixteenth century affected the nature of cartographic materials and their circulation in East Asia. This has led us to explore relationships between social, political, and intellectual trends of the time and the production and use of maps, revealing rich seams in East Asian cultural history that transcend borders.

In contrast to the roughly contemporary explosion of cartography in some parts of Europe – where new technologies meshed with an age of discovery and the conceptual shifts of the Enlightenment to produce not only the expectation that maps should be constantly updated, but a veritable avalanche of new data with which to do so – map publishing in China up to the mid-Ming existed in the social context of an examination system which played a large role in shaping the evolution of a cartographic canon. Aside from the maps found in gazetteers, a large proportion of early printed maps (as well as maps engraved on stelae erected at school grounds) served as study aids for the Classics and dynastic histories. To contrast this with the situation in Europe, we might imagine what the results would have been if the great cartographers of the 1500s had focused primarily on reconstructions of biblical geography and the shifting borders of post-Roman political entities, rather than piecing together the contours of a world in the process of colonization.

The examination system was therefore a more important driving force for the production of Chinese maps circulating in the public realm than was the territorial imperative or the other processes of early modern state formation. The subject matter of *printed* maps in late imperial China has a strong tendency toward the didactic rather than the utilitarian. In general, a didactic map was more likely to be needed in multiple examples. Utilitarian maps, such as nautical handbooks for sailing along the coast, were often privileged documents that were copied by hand and not mass-produced for general circulation. Published route books from this period that were

Akin, Alexander, *East Asian Cartographic Print Culture: The Late Ming Publishing Boom and its Trans-Regional Connections*. Taylor & Francis Group, 2021
DOI: 10.5117/9789463726122_CONCL

meant for real use in navigating road networks relied on text rather than images to provide the level of detail that made them worthwhile.

The exposure of the ordinary civilian to maps appears to have increased significantly after the middle of the sixteenth century, as cartographic representations entered the popular literate consciousness to an unprecedented degree. Maps that had existed before in limited manuscript versions, or earlier works that had nearly disappeared, found renewed life through printing. By the end of the 1600s, the typical household almanac included a simple orienting map as one of its essential illustrations, and in works like the *Tushu bian*, the distillation of the entire cartographic repertoire – from depictions of the dawn of civilization to maps of the Grand Canal and even miniature copies of the newly introduced Jesuit world maps – could be found in one publication.

In sharp contrast to the changes taking place around the same time in the publishing centers of Western Europe, however, this cartographic florescence took place without any dramatic changes in either printing technology or the methods used for cartographic drafting. The technologies used in most mapping, both practical and conceptual, were by this time quite old, though sometimes quite advanced. Few new surveys took place above the local level; new maps were typically drawn on the basis of old maps and textual data.

Late Ming publishing exemplified that old adage, *Tianxia wenzhang yi da chao* (天下文章一大抄), which might be loosely translated as 'All writings under heaven are copied from somewhere else.' Some of the maps from this period served as prototypes for later copies, reused for as long as their social function persisted. Classic works from the Song dynasty, supplemented by a number of pathbreaking works published during the Ming, formed the core reservoir from which the vast majority of later printed maps were copied, or their methodologies inspired. In the process of transcription, engraving and publication, errors were frequently introduced, often making printed maps less accurate than the originals upon which they were based. Despite these shortcomings inherent in the means of production, however, the proliferation of cartographic texts in the late Ming facilitated contact with a diverse array of contending cartographies. The changes we see in the late Ming are therefore not in the techniques used to create maps but rather in the ways that maps were used. The expansion of publishing made a greater variety of cartographic materials available to the public while simultaneously widening the gap between the most widely available maps, those published in books and on paper sheets, and the most technically sophisticated.

Ming texts on military affairs reflected contemporary events, first focusing primarily on the Mongols and the northern frontier, then on Japan and the sites of *wokou* raids, later on the Jurchens in the northeast. Each change in regional focus created a miniature cartographic tradition of its own. In later works on the same topics, we have seen that publishers often borrowed the same maps to illustrate their own books. For example, Zheng Ruozeng's maps of Japan appeared in a number of later treatises and encyclopedias, and were even exported to and reprinted in Japan itself to help understand how others saw the country.

The late Ming also saw a dramatic increase in the proportion of local administrative gazetteers that included maps. The maps in a county gazetteer could attract notice from superiors at the prefectural level and get reprinted in the gazetteer for the prefecture. Meanwhile, we have seen concepts familiar at the national level, such as the historical time series popularized through the *Lidai dili zhizhang tu*, or the grid network used in the *Guang yutu*, applied at the level of the local gazetteer. Influence could move up or down the administrative hierarchy, with somewhat haphazard or idiosyncratic results.

In my introduction I noted that Kai-wing Chow's notion of the late Ming 'literary public sphere,' a community of scholars linked together by the sharing of texts, also has cartographic repercussions, for the publishing boom fostered a 'community of readers' with a cartographic consciousness. Late Ming publications about military affairs and frontier security, often composed by private authors working from published memorials and personal experience, comprise one arena in which the role of cartography within the formation of public discourse is particularly important. As we have seen, the later editions of works like the *Guang yutu* reflect this discourse. Later printings came to include newly added maps taken from other books that had recently appeared, even as maps from the *Guang yutu* were in turn copied in new works by other authors.

One of the few truly novel additions to late Ming cartography was the introduction of a different worldview by Catholic missionaries in the late sixteenth and early seventeenth century. After showing that this was not the first perspective on the scene to 'decenter' China, I have argued for a Ming-centric perspective on the Jesuit contribution to Ming cartography, noting that the Jesuit adoption of a strategy of cartographic evangelism was the result of their encounter with an already burgeoning circulation of maps in locally printed works. It is fruitful to consider the Jesuits as participants in the late Ming publishing boom, for despite their introduction of new concepts, they used the same printing technology and, often, the same woodblock carvers who worked for other local patrons. Once published, their works entered the fray of late Ming intellectual discourse; they essentially became immigrant contributors to an ongoing domestic conversation. As we have seen, Japanese readers of Jesuit works from the Ming could even mistake their authors for men of China.

Local reactions to Jesuit works were quite variable, from the 'inclusive' tack of even non-converts such as Zhang Huang, Wang Qi, and Wang Siyi, to the defensive posture of Chen Zushou. Future research in Jesuit-Chinese contact and influence must outline something more than a late Ming Response to the West: Individual authors reveal distinctive and often complex reactions to both long-developed indigenous traditions and newly imported Jesuit theories and data, both of which were becoming available to an unprecedentedly broad audience at the time. The more we learn about engagement with imported geographical thought in this period, the more urgent it becomes to problematize the entire paradigm of contrastive Eastern and Western cartographies.

Nor were these developments confined to the Ming. The expansion of the publishing industry, and the medium of classical Chinese as a common written language, facilitated regional trade in texts and their republication in neighboring states, and this phenomenon resulted in significant sharing of both text and images between cartographic works published in different countries. Despite the much longer persistence of primarily manuscript cartographic practice in Chosŏn Korea, printed maps from the Ming influenced local map culture there, with writers like Ŏ Sukkwŏn proudly citing loyalty to Ming models. The perpetuation of cartographic diversity in Chosŏn followed a different pattern than in the heavily commercialized world of Ming publishing; the (admittedly incomplete) record suggests that Chosŏn Korea did not witness anything approaching the printed cartographic diversity of its more powerful neighbor until the nineteenth century. Nevertheless, it is clear that even in the absence of a local 'publishing boom' until a relatively late date, Ming publications had significant reverberations in literate Chosŏn society.

The movement of geographical and cartographic material was not a unidirectional flow from the Ming to its neighbors, as exemplified by the *Chaoxian tushuo*'s importation and reprinting of a Korean miniature atlas in China. Late Ming publishing centers should be considered as nodes of cartographic and geographic synthesis that were connected within a larger web. Even in periods when security concerns prompted attempts to staunch this flow, maps had ways of moving across borders.

For reasons elaborated in Chapter Five, fewer fixed cartographic preconceptions inhibited Japanese scholars from considering new paradigms. We have reviewed several possible reasons for the greater influence of European cartography in Japan than elsewhere in East Asia. All the same, conservatism was a hallmark of Japanese cartography too. Just as reprints of older books and images were part of the Ming boom, so Japanese printers also took up a number of stable and familiar tropes, such as the simplistic Gyōki-style map, making them far more accessible without necessarily updating them, and Buddhist geographical conceptions endured long into the era of contact with the broader world.

Print culture in this period also contributed to the standardization of images of states and their provinces into instantly recognizable forms. I have argued that this phenomenon offers contrasts with Benedict Anderson's thesis, based primarily on the very different experience of Southeast Asia, that European colonialism and its attendant technological and administrative handmaidens (including print capitalism) were responsible for spreading the notion of states having a cartographic 'footprint.' This thesis fits uncomfortably not only with Chinese data, but with developments in Chosŏn Korea and Japan. The *Sŭngnam*, a national gazetteer styled after that of the Ming, was illustrated with basic provincial maps that continued to serve as the main style of Korean self-representation for at least three hundred years. The fact that these maps became familiar, stereotypical images despite the lack of widespread commercial printing poses a challenge to the linkage of a recognizable national 'footprint' to print capitalism. Any narrative of the development of cartography that emphasizes the role of maps in the process of Western colonization, setting in indelible ink rapacious claims upon the earth, is one that must take better account of the cartographic record in East Asia and its divergences from this model.

To take the last point one step further, those of us who work with East Asian cartography must do our part to build a greater storehouse of works in Western languages on the multifarious manifestations of mapping in this part of the world. We have a role to play in making East Asian cartographic history a more accessible part of world history, no longer an "exception to the pattern" or a special case, but rather an integral part of the pattern itself.

Appendices

Appendix 1

Sources cited in the compilation of the *Chouhai tubian,* reflecting Zheng Ruozeng's access to both published works and military maps used by relevant officials. Most of the latter are believed to be lost.

皇明地理十六圖 *Huang ming dili shiliu tu* (Imperial Ming geography: 16 maps). 南京都察院本 Volume held by the Nanjing Chief Surveillance Bureau.

大明輿地指掌圖 *Da Ming yudi zhizhang tu* (Convenient atlas of the Great Ming). 大學士桂萼 Grand Academician Gui E.[1]

歷代地理指掌圖 *Lidai dili zhizhang tu* (Convenient historical atlas). 宋學士蘇軾都御史唐順之 (Song Grand Academician Su Shi;[2] [reprinted by or held by?] Censor-in-Chief Tang Shunzhi.[3]

廣輿圖 *Guang yutu* (Expanded Atlas) 壯元羅洪先 *Zhuangyuan* [top-ranked *jinshi*] Luo Hongxian.

沿海七邊圖 *Yanhai qibian tu* (Map of the coast and seven frontiers). 侍郎錢邦彥 Vice Minister Qian Bangyan.

古今形勝圖 *Gujin xingsheng tu* (Map of ancient and present topological advantages). 都御史喻時 Censor-in-Chief Yu Shi.

天下圖誌 *Tianxia tuzhi* (Illustrated gazetteer of All Under Heaven). 閩廖世昭 Liao Shizhao of Fujian.

分野輿圖 *Fenye yutu* (Map of correlations between the Heavens and Earth). 星源游濟川 You Jichuan of Xingyuan.

嶺海輿圖 *Linghai yutu* (Map of mountains and seas) 御史姚虞 Censor Yao Yu.[4]

八閩五十一圖 *Ba Min wushi yi tu* (Fifty-one maps of the eight Min [Fujian]).

福建五澳圖 Fujian wu ao tu (Map of the five harbors of Fujian).

1 Translations of titles are, where possible, taken from Hucker's *Dictionary of Official Titles in Imperial China*. Gui presented the emperor a seventeen-part map with two *juan* of textual notes in 1529, its title recorded as 輿地指掌圖 *Yudi zhizhang tu*. (Chou Tai-chi, 'Kuei O,' p. 758). This is probably the same map or a revision of it.

2 Now attributed to Shui Anli, this text (discussed in Chapter Two) was long attributed by many to Su Shi.

3 Tang Shunzhi (1507-1560) was a friend of Luo Hongxian and the two men were dismissed from office at the same time in 1541 for writing similar memorials offending the emperor. Tang exhibited some interest in mathematics and served in various military posts, but is best known as an eccentric though talented essayist. (Huang, 'T'ang Shun-chih'). The *Lidai dili zhizhang tu* is not listed by Huang among his works; it is possible that instead of revising or republishing it he simply was the source of the copy used for reference.

4 A version of this work has been republished: Yao Yü 姚虞, *Linghai yutu* 嶺海輿圖.

兩浙海邊圖 *Liang zhe haibian tu* (Coastal map of Zhejiang). 兵部郎郭仁 Guo Ren, Gentleman of the Ministry of War.

浙東海邊圖 *Zhedong haibian tu* (Map of the Zhedong Coast). 都御史周倫 Censor-in-Chief Zhou Lun.

浙東海邊圖 *Zhedong haibian tu* (Map of the Zhedong Coast). 太守秦汴 Prefect Qin Bian.

浙海圖 *Zhe hai tu* (Map of the Zhe[jiang] Sea[coast]). 總兵俞大猷 Regional Commander Yu Dayou.[5]

浙海圖 *Zhe hai tu* (Map of the Zhe[jiang] Sea[coast]). 總兵盧鏜 Regional Commander Lu Tang.[6]

浙海圖 *Zhe hai tu* (Map of the Zhe[jiang] Sea[coast]). 都司黎秀 Li Xiu of the Regional Military Commission.

蘇松海邊圖 *Susong haibian tu* (Map of the coast at Susong). 把總X揮陳習 (Squad Leader [*obscure character*] Hui Chenxi).

三吳水利圖 Sanwu shui li tu (Map of the waterways of the Sanwu area). 長洲文嘉 Wen Jia of Changzhou.

淮楊圖 Huaiyang tu (Map of Huaiyang). 都御史唐順之 Censor-in-Chief Tang Shunzhi.

5 James F. Millinger describes Yu as 'one of the most important military commanders of the mid-16th century.' His writings, including maps, were first published in 1565 under the title *Zhengqi tang ji* 正氣堂集. (Millinger, 'Yü Ta-yu.')

6 A military leader for many years on the southeastern seaboard, Lu (c. 1520-1570) was known to contemporaries for his suppression of piracy and capture or execution of prominent coastal raiders. His record was marred by the execution of several 'pirates' who turned out to be legitimate Malaccan merchants, a mistake for which he was almost executed. Beginning in 1556 he served under Hu Zongxian, one of the compilers of the *Chouhai tubian*. (Wiethoff, 'Lu T'ang.')

Appendix 2

The *fanli* section of the *Chongzhen Jiaxing xianzhi* indicates that Jiaxing County did not previously have its own gazetteer. The following books were listed as sources for its compilation. [Clarification provided in square brackets.] They illustrate the breadth of both classical and local historical resources that could be drawn upon in this part of the empire.

春秋 Chunqiu [*Spring and Autumn Annals*, traditionally attributed to Confucius].
吳越春秋 Wu Yue Chunqiu [Attributed to Zhao Ye, Eastern Han].
越絶書 Yuejue Shu [Attributed to Yuan Kang, Eastern Han].
浙江通志 Zhejiang tongzhi [1561].
元嘉禾志 Yuan Jiahe zhi [Jiahe is a former name for Jiaxing. Compiled by Xu Shuo in the Zhiyuan era].
國朝弘治柳志 Guochao Hongzhi Liu zhi [*Jiaxing fuzhi*, comp. Liu Yan, 1493].
嘉興趙志 Jiajing Zhao zhi [*Jiaxing fu tuji*, comp. Zhao Wenhua, 1549].
萬曆沈志 Wanli Shen zhi [*Jiaxing fuzhi*, comp. Shen Yaozhong, 1600].
秀水黄志 Xiushui Huang zhi [*Xiushui xianzhi*, comp. Huang Hongxian, 1596].

Appendix 3

Translation of Ricci's preface to the world map (Chinese text follows)

The land and the sea are basically round in shape, and together they form a globe, which is situated in the celestial sphere like the yolk of a chicken's egg is suspended in the white. It has been said that the earth is '*fang* (方),' but this refers to its being fixed and immobile, not to a [square] shape.

The heavens envelop the earth and they mutually correspond. Therefore the heavens have two poles, south and north, and the earth does as well. The heavens are divided into 360 degrees, and the earth is too. In the middle of the heavens lies the [celestial] equator. South from the equator 23.5 degrees lies the Tropic of Capricorn, while 23.5 degrees to the north of the equator lies the Tropic of Cancer. Thus, the Middle Kingdom is north of the Tropic of Cancer.

When the sun moves over the equator, day and night are equal. When it moves south of the equator, the day is shorter. When it moves north of the equator, the day is longer. Thus the celestial sphere has an equator in the middle, and two circles of short day (Tropic of Capricorn) and long day (Tropic of Cancer) to the south and north, determining the way the sun circumnavigates the terrestrial globe in its orbit. Likewise, the globe of earth has three corresponding lines. However, the heavens are deep and vast as they envelop the earth and extend outward, while the earth, being situated within the heavens, is very small and constricted. In this respect alone they differ.

Investigation shows that for someone walking due north, each 250 *li* traveled brings the north pole higher by one degree and the south pole lower by one degree. Walking due south, each 250 *li* traveled brings the north pole lower by one degree and the south pole higher by one degree. This proves not only that the form of the earth is round, but furthermore that 250 *li* equal one degree. Thus one circuit of the earth, whether going east-west or north-south, in fact adds up to 90,000 *li*. Whether going north-south or east-west, they are equivalent with no objective difference.

Now, the earth has a thickness (diameter) of 28,636.36 *li*. It is inhabited on all sides, whether above or below. It is a sphere, which has no 'up' or 'down' to begin with. Is there anywhere in the world where one does not look up to see the sky? Absolutely everywhere under heaven, wherever the feet plant themselves is down, and wherever the head points, is up. Those who say that one would point up or down based on location are wrong.

Now, when I sailed by sea from the West to the Middle Kingdom I arrived at the equator and saw that the two poles, north and south, were both on the horizon with no difference in elevation on either side. Continuing to the south, when I passed the Cape of Good Hope I saw that the south [celestial] pole was 36 degrees above the horizon. Thus the Middle Kingdom and the Cape of Good Hope correspond in latitude, above and below. When I was there I saw the sky above me, I never saw it below me. Thus I say it is worthy of belief that the Earth is round and inhabited on all sides.

Using geodesy (*tianshi* 天勢) to divide the mountains and seas, there are five bands [i.e., climes] from north to south. The first is between the Tropic of Capricorn and the Tropic of Cancer. This is a very hot band, because it is close to the sun's path. The second is within the Arctic circle, and the third is within the Antarctic circle. These two places are very cold bands, because they are far from the sun's path. The fourth is between the Arctic Circle and the Tropic of Cancer, and the fifth is between the Antarctic Circle and the Tropic of Capricorn. These two are called the 'perfect' bands because they are neither too cold nor too hot, for the sun is neither too far nor too near.

Dividing the earth according to land forms, it is comprised of five large continents: Europe, Libya [i.e., Africa], Asia, North and South America, and Magellanica.[7]

As for Europe: On the south it borders the Mediterranean, on the north its frontiers are Greenland and the Ice Sea, on the east it borders the Tanais [the River Don] and the Sea of Meotides [Sea of Azov], and on the west it borders the Atlantic.

As for Libya: To the south it extends to the Cape of Good Hope, on the north it borders the Mediterranean, to the east it reaches the Western Red Sea and San Lorenzo [Madagascar], and to the west the Ocean Sea ['Oceano']. This continent is only connected to Asia by a thin isthmus under the Holy Land, and is otherwise surrounded on all sides by seas.

As for Asia: To the south it reaches Sumatra, Luzon and other islands, on the north it reaches Nova Zembla[8] and the North Sea, on the east it extends

7 Magellanica is the antipodal continent. The notion that there was a great southern continent, or antipode, balancing the northern hemisphere's land masses began long before any missions of exploration had actually visited the region. When various sailors spotted islands in the far southern seas during the fifteenth and sixteenth centuries, these were often interpreted as outcroppings of the great southern continent. Mercator included the unknown land in his famous 1569 projection, which was borrowed by Ortelius and thus found its way onto Matteo Ricci's maps. See Tooley, *Early Antarctica: a glance at the beginnings of cartographic representation of the South Polar regions* for a historical overview.

8 An island discovered by traders seeking an Arctic route to China in the 1550s.

to Japan and the Great Ming Sea, and to the west it reaches the Tanais [the River Don], the Sea of Meotides [Sea of Azov], the Western Red Sea and the Lesser Western Sea [Indian Ocean].

As for America: It is surrounded on all sides by seas; a thin isthmus of land connects the north and south.

As for Magellanica: It is completely situated in the south; one can only see the [celestial] south pole while the [celestial] north pole is hidden. It has not yet been explored and for this reason I dare not fix [its borders], only the northern edge, Greater and Lesser Java,[9] and the Straits of Magellan.

These continents are differentiated with color to help distinguish them. It would be difficult to do so with each individual country, however, as they are so numerous. Generally speaking, each continent has over one hundred countries. I originally wanted to make a globe, but although changing it to [flat] map form was difficult, I had no choice but to transform the spherical into a planar map drawn inside of circles. In order to know the real form, one must link the western and eastern seas as one.

Originally I wanted to show lines of latitude and longitude at each degree, but to reduce confusion I have made squares of ten degrees [to a side]. By means of this it is possible to place each country in its proper place.

The east-west lines of latitude measure the distance on the earth from the equator, counting northward to the North Pole and southward to the South Pole. The north-south running lines of longitude measure the breadth of the earth, taking the Fortunate Islands as the meridian point, running in intervals of ten degrees until, at 360 degrees, one reaches the starting point again. For example, if observations show that Nanjing is 32 degrees above the equator and 128 degrees east of the Fortunate Islands, we can place it in its proper position [on the map].

All countries north of the equator are governed by the north pole and are thus in the northern hemisphere, while those countries south of the equator are governed by the south pole and are thus in the southern hemisphere. Thus the folly of the Buddhists is evident in their claim that the Middle Kingdom is in southern Jambudvīpa, and in their calculations of the altitude of Mount Sumeru.[10]

The lines of latitude also serve to express the altitude of the celestial poles. The number of degrees from the equator and the number of degrees by which

9 Note that the great southern continent was thought to extend as far north as these Pacific islands.

10 Buddhist cosmology held that the known world was centered on an enormous continent, Jambudvīpa, and that an axial mountain called Sumeru was of enormous height.

the pole rises above the horizon are equal. But in the south one counts the degrees by which the south pole rises above horizon, in the north, one counts the degrees by which the north pole rises above horizon. Thus, when it is seen that the capital is 40 degrees north of the equator, one knows that the elevation of the north [celestial] pole is 40 degrees [as observed from] the capital. When it is seen that the Cape of Good Hope is 36 degrees south of the equator, one knows that the elevation of the south [celestial] pole is 36 degrees [as observed from] the Cape of Good Hope. All places of the same latitude have the same [celestial] polar altitude. It follows that the coldness or heat of the four seasons is of equal intensity [in all places of that latitude]. If two places are equidistant from the equator but one is south of it and one is north, their seasons and the length of the days are equal, but temporally reversed, so that summer here is winter there. The length of night or day varies according to distance from the equator. I have fashioned at the edge of the map a chart, marked at every 5-degree interval; from the length of the day and night, whether in the east or west, above or below, in any given case you can tell the precise distance from the equator.

A line of longitude indicates the amount of time that separates two places. The sun makes one revolution in a day, so in each *chen* [period of two hours] it moves 30 degrees. Thus two places separated by 30 degrees are one *chen* apart. Thus if we see that the Jurchens are 140 degrees from the Fortunate Islands, while Burma is 110 degrees, then clearly we see that they are separated by one *chen*. So, when for the Jurchens it is the *mao chen* (5-7 a.m.), in Burma it is the *yin chen* (3-5 a.m.), and so on.

If two places differ by six *chen*, their day and night are opposite. If two people in such a situation are at the same number of degrees from the equator, but one south and one north, then they have the soles of their feet facing each other. Thus Nanjing is 32 degrees north of the equator and 128 degrees from the Fortunate Islands, while Mapazo[11] in South America is 32 degrees south of the equator and 308 degrees from the Fortunate Islands, so people in Nanjing and Mapazo are walking with the soles of their feet facing one another. From this we see that two places of the same longitude will witness a solar or lunar eclipse at the same *chen* and the same hour.

This is the general outline; the particulars can be found on the map.

Matteo Ricci

11 In the Amazon.

地與海本是圓形，而合為一球,居天球之中，誠如雞子黃在青內。有謂地為方者，乃語其定而不移之性,非語其形體也。天既包地，則彼此相應，故天有南北二極，地亦有之。天分三百六十度，地亦同之。天中有赤道，自赤道而南二十三度半為南道，赤道而北二十三度半為北道。按中國在赤道之北。日行赤道,則晝夜平；行南道則晝短；行北道則晝長。故天球有晝夜平圈列於中,晝短晝長二圈列於南北,以著日行之界。地球亦設三圈對於下焉。但天包地外為甚大，其度廣；地處天中為甚小,其度狹。此其差異者耳。查得直行北方者，每路二百五十里，覺北極出高一度，南極入低一度。直行南方者,每路二百五十里，覺北極入低一度，南極出高一度。則不特審地形果圜。而並徵地之每一度廣二百五十里。則地之東西南北各一週有九萬里實數也。是南北與東西數相等,而不容異也。夫地厚二萬八千六百三十六里零百分里之三十六分,上下四旁皆生齒所居,渾淪一球，原無上下。蓋在天之內，何瞻非天，總六合內，凡足所佇即為下，凡首所向即為上；其專以身之所居分上下者，未然也。且予自大西浮海入中國，至晝夜平線，已見南北二極皆在平地，略無高低；道轉而南過大浪山，已見南極出地三十六度，則大浪山與中國上下相為對待矣。而吾彼時只仰天在上，未視之在下也。故謂地形圓而周圍皆生齒者，信然矣。以天勢分山海，自北而南為五帶；一在晝長,晝短二圈之間,其地甚熱,帶近日輪故也；二在北極圈之內；三在南極圈之內，此二處地居甚冷帶遠日輪故也；四在北極晝長二圈之間，五在南極晝短二圈之間，此二地皆謂之正帶，不甚冷熱，日輪不遠不近故也。又以地勢分輿地為五大州：曰歐邏巴，曰利未亞，曰亞細亞，曰南北亞墨利加，曰墨瓦蠟泥加。若歐邏巴者，南至地中海，北至臥蘭的亞及冰海，東至大乃河、墨何的湖大海，西至大西洋。若利未亞者，南至大浪山，北至地中海，東至西紅海，仙勞冷祖島，西至河摺亞諾滄。即此州只以聖地之下微路與亞細亞相聯，其餘全為四海所圍。若亞細亞者，南至蘇門荅臘，呂宋等島，北至新增，白臘及北海，東至日本島，大明海，西至大乃河，墨阿的湖大海，西紅海，小西洋。若亞墨利加者，全為四海所圍，南北以微地相聯。若墨瓦蠟泥加者，盡在南方。惟見南極出地，而北極恆藏焉。其界未審何如，故未敢訂之。惟其北邊，與大小瓜哇及墨瓦蠟泥峽為境也。其各州之界，當以五色別之，令其便覽。各國繁夥難悉，大約各州俱有百余國。原宜作圓球，以其入圖不便，不得不易圓為平，反圈為線耳。欲知其形，必須相合，連東西二海為一片可也。其經緯線，本宜每度畫之，今且惟每十度為一方，以免雜亂。依是可分置各國於其所。東西緯線天下之長，自晝夜平線為中而起，上數至北極，下數至南極。南北經線數天下之寬，自福島起為一十度，至三百六十度復相接焉。試如察得南京離中線以上三十二度，離福島以東一百二十八度，則安之於其所也。凡地在中線以上至北極，則實為北方；凡在中線以下則實為南方焉。釋氏謂中國在南瞻部洲，並計須彌山出入地數，其謬可知也。又用緯線以著各極出地幾何。蓋地離晝夜平線度數，與極出地度數相等。但在南方則著南極出地之數，在北方則著北極出地之數也。故視京師隔中線以北四十度，則知京師北極高四十度也；視大浪山隔中線以南三十六度，則知大浪山南極高三十六度也。凡同緯之地，其極出地數同，則四季寒暑同態焉。若兩處離中線度數相同，但一離于南一離於北，其四季並晝夜刻數均同，惟時相反，此之夏為彼之冬耳。其長晝，長夜，離中線愈遠，則其長愈多。余為式，以記于圖邊，每五度其晝夜長何如，則東西上下隔中線數一，則皆可通用焉。用經線以定兩處相離幾何辰也。蓋日輪一日作一週，則每辰行三十度，而兩處相違三十度，並謂差一辰。故視女直島離福島一百四十度，而緬甸離一百一十度，則明女直於緬甸差一辰，而凡女直為卯時，緬方為寅時也。其餘倣是焉。設差六辰,則兩處晝夜相反焉。如所離中線度數又同，而差南北，則兩地人對足底反行，故南京離中線以北三十二度，離福島一百二十八度，而南亞墨利加之瑪八作離中線以南三十二度，離福島三百又零八度，則南京於瑪八作人相反足底行矣。從此可曉同經線處並同辰而同時見日月蝕矣。此其大略也，其詳則備於圖云。

利瑪竇撰。

Appendix 4

Translation of the *Sancai tuhui's* adaptation of the Ricci preface (Chinese text follows)

The land and the sea are basically round in shape, and together they form a globe, which is situated in the celestial sphere like the yolk of a chicken's egg is suspended in the white. It has been said that the earth is '*fang* (方),' but this refers to its being fixed and immobile, not to a [square] shape.

The heavens envelop the earth and they mutually correspond. Therefore the heavens have two poles, south and north, and the earth does as well. The heavens are divided into 360 degrees, and the earth is too. In the middle of the heavens lies the equator. South from the equator 23.5 degrees lies [the Tropic of Capricorn, while 23.5 degrees to the north of the equator lies][12] the Tropic of Cancer. Thus, the Middle Kingdom is north of the Tropic of Cancer.

When the sun moves over the equator, day and night are equal. When it moves south of the equator, the day is shorter. When it moves north of the equator, the day is longer. Thus the celestial sphere has an equator in the middle, and two circles of short day (Tropic of Capricorn) and long day (Tropic of Cancer) to the south and north, determining the way the sun circumnavigates the terrestrial globe in its orbit. Likewise, the globe of earth has three corresponding lines.

Investigation shows that for someone walking due north, each 250 *li* traveled brings the north pole higher by one degree and the south pole lower by one degree. For one walking due south, each 250 *li* traveled brings the north pole lower by one degree and the south pole higher by one degree. This proves not only that the form of the earth is round, but furthermore that 250 *li* equal one degree. Thus one circuit of the earth, whether going east-west or north-south, really adds up to 90,000 *li*. Whether going north-south or east-west, they are equivalent with no objective difference.

Now, the earth has a thickness (diameter) of 28,636.36 *li*.[13] It is inhabited on all sides, whether above or below. To put it simplistically, a sphere originally has no up or down. Is there anywhere in the world where one does not look up to see the sky? Absolutely everywhere under heaven, wherever

12 The words inside the brackets are from the Ricci original, but have been omitted in a copyist's error.

13 The word 天 at the beginning of this sentence is an error for 夫, which appears in Ricci's version.

the feet plant themselves is down, and wherever the head points is up. It is impossible to distinguish up or down based on where one is. Now, when one sails by sea from the West to the Middle Kingdom one arrives at the equator and sees that the two poles, north and south, are both on the horizon with no difference in elevation on either side. Continuing to the south, when one passes the Cape of Good Hope one sees that the south pole (of the heavens) is 32 degrees above the horizon.[14] Thus the Middle Kingdom and the Cape of Good Hope correspond in latitude, above and below.[15]

Using geodesy (*tianshi* 天勢) to divide the mountains and seas, there are five bands from north to south. The first is between the Tropic of Capricorn and the Tropic of Cancer. This is a very hot band, because it is close to the sun's path. The second is within the Arctic circle, and the third is within the Antarctic circle. These two places are very cold bands, because they are far from the sun's path. The fourth is between the Arctic Circle and the Tropic of Cancer, and the fifth is between the Antarctic Circle and the Tropic of Capricorn. These two are called the 'perfect' bands because they are neither too cold nor too hot, for the sun is neither too far nor too near.

Dividing the earth according to landforms, it is comprised of six large continents: Europe, Libya [i.e., Africa], Asia, North America, South America, and Magellanica.[16]

Europe borders the Mediterranean on the south, on the north its frontiers are Greenland and the Ice Sea, on the east it borders the Tanais [the River Don] and the Sea of Meotides [Sea of Azov], and on the west it borders the Atlantic.

Libya extends south to the Cape of Good Hope, on the north it borders the Mediterranean, to the east it reaches the Western Red Sea[17] and San Lorenzo [Madagascar], and to the west the Ocean Sea [Mare Oceano]. This continent is only connected to Asia by a thin isthmus under the Holy Land, and is otherwise surrounded on all sides by seas.

Asia reaches Sumatra, Luzon, and other islands in the south, on the north it reaches Nova Zembla and the North Sea, on the east it extends to Japan and the Great Ming Sea, and to the west it reaches the Tanais [the River Don], the Sea of Meotides [Sea of Azov], the Western Red Sea and the Lesser Western Sea [Indian Ocean].

14 This is 36 degrees in Ricci's version.

15 Ricci's line 'When I was there I saw the sky above me, I never saw it below me. Thus I say it is worthy of belief that the Earth is round and inhabited on all sides' has been removed.

16 Note that there are six continents rather than Ricci's five, because the Americas have been split into two continents.

17 紅 is replaced with 江, an error also seen in the Japanese *Sanzai zue* version.

The two Americas are surrounded on all sides by seas; a thin isthmus of land connects the north and south.

Magellanica is completely in the south; one can only see the [celestial] south pole while the [celestial] north pole is always hidden. It has not yet been explored and for this reason I dare not fix [its borders], only the northern edge, Greater and Lesser Java, and the Straits of Magellan.

Generally speaking, each continent has over one hundred countries.[18]

The spherical globe has been rendered into a flat circle. The western and eastern seas are [actually] joined together as one.

Squares of ten degrees [each way] are made of latitude and longitude lines. The east-west lines of latitude measure the distance on the earth from the equator, counting northward to the North Pole and southward to the South Pole. The north-south running lines of longitude measure the breadth of the earth, taking the Fortunate Islands as the meridian point, running in intervals of ten degrees until, at 360 degrees, one reaches the starting point again.[19]

All countries north of the equator are governed by the north pole and are thus in the northern hemisphere, while those countries south of the equator are governed by the south pole and are thus in the southern hemisphere. The Buddhists' talk of southern Jambudvīpa and their calculations of the altitude of Mount Sumeru are foolish.[20]

The lines of latitude also serve to express the altitude of the celestial poles. The number of degrees from the equator and the number of degrees by which the pole rises above the horizon are equal. But in the south one counts the degrees by which the south pole rises above horizon, and in the north it is also thus. Thus, when it is seen that the capital is 40 degrees north of the equator, one knows that the elevation of the north [celestial] pole is 40 degrees [as observed from] the capital. Nanjing is 32 degrees north of the equator and 130 degrees east of the Fortunate Islands.[21] When it is seen that the Cape of Good Hope is 36 degrees south of the equator, one knows that the elevation of the south [celestial] pole is 36 degrees [as observed from] the Cape of Good Hope. All places of the same latitude have the same [celestial] polar altitude. Thus, the coldness or heat of the four seasons is of equal intensity [in all places of that latitude]. If two places are equidistant

18 The note about differentiating the continents with color has been removed, as the book was printed in black ink and not colored by hand.

19 The example of Nanjing's location is omitted.

20 The reference to China's location on the Buddhist continent is omitted.

21 This line has been interpolated from further down in the passage, and the longitudinal figure has been rounded upwards.

from the equator but one is south of it and one is north, their seasons and the length of the days are equal, but temporally reversed, so that summer here is winter there. Two places of the same longitude will witness a solar or lunar eclipse at the same *chen* and the same hour. The time of day is the same at any point along the same line of longitude, regardless of its distance from the equator.

A line of longitude indicates the amount of time that separates two places. The sun makes one revolution in a day, so in each *chen* [period of two hours] it moves 30 degrees. Thus two places separated by 30 degrees are one *chen* apart. Thus if we see that the Jurchens are 140 degrees from the Fortunate Islands, while Burma is 110 degrees, then clearly we see that they are separated by one chen. So, when for the Jurchens it is the *mao* 卯 *chen* (5-7 a.m.), in Burma it is the *yin* 寅 *chen* (3-5 a.m.), and so on.

For the details see the full map; this is [merely] a sketch version for reference.

地與海本是圓形，而仝為一球，居天球之中，如鷄子黃在青内。有謂地為方者，乃語其定而不移之性，非語其形體也。天既包地，則彼此相應，故天有南北二極，地亦有之。天分三百六十度，地亦同之。天中有赤道，自赤道而南二十三度半為北道，據中国在北道之北。日行赤道則晝夜平，行南道則晝短，行北則晝長。故天球有晝夜平圈列于中，晝短晝長二圈列于南北以著日行之界。地球亦有三圈對于下也。

查得直行北方者，每路二百五十里覺北極出高一度，南極入低一度。直行南方者，每路二百五十里覺北極入低一度，南極出高一度。則不特審地形果圓，而並徵地每度廣二百五十里。則地之東西南北各一週有九萬里實數也。是南北與東西數相等不異。天地厚二萬八千六百三十六里零三十六丈。

上下四旁皆生齒所居。渾淪一球原無上下。蓋在天之内，何瞻非天，總六合内，凡足所竚即為下，凡首所向即為上，不專以身之所居分上下也。予自太西浮海入中國，至晝夜平線，已見南北二極皆在平地，畧無高低，道轉而南過大浪山，已見南極出地三十二度，則大浪山與中國上下相為對待矣。

以天勢分山海，自北方而南為五帶。一在晝長，晝短二圈之間，其地甚熱，帶近日輪故也；二在北極圈之内，三在南極圈之内此，二處地俱甚冷帶，遠日輪故也；四在北極晝長二圈之間，五在南極晝短二圈之間，此二地皆謂之正帶，不甚冷熱，日輪不遠不近故也。又以地勢分輿地為六大州：曰歐邏巴，曰利未亞，曰亞細亞，曰北亞墨利加，曰南亞墨利加，曰墨瓦臘泥加。

歐邏巴南至地中海，北至臥蘭的亞及冰海，東至大乃河，墨河的湖大海，西至大西洋。利未亞南至大浪山，北至地中海，東至西江海，仙勞泠祖島，西至河摺亞諾滄。即此州只以聖地之下微路與亞細亞相聯，其餘全為四海所圍。亞細亞南至沙馬大臘，呂宋等島，北至新增，白臘及北海，東至日本島，大明海，西至大乃河，墨河的湖大海，西江海，小西洋。 南北亞墨利加全為四海所圍，南北以微地相聯。墨瓦臘泥加盡在南方。惟見南極出地，而北極常藏焉。其界未審何如，故不敢訂之。惟其北邊，與大小爪哇及墨瓦臘泥峽為境也。大約各州有百餘國。

圓球平圈為線合連東西二海形為一方。每十度一方為經緯線，東西緯線數天下之長，自晝夜平線為中而起，上數至北極，下數至南極。南北經線數天下之寬，自福島起為一十度，三百六十度復相接。 凡地在中線以上主北極則實爲北方，中線以下主南極實爲南方也。釋氏謂南贍部洲，並計須彌山出入地數，謬也。

又用緯線以着各極出地幾何。蓋地離晝夜平線度數，與極出地度數相等。但在南則若南極出地之數，在北方亦然。故視北京隔中線以北四十度，則知京師北極高四十度也。南京離中線以上三十二度，離福島以東一百三十度也。視大浪山隔中線以南三十六度，則知大浪山南極高三十六度也。凡同緯之地，其極出地數同，則四季寒暑同態焉。若兩處離中線度數相同，但一離南一離北，其四季晝夜刻數均同，惟時相反，耳此之夏為彼之冬也。

同經線處並同辰而同時見日月蝕矣。晝夜時辰遠近以中線度之不差。蓋日輪一日作一週，則每辰行三十度，而兩處相違三十度，謂差一辰。故女直離福島一百四十度而緬國離一百一十度，則女直于緬國差一辰，凡女直之卯為緬之寅時也。餘倣是詳見全圖玆畧載以覽歿。

自古帝王之御世者必一統天下而後為盛。義農以上疆理之制，世遠莫之詳矣。其見諸載籍者，謂黃帝畫野分州，得百里之國萬區。帝嚳創制九州，統領萬國。堯遭洪水，天下分絶為十二州，使禹平治水土，更制九州列五服。禹繼唐虞之盛，塗山之會，執玉帛者萬國，而四百年間，遞相兼併。逮商受命，其能存者纔三千餘國，亦為九州，分統天下。及周克商，尚有千八百國，而分天下為九畿。至成王時，仍曰九州。厥後諸侯相呑，列國耗盡，陵夷至於戰國，天下分而為七。秦並六國，罷侯置守，分天下為四十郡。漢因秦制，加置郡國。武帝攘胡開越，四履彌廣，分天下為十三州郡，皆置刺史。既而三國鼎峙。至晉始合為一置州凡十有九。未幾，南北分裂，至隋復合為一，盡廢郡為州。唐承隋後，置州愈多。貞觀初，分天下為十道，開元中又增至十五道。宋承五季，削平偏據，至道末，分天下為十五路，宣和中，又增至二十六路。元氏以夷狄入主華夏，内立中書省一以領。雖過於前，而東南島夷則未盡附，惟我皇明誕膺天命統一華夷。幅員之廣東盡遼左，西極流沙，南越海表，北抵沙漠。四極八荒靡不來庭。而疆理之制則以京畿府州直隸六部，天下分爲十三步政司，曰山西，曰山東，曰河南，曰陝西，曰浙江，曰江西，曰湖廣，曰四川，曰福建，曰廣東，曰廣西，曰雲南，曰貴州，以統諸府州縣。而都司衛所則錯置於其間，以為防禦。總之為府一百四十九，為州一百二十八，位縣一千一百五，而邊陲之地，都司衛所及宣慰，招討，宣撫，安撫等司，與夫四夷受官封，執臣禮者，皆以次具載於志焉。顧昔周官詔觀事則有志，詔地事則有圖。故今復為圖分置於兩畿各布政司之前，又為天下總圖於首。披圖而觀，庶天下疆域之大了然在目，而我皇明一統之盛冠乎古今者垂之萬世有足徵云。

Appendix 5

Comparison of categories of information given in the *Da Ming yitong zhi* and the *Sŭngnam*

Note that the *Da Ming yitong zhi* is divided into prefectures, while the *Sŭngnam* is divided into much smaller counties. Thus any given *Sŭngnam* entry is unlikely to include all the features listed. Both the *Da Ming yitong zhi* and the *Sŭngnam* also list more specialized features for atypical locales, which are not included here. From this list can be grasped at a glance the *Sŭngnam*'s borrowing of an administrative focus from the *Da Ming yitong zhi*, while not entirely shedding the literary subject matter of its earlier version influenced by the *Fangyu shenglan*.

Da Ming yitong zhi

健置沿革 Establishment of and changes in administrative boundaries
郡名 Names of commanderies
形勝 Topography (quotes from literary and administrative works)
風俗 Local customs, cultural traits (ditto)
山川 Mountains and rivers
土產 Local products
公署 Official offices
學校 Schools
書院 Academies
宮室 Mansions and prominent buildings
關梁 Passes and bridges
寺觀 Buddhist and Daoist temples
祠廟 Ancestral halls, Confucian temples
陵墓 Tombs
古蹟 Ruins
名官 Biographies of famous officials
流寓 Sojourners from elsewhere
人物 Biographies of local personages
列女 Biographies of virtuous women
仙釋 Biographies of Daoist and Buddhist notables

Sŭngnam

健置沿革 Establishment of and changes in administrative boundaries, with subdivision called 官員 (listing official positions)
郡名 Names of commanderies
姓氏 Prominent surnames
形勝 Topography (quotes from literary and administrative works)
風俗 Local customs, cultural traits (ditto)
山川 Mountains and rivers
土產 Local products
城郭 Walled fortifications
烽燧 Beacon stations
宮室 Mansions and prominent buildings
樓亭 Towers and pavilions
學校 Schools

驛院 Post stations
橋梁 Bridges
佛宇 Buddhist temples
祠廟 Ancestral halls, Confucian temples
塚墓 / 陵墓 Tombs
古蹟 Ruins
名官 Biographies of famous officials
人物 Biographies of local personages
寓居 Resident sojourners (sometimes as subcategory of 人物)
列女 Biographies of virtuous women
題詠 Verses on themes

Bibliography

Note: In some cases I have used Romanization inconsistent with standard systems when representing the names of Chinese, Korean, or Japanese scholars who use a different Romanization in their own English publications; I have adhered to their own use in English.

Primary sources

Ai Nanying 艾南英. *Yugong tuzhu* 禹貢圖註, Daoguang 11. Facsimile edition in *Lidai Yugong wenxian jicheng* 歷代禹貢文獻集成 (7 volumes), vol. 3: 1017-1062. Xi'an: Xi'an ditu chubanshe, 2006.

Aleni, Giulio. *Zhifang wai ji* 職方外紀. [China: s.n.], Chongzhen period.

Bing bu (Board of War) 兵部. *Jiubian tushuo* 九邊圖說. S.n.: Longqing year 3 [1569].

Chen Shikai 陳師凱. *Shu Cai shi zhuanpang tong* 書蔡氏傳旁通. Wenyange facsimile edition. [Taibei]: Taiwan shangwu yinshuguan, [1983].

Chen Yi 陳沂. *Jinling gujin tukao* 金陵古今圖考. S.n., Zhengde 10 [1515].

Chen Yuanjing 陳元靚, *Shilin guangji* 事林廣記. S.n., Yuan period.

Chen Zilong 陳子龍. Preface to Wu Guofu 吳國輔, *Gujin yuditu* 古今輿地圖, Chongzhen 11 (1638). Facsimile edition with title as *Jingu yuditu* 今古輿地圖. In *Siku jinhui shu congkan bubian* 四庫禁燬書叢刊補編, Shi bu 史部 (311 volumes), vol. 26. Beijing: Beijing chubanshe, 2005.

Chen Zushou 陳組綬. *Huang Ming zhifang ditu* 皇明職方地圖. S.n., Chongzhen 9 [1636].

Cheng Dachang 程大昌. *Yugong lun, shanchuan dilitu* 禹貢論山川地理圖. Facsimile of Song edition. Beijing: Zhonghua shuju 中華書局: 1985.

Chosŏn wangjo sillok 朝鮮王朝實錄. Seoul: T'amgudang, 1984.

Daoxuan 道宣, with introduction by Fan Xiangyong 范祥雍. *Shijia fangzhi* 释迦方誌. Beijing: Zhonghua shuju 中華書局, 1983.

Fan Lai 范淶, and Zhang Huang 章潢. *Wanli xin xiu Nanchang fuzhi* 萬曆新修南昌府志. Facsimile of Wanli 16 edition. In *Riben cang Zhongguo hanjian difangzhi congkan* 日本藏中國罕見地方志叢刊 (32 volumes), vol. 6. Beijing: Shumu wenxian chubanshe 書目文獻出版社, [1990-1991].

Fan Zhongyi 范中義. *Qi Jiguang bingfa* 戚繼光兵法. Beijing: Shishi Chubanshe, 1998.

Fu Yin 傅寅. *Yugong shuoduan* 禹貢說斷. Wenyange facsimile edition. [Taibei]: Taiwan shangwu yinshuguan 臺灣商務印書館, [1983].

Han Paek-kyŏm 韓百謙. *Tongguk chiriji* 東國地理誌. Facsimile edition. Seoul: Ilchogak, 1982.

Hong Mai 洪邁. *Rong zhai sui bi* 容齋隨筆. Shanghai: Shanghai guji chubanshe, 1978.

Hu Yaoguang 許瑤光, and Wu Yangxian 吳仰賢, eds. *Jiaxing fuzhi* 嘉興府志. Jiaxing: Guangxu 4 [1878].

Hu Zhenheng 胡震亨, ed. *Haiyan xian tujing* 海鹽縣圖經. Haiyan: Tianqi 天啓 2 [1622].

Huang Bian 黄汴. *Yitong lucheng tuji* 一統路程圖記. Facsimile reprint in *Siku quanshu cunmu congshu* 四庫全書存目叢書, Shi bu 史部 (292 volumes), vol. 166: 481-562. Jinan: Qi Lu shushe chubanshe, 1997.

Huang Jinsheng 黄金聲, ed. *Daoguang Jinhua xianzhi* 道光金華縣志. S.n.: Daoguang 3 [1823].

Huanyu tongqu 寰宇通衢. Facsimile reprint in *Siku quanshu cunmu congshu* 四庫全書存目叢書, Shi bu 史部 (292 volumes), vol. 166: 164-237. Jinan: Qi Lu shushe chubanshe, 1997.

Ishikawa Ryūsen 石川流宣. *Bankoku sōkaizu* 萬國總界圖. Edo: Suhara Mohē, Hōei 5 [1708].

Ji Zengyun 嵇曾筠, ed. *Zhejiang tongzhi* 浙江通志. Facsimile reprint in *Yingyin Wenyuange Siku quanshu*, vols 519-526. Taibei: Taiwan shangwu yinshuguan, 1983.

Jueluo Shilin 覺羅石麟, and Chu Dawen 儲大文, eds. *Shanxi Tongzhi* 山西通志. Facsimile reprint in *Yingyin Wenyuange Siku quanshu*, vols 542-550. Taibei: Taiwan shangwu yinshuguan, 1983.

Kim, Chŏng-ho 金正浩. *Ch'ŏnggudo* 青邱圖. Seoul: Minjok Munhwa Ch'ujinhoe, 민족문화추진회, 1971.

—. *Taedong yŏjido wŏndo* 大東輿地圖原圖. Seoul: Kyŏnghŭi Taehakkyo Pusŏl Han'guk Chŏnt'ong Munhwa Yŏn'guso 慶熙大學校附設韓國傳統文化研究所, 1977.

Li Chengxun 李承勛. *Chaoxian tushuo* 朝鮮圖說. S.n., Autumn of 1600 (萬曆庚子季秋).

—. *Jixiao xinshu* 紀效新書. Republication of Qi Jiguang's edition, Wanli 16 [1588].

—. *Mingjian ji* 名劍記. In the *Shuo fu* 說郛, vol. 158. Shunzhi year 3 [1646], Xu *juan* 36.

—. *Xu Mingma ji* 續名馬記. Facsimile of Ming edition, in *Xuxiu Siku quanshu* 續修四庫全書, Zi bu (370 volumes), vol. 1119: 319-338. Shanghai: Shanghai guji chubanshe: [1995-1999].

Li Lian.李濂. *Bianjing yiji zhi* 汴京遺蹟志. Facsimile of the Wenyange edition. [Taibei]: Taiwan shangwu yinshuguan 臺灣商務印書館, [1983].

Li Xian 李賢, ed. *Da Ming yitong zhi* 大明一統志. Nei fu, Ming Shuntian 5 [1461].

—. *Da Ming yitong zhi* 大明一統志. Facsimile edition. Xi'an: San Qin chubanshe 三秦出版社, 1990.

Li Yangong 李言恭. *Riben kao* 日本考. Facsimile reprint in *Siku quanshu cunmu congshu* 四庫全書存目叢書, Shi bu 史部 (292 volumes), vol. 255: 485-569. Jinan: Qi Lu shushe chubanshe, 1997.

Liu Xu 劉昫, ed. *Jiu Tang shu* 舊唐書. Beijing: Zhonghua shuju 中華書局, 1975.

Liu Yan 柳琰, compiler. *Jiaxing fuzhi* 嘉興府志, Hongzhi 5 (1493). Facsimile reprint in *Siku quanshu cunmu congshu* 四庫全書存目叢書, Shi bu 史部 (292 volumes), vol. 179: 1-426. Jinan: Qi Lu shushe chubanshe, 1997.

Luo Hongxian 羅洪先. *Guang yutu* 廣輿圖. [China]: Han Jun'en, Du Si; Ming Jiajing 45 [1566].

—. *Guang yutu* 廣輿圖. Facsimile reprint of 1799 edition. Taibei: Xuehai chubanshe 學海出版社, 1969.

—. *Guang yutu* 廣輿圖. Facsimile reprint of undated Jiajing edition in *Xuxiu siku quanshu* 續修四庫全書, Shi bu 史部 (670 volumes), vol. 586: 411-528. Shanghai: Shanghai guji chubanshe, 2002.

Luo Kai 羅炌. *Chongzhen Jiaxing xianzhi* 崇禎嘉興縣志. Facsimile of 1637 edition. Beijing: Shumu wenxian chubanshe 書目文獻出版社, 1991.

Mao Huang 毛晃. *Yugong zhinan* 禹貢指南. Wenyange facsimile edition. [Taibei]: Taiwan shangwu yinshuguan, [1983].

Mao Ruizheng 茅瑞徵. *Dongyi kaolue* 東夷考略. Facsimile edition. Taibei: Guangwen shuju, 1977.

—. *Yugong huishu* 禹貢匯疏. Facsimile reprint in *Lidai Yugong wenxian jicheng* 歷代禹貢文獻集成 (7 volumes), vol. 2: 465-762. Xi'an: Xi'an ditu chubanshe, 2006.

Matsushita Kenrin 松下見林. *Ishō Nihon den* 異稱日本傳. Sesshū 攝州: Morita Shōtarō 毛利田庄太郎, Genroku 6 [1693].

Ni Yue 倪岳. *Qingxi mangao* 青谿漫稿. Facsimile of Wenyange edition. [Taibei]: Taiwan shangwu yinshuguan 臺灣商務印書館, [1983].

No Sasin 盧思愼, ed. *Sinjŭng Tongguk yŏji sŭngnam* 新增東國輿地勝覽. Seoul: Tongguk Munhwasa facsimile edition, Tan'gi 4291 [1958].

—. *Sinjŭng Tongguk yŏji sŭngnam.* P'yŏngyang: Chosŏn Minjujuŭi Inmin Konghwaguk Kwahagwŏn Ch'ulp'ansa, 1959.

—. *Sinjŭng Tongguk yŏji sŭngnam.* Hansŏng, Mallyŏk 39 [1611].

Pan Shaoyi 潘紹詒 and Zhou Rongchun 周榮椿, eds. *Guangxu Chuzhou fuzhi* 光緒處州府志. Guangxu year 3 [1877].

Peng Dayi 彭大翼. *Shantang sikao* 山堂肆考. Facsimile of Wenyange edition. [Taibei]: Taiwan shangwu yinshuguan 臺灣商務印書館, [1983].

Purchas, Samuel. *Hakluytus Posthumus, or, Purchas his Pilgrimes: Contayning a History of the World in Sea Voyages and Lande Travells by Englishmen and Others.* Facsimile of 1625 edition. Glasgow: J. MacLehose and Sons, 1905-1907.

Qian Yikai 錢以塏. *Jiaxing fuzhi* 嘉興府志. Kangxi 60 [1721].

Ren Chao 仁潮, *Fajie anli tu* 法界安立圖. Facsimile edition of undated Qing copy. Taibei: Xin Wenfeng chuban gufen youxian gongsi 新文豐出版股份有限公司, 1977.

Shui Anli 稅安禮. *Lidai dili zhizhang tu* 歷代地理指掌圖. Facsimile edition of Song edition held at the Tōyō Bunko, published as *Songben lidai dili Zhizhang tu* 宋

本歷代地理指掌圖 with prefaces by Tan Qixiang and Cao Wanru. Shanghai: Shanghai guji chubanshe 上海古籍出版社, 1989.

Sim Paek-kang, ed. *Tangun ko kirok sagong*. Seoul: Minjok Munhwa Yŏn'guwŏn, 2001.

Terajima Ryōan 寺島良安. *Wa-Kan sansai zue* 和漢三才圖會. Facsimile of 1713 edition. Tokyo: Yoshikawa Kōbunkan 吉川弘文館, Meiji 39 [1906].

Wang Guanglu 王光魯. *Yueshi yueshu* 閱史約書. Facsimile reprint of 1643 edition in *Siku quanshu cunmu congshu* 四庫全書存目叢書, Shi bu 史部 (292 volumes), vol. 32: 741-898. Jinan: Qi Lu shushe chubanshe, 1997.

Wang Qi 王圻, and Wang Siyi 王思義. ***Sancai tuhui*** 三才圖繪. [China]: Wang Erbin, Chongzhen reign (1628-1644).

—. *Sancai tuhui*. Facsimile of c. 1609 edition. Shanghai: Shanghai guji chubanshe 上海古籍出版社, 1988.

Wang Shu 王澍. *Yugong pu* 禹貢譜. Kangxi 46 [1707]. Facsimile reprint in *Lidai Yugong wenxian jicheng* 歷代禹貢文獻集成 (7 volumes), vol. 3: 1063-1103. Xi'an: Xi'an ditu chubanshe, 2006.

Wang Tingzeng 王廷曾, ed. *Yiwu xianzhi* 義烏縣志. Yiwu: 1692.

Wang Yingdian 王應電. *Zhouli zhuan* 周禮傳. Facsimile of Wenyange edition. [Taibei]: Taiwan shangwu yinshuguan 臺灣商務印書館, [1983].

Xia Yunyi 夏允彝. *Yugong gujin hezhu* 禹貢古今合注. Facsimile reprint in *Lidai Yugong wenxian jicheng* 歷代禹貢文獻集成 (7 volumes), vol. 2: 763-922. Xi'an: Xi'an ditu chubanshe, 2006.

Xu Qilong 徐企龍. *Wuche wanbao quanshu* 五車萬寶全書. Chūgoku nichiyō ruisho shūsei 中国日用類書集成, vols 8-9. Tōkyō: Kyūko Shoin, Heisei 13 [2001].

Xu Sanyou 徐三友. *Wuche bajin* 五車拔錦. Chūgoku nichiyō ruisho shūsei 中国日用類書集成, vols 1-2. Tōkyō: Kyūko Shoin, Heisei 11 [1999].

Yao Yü 姚虞. *Linghai yutu* 嶺海輿圖. Congshu jicheng chubian 叢書集成初編. Shanghai: Shangwu yinshuguan, Minguo 26 [1937].

Yao Zhong 堯中, ed. *Jiaxing fuzhi* 嘉興府志. Ming Wanli 萬曆 gengzi 庚子 [year 28, 1600].

Yi Ŭr-ho 李乙浩, ed. *Haedong chegukki / Kanyang nok* 海東諸國記 / 看羊錄. Seoul: Taeyang Sŏchŏk, 1972.

Yu Shangling 于尚齡. *Jiaxing fuzhi* 嘉興府志. Daoguang year 20 [1840].

Yu Xiangdou 余象斗. *Santai wanyong zhenzong* 三台萬用正宗. Chūgoku nichiyō ruisho shūsei 中国日用類書集成, vols 3-5. Tōkyō: Kyūko Shoin, Heisei 12 [2000].

Yue Shi 樂史. *Taiping huanyu ji* 太平寰宇記. Hangzhou: Gujiu shudian, 1982.

Zhang Ao 張鏊. *Jiaoli jiaoping shilue* 交黎剿平事略. Facsimile reprint of 1551 edition in *Zhongguo bianjiang shi di congshu chubian* 中國邊彊史地叢書初編 (12 volumes), vol. 5: 875-1336. Taibei, [s.n.], 1969.

Zhang Cheng 張誠, and Wang Heng 王恒, eds. *Pinghu xianzhi* 平湖縣志. Qianlong 55 [1790].

Zhang Huang 章潢. *Tushu bian* 圖書編. Facsimile of Wanli 41 [1613] edition. Taibei: Chengwen chubanshe 成文出版社, 1971.

Zhang Jian 張鑑. *Xi Xia jishi benmo* 西夏紀事本末. Facsimile edition. Taibei: Wenhai chubanshe 文海出版社, 1981.

Zhang Tianfu 張天復. *Huangyu kao* 皇輿考. s.n., Jiajing 36 [1557].

—. *Huangyu kao* 皇輿考. Facsimile of Ming Wanli 16 [1588] edition. Xuanlantang congshu 玄覽堂叢書, Chu ji, vols 6-7. Taibei: Guoli Zhongyang tushuguan 國立中央圖書館, Zhengzhong shuju yinxing 正中書局印行, 1981.

Zhao Weiyu 趙惟崳, ed. *Jiaxing xianzhi* 嘉興縣志. Guangxu 34 (1908). Facsimile edition in *Zhongguo difangzhi jicheng: Zhejiang fuxianzhi ji* 中國地方志集成浙江府縣志輯, vol. 15. Shanghai: Shanghai shudian, 1993.

Zhao Ying 趙瀛, and Zhao Wenhua 趙文華, eds. *Jiaxing fu tuji* 嘉興府圖記, Jiajing 28 (1549). Facsimile reprint in *Siku quanshu cunmu congshu* 四庫全書存目叢書, Shi bu 史部 (292 volumes), vol. 191: 302-540. Jinan: Qi Lu shushe chubanshe, 1997.

Zheng Qiao 鄭樵. *Tongzhi* 通志. Facsimile edition. Taibei: Taiwan shangwu yinshuguan 臺灣商務印書館, 1983.

—. *Tongzhi lue* 通志畧. Facsimile edition. Shanghai: Zhonghua shuju 中華書局, [1936?].

Zheng Ruozeng 鄭若曾. *Chouhai tubian* 籌海圖編. Publisher Hu Weiji 胡惟極. Tianqi 甲子 [year 4, 1624].

—. *Chouhai tubian* 籌海圖編. Wenyange facsimile edition. [Taibei]: Taiwan shangwu yinshuguan 臺灣商務印書館, [1983].

—. *Chaoxian tushuo* 朝鲜圖說. In *Chaoxian shiliao huibian* 朝鲜史料汇编 (20 volumes), vol. 19: 337-434. Beijing: Quanguo tushuguan wenxian suowei fuzhi zhongxin, 2004.

—. *Riben tuzuan* 日本圖纂. In *Nihon kigo no kenkyū* 日本寄語の研究. Kyoto: Kyōto Daigaku Kokubun Gakkai 京都大學國文學會, Showa 40 [1965].

—. *Zheng Kaiyang zazhu* 鄭開陽雜著. Facsimile of manuscript copy. Nanjing: Guoxue tushuguan Taofeng lou 國學圖書館陶風樓, 壬申 [1932].

Zhipan 志磐. *Fozu tongji* 佛祖統紀. Facsimile edition of a Ming reprint, in *Xuxiu Siku quanshu* 續修四庫全書, Zi bu 子部, Zong jiao lei 宗教類, vol. 1287. Shanghai: Shanghai guji chubanshe, [between 1995 and 1999].

Secondary sources

Ahmad, Nafis. *Muslim Contribution to Geography*. Lahore: M. Ashraf, [1947].

Akin, Alexander. 'The *Jing Xing Ji* of Du Huan: Notes on the West by a Chinese Prisoner of War.' *Harvard Middle Eastern and Islamic Review* 5 (1999-2000): 77-102.

—, with David Mumford. 'Yu Laid out the Lands: Georeferencing the Chinese *Yujitu* [Map of the Tracks of Yu] of 1136.' *Cartography and Geographic Information Science* 3, no. 3 (March 2013): 154-169.

Amelung, Iwo. 'New Maps for the Modernizing State: Western Cartographic Knowledge and Its Application in 19th and 20th Century China.' In *Graphics and Text in the Production of Technical Knowledge in China, the Warp and the Weft*, edited by Francesca Bray, Georges Métailié and Vera Dorofeeva-Lichtman, 685-726. Leiden: Brill, 2007.

Anderson, Benedict. *Imagined Communities*. Revised edition. New York: Verso, 1991.

Ayusawa Shintarō. 'Geography and Japanese Knowledge of World Geography.' *Monumenta Nipponica* 19, no. 3/4 (1964): 275-294.

—. 'The Types of World Map Made in Japan's Age of National Isolation.' *Imago Mundi* 10 (1953): 123-127.

—. *Mateo Ritchi no sekai zu ni kansuru shiteki kenkyū* マテオ「リッチの世界図に関する史的研究 (A historical study of Matteo Ricci's world map). Yokohama: Yokohama Shiritsu Daigaku, 1953.

Bae Woo Sung 배우성. 'Yet chido wa segye' 옛지도와 세계. In *Uri yet chido wa kŭ arŭmdaum* 우리 옛지도와 그 아름다움, edited by Han Yŏng-u, An Hwi-jun, Pae, U-sŏng, 171-182. Seoul: Hyohyŏng Ch'ulp'an, 1999.

—. 'Worldviews & Cartography.' In *The Artistry of Early Korean Cartography*, edited by Alexander Akin, 91-129. Larkspur, CA: Tamal Vista Press, 2008.

Bartholomew, James R. *The Formation of Science in Japan: Building a Research Tradition*. New Haven, CT: Yale University Press, 1989.

Beans, George Harry. *A List of Japanese Maps of the Tokugawa Era*. Jenkintown, PA: Tall Tree Library, 1951.

Bennett, Steven J. 'Patterns of the Sky and Earth: A Chinese Science of Applied Cosmology.' *Chinese Science* 3 (1978): 1-26.

Bernard, Henri. *Matteo Ricci's Scientific Contribution to China*. Translated by Edward Chalmers Werner. Peiping: H. Vetch, 1935.

Berry, Mary Elizabeth. *Hideyoshi*. Cambridge, MA: Council on East Asian Studies; Harvard University Press, 1989.

—. *Japan in Print: Information and Nation in the Early Modern Period*. Berkeley: University of California Press, 2006.

Black, Jeremy. *Maps and History: Constructing Images of the Past*. New Haven, CT: Yale University Press, 1997.

Bol, Peter K. 'The Rise of Local History: History, Geography, and Culture in Southern Song and Yuan Wuzhou.' *Harvard Journal of Asiatic Studies* 61, no. 1 (June 2001): 37-76.

—. 'The "Localist Turn" and "Local Identity" in Later Imperial China.' *Late Imperial China* 24, no. 2 (December 2003): 1-50.

—. 'Creating a GIS for the History of China.' In *Placing History: How Maps, Spatial Data, and GIS Are Changing Historical Scholarship*, edited by Anne Kelly Knowles and Amy Hillier, 25-57. Redlands, CA: ESRI Press, 2007.

Boxer, Charles Ralph. *The Christian Century in Japan: 1549-1650*. Berkeley: University of California Press, 1951.

Brockey, Liam Matthew. *Journey to the East: The Jesuit Mission to China, 1579-1724*. Cambridge, MA: Harvard University Press, 2007.

Brokaw, Cynthia J., and Kai-wing Chow, eds. *Printing and Book Culture in Late Imperial China*. Berkeley: University of California Press, 2005.

Brook, Timothy. *The Chinese State in Ming Society*. London and New York: RoutledgeCurzon, 2005.

—. 'Communications and Commerce.' In *Cambridge History of China*, vol. 8, edited by D. Twitchett and F. Mote, 579-707. Cambridge: Cambridge University Press, 1998.

—. 'Edifying Knowledge: The Building of School Libraries in Ming China.' *Late Imperial China* 17, no. 1 (June 1996): 93-119.

—. *Geographical Sources of Ming-Qing History*. Second edition. Ann Arbor: Center for Chinese Studies, University of Michigan, 2002.

—. 'Guides for Vexed Travelers: Route Books in the Ming and Qing.' In three parts: *Ch'ing-shih wen-t'i* 4, no. 5 (June 1981): 32-76; 4, no. 6 (December 1981): 130-140; 4, no. 8 (December 1982): 96-109.

—. 'Mapping Knowledge in the Sixteenth Century: The Gazetteer Cartography of Ye Chunji.' *Gest Library Journal* 7, no. 1 (Spring 1994): 5-32.

—. *Mr Selden's Map of China: Decoding the Secrets of a Vanished Cartographer* Toronto: House of Anansi Press, 2013.

Brown, Philip C. 'Never the Twain Shall Meet: Land Survey Techniques in Tokugawa Japan.' *Chinese Science* 9 (1989): 53-79.

Cao Wanru, ed. *Atlas of Ancient Maps in China* 中國古代地圖集*: From the Warring States Period to the Yuan Dynasty*. Beijing: Wenwu chubanshe 文物出版社 1990.

—. *Atlas of Ancient Maps in China* 中國古代地圖集*: The Ming Dynasty 1368-1644*. Beijing: Wenwu chubanshe 文物出版社 1994.

Cary-Elwes, Columba. *China and the Cross: A Survey of Missionary History*. New York: P.J. Kennedy and Sons, 1957.

Chan, Albert. 'Johann Adam Schall in the Pei-yu lu of T'an Ch'ien and the Eyes of His Contemporaries.' In *Western Learning and Christianity in China: The Contribution and Impact of Johann Adam Schall von Bell, 1592-1666*, edited by Roman Malek, 273-302. Sankt Augustin, Germany: China-Zentrum and Monumenta Serica Institute; Nettetal, Germany: Steyler Verlag, 1998.

Chang Kuei-sheng. 'Cheng Ho.' In *Dictionary of Ming Biography*, edited by L. Carrington Goodrich and Chaoying Fang, 194-200. New York: Columbia University Press, 1976.

Ch'en, Kenneth. 'A Possible Source for Ricci's Notices on Regions Near China.' *T'oung Pao* 34 (1938): 179-190.

Chen Zhengxiang 陳正祥. *Zhongguo ditu xue shi* 中國地圖學史. Hong Kong: Shangwu yinshuguan, Xianggang fen guan, 1979.

Chia, Lucille. *Printing for Profit: The Commercial Publishers of Jianyang, Fujian (11th-17th Centuries)*. Cambridge, MA: Harvard University Press, 2002.

—. 'Mashaben: Commercial Publishing from the Song to the Ming.' In *The Song-Yuan-Ming Transition in Chinese History*, edited by Paul Jakov Smith and Richard von Glahn, 284-328. Cambridge, MA: Harvard University Asia Center, 2003.

Choe Suh Myun. 'Tokdo in Old Maps.' *Korea Observer* 29, no. 1 (Spring 1998): 187-203.

Chou Tai-chi. 'Kuei O.' In *Dictionary of Ming Biography*, edited by L. Carrington Goodrich and Chaoying Fang, 756-759. New York: Columbia University Press, 1976.

Chow, Kai-Wing. 'Writing for Success: Printing, Examinations, and Intellectual Change in Late Ming China.' *Late Imperial China* 17, no. 1 (June 1996): 120-157.

—. *Publishing, Culture, and Power in Early Modern China*. Stanford, CA: Stanford University Press, 2004.

Cieslik, Hubert. 'The Case of Christovão Ferreira.' *Monumenta Nipponica* 29, no. 1 (Spring 1974): 1-54.

Clunas, Craig. *Superfluous Things: Material Culture and Social Status in Early Modern China*. Urbana: University of Illinois Press, 1991.

Cooper, Michael. *Rodrigues the Interpreter: An Early Jesuit in Japan and China*. New York and Tokyo: Weatherhill, 1974.

Cortazzi, Hugh. *Isles of Gold: Antique Maps of Japan*. New York: Weatherhill, 1983.

Cunningham, William, *The Cosmographical Glasse*. Amsterdam, Theatrum Orbis Terrarum; New York, Da Capo Press: 1968.

Dardess, John W. 'Did the Mongols Matter? Territory, Power and the Intelligentsia in China from the Northern Song to the Early Ming.' In *The Song-Yuan-Ming Transition in Chinese History*, edited by Paul Jakov Smith and Richard von Glahn, 111-134. Cambridge, MA: Harvard University Asia Center, 2003.

Day, John D. 'The Search for the Origins of the Chinese Manuscript Copies of Matteo Ricci's Maps.' *Imago Mundi* 47 (1995): 94-117.

D'Elia, Pasquale M. *Il mappamondo cinese del p. Matteo Ricci, S.I. (terza edizione, Pechino, 1602) conservato presso la Biblioteca vaticana. Commentato, tradotto e annotato dal P. Pasquale M. D'Elia, S.I.; con XXX tavole geografiche e 16 illustrazioni fuori testo.* Vatican City: Biblioteca Apostolica Vaticana, 1938.

—. *Fonti ricciane; documenti originali concernenti Matteo Ricci e la storia delle prime relazioni tra l'Europa e la Cina (1579-1615) ed. e commentati de Pasquale M. d'Elia sotto il patrocinio della Reale accademia d'Italia*. Rome, Libreria dello Stato, 1942-1949.

—. 'Recent Discoveries and New Studies (1938-60) on the World Map in Chinese of Father Matteo Ricci S.I.' *Monumenta Serica* 20 (1961): 82-164.

De Weerdt, Hilde. 'The Cultural Logics of Map Reading: Text, Time and Space in Printed Maps of the Song Empire,' In *Knowledge and Text Production in an Age of Print: China, 900-1400*, edited by Lucille Chia and Hilde De Weerdt, 239-270. Leiden: Brill, 2011.

—. 'Maps and Memory: Readings of Cartography in Twelfth- and Thirteenth-Century Song China.' *Imago Mundi: International Journal for the History of Cartography* 61, no. 2 (2009): 145-167.

Debergh, Minako. 'La Carte du Monde du P. Matteo Ricci (1602) et sa version Coréenne (1708) conservée à Osaka.' *Journal Asiatique* 274, no. 3-4 (1986): 417-454.

Dennis, Joseph R. *Writing, Publishing, and Reading Local Gazetteers in Imperial China, 1100-1700*. Cambridge, MA: Harvard University Asia Center, 2015.

Destombes, Marcel. 'Une Carte Chinoise du XVIe Siècle Découverte a la Bibliothèque Nationale.' *Journal Asiatique* 262 (1974): 193-212.

Dudink, Adrian. 'The Inventories of the Jesuit House at Nanking Made Up During the Persecution of 1616-1617 (Shen Que, Nangong shudu, 1620).' In *Western Humanistic Culture Presented to China by Jesuit Missionaries (XVII-XVIII centuries), Proceedings of the Conference Held in Rome, October 25-27, 1993*, edited by Federico Masini, 121-157. Rome: Institutum Historicum S.I., 1996.

Dunne, George H. *Generation of Giants: The Story of the Jesuits in China in the Last Decades of the Ming Dynasty*. Notre Dame, IN: University of Notre Dame Press, 1962.

Edney, Matthew. 'Theory and the History of Cartography.' *Imago Mundi* 48 (1996): 185-191.

Elison, George. *Deus Destroyed: The Image of Christianity in Early Modern Japan*. Cambridge, MA: Council on East Asian Studies, Harvard University, 1988.

Elliott, James. *The City in Maps: Urban Mapping to 1900*. London: British Library, 1987.

Elman, Benjamin A. 'Comparing and Classifying: Ming Dynasty Compendia and Encyclopedias (*Leishu*).' 'Qu'était-ce qu'écrire une encyclopédie en Chine? / What did it mean to write an Encyclopaedia in China?' edited by Florence Bretelle-Establet and Karine Chemia. Special issue, *Extrême-Orient, Extrême-Occident* hors série (2007): 131-157.

—. 'Geographical Research in the Ming-Ch'ing Period.' *Monumenta Serica* 35 (1981-1983): 1-18.

—. 'Ming-Qing Border Defence, the Inward Turn of Chinese Cartography, and Qing Expansion in Central Asia in the Eighteenth Century.' In *The Chinese State at the Borders*, edited by Diana Lary, 29-56. Vancouver: UBC Press, 2007.

—. *On Their Own Terms: Science in China, 1550-1900*. Cambridge, MA: Harvard University Press, 2005.

Fang, Chaoying and L. Carrington Goodrich. 'Kuo Tzu-chang.' In *Dictionary of Ming Biography*, edited by L. Carrington Goodrich and Chaoying Fang, 775-777. New York: Columbia University Press, 1976.

Fang, Lienche Tu. 'Shen Ch'ueh.' In *Dictionary of Ming Biography*, edited by L. Carrington Goodrich and Chaoying Fang, 1177-1179. New York: Columbia University Press, 1976.

Feng Tianyu, 'Some Late Ming Western Technical Terms and Their Implications for Cultural Transformation: "Brain," "Geometry," "Globe," and "Cathay."' *Frontiers of History in China* 1, no. 1 (January 2006): 47-60.

Foss, Theodore N. 'A Western Interpretation of China: Jesuit Cartography.' In *East Meets West: The Jesuits in China, 1582-1773*, edited by Charles E. Ronan and Bonnie B.C. Oh, 209-251. Chicago: Loyola University Press, 1988.

Fracasso, Riccardo. 'Shan hai ching.' In *Early Chinese Texts: A Bibliographical Guide*, edited by Michael Loewe, 357-367. Berkeley, CA: Society for the Study of Early China and Institute of East Asian Studies, University of California, 1993.

Franke, Wolfgang. *An Introduction to the Sources of Ming History.* Kuala Lumpur: University of Malaya Press, 1968.

—. 'Ricci, Matteo.' In *Dictionary of Ming Biography*, edited by L. Carrington Goodrich and Chaoying Fang, 1137-1144. New York: Columbia University Press, 1976.

Fuchs, Walter. *The 'Mongol Atlas' of China by Chu Ssu-Pen and the Kuang-Yü-Tu.* Peiping: The Catholic University Press, 1946.

Gernet, Jacques. *China and the Christian Impact: A Conflict of Cultures.* Translated by Janet Lloyd. Cambridge and New York: Cambridge University Press, 1985.

—. 'Christian and Chinese Visions of the World in the Seventeenth Century.' *Chinese Science* 4 (1980): 1-17.

—. 'Space and Time: Science and Religion in the Encounter Between China and Europe.' *Chinese Science* 11 (1993-1994): 93-102.

Giles, Lionel. 'Translations from the Chinese World Map of Father Ricci.' *The Geographical Journal* 52, no. 6 (December 1918): 367-385, and 53, no. 1 (January 1919): 19-30.

Goodall, John A. *Heaven and Earth: 120 Album Leaves From a Ming Encyclopedia, San-Ts'ai T'u-hui, 1610.* London: Lund Humphries, 1979.

Goodrich, L. Carrington, and Chaoying Fang, eds. *Dictionary of Ming Biography.* New York: Columbia University Press, 1976.

—. 'Cattaneo, Lazzaro.' In *Dictionary of Ming Biography*, edited by L. Carrington Goodrich and Chaoying Fang, 31-33. New York: Columbia University Press, 1976.

Goodrich, L. Carrington, and C.N. Tay. 'Chang Huang.' In *Dictionary of Ming Biography*, edited by L. Carrington Goodrich and Chaoying Fang, 83-85. New York: Columbia University Press, 1976.

Hales, Dell R. 'Yü Hsiang-tou.' In *Dictionary of Ming Biography*, edited by L. Carrington Goodrich and Chaoying Fang, 1612-1614. New York: Columbia University Press, 1976.

Hamel, Hendrik. *Hamel's Journal and a Description of the Kingdom of Korea, 1653-1666* (revised edition). Translated by Br. Jean-Paul Buys. Seoul: Royal Asiatic Society, Korea Branch, 1998.

Han Yŏng-u 한영우, An Hwi-jun 안휘준, and Pae U-sŏng 배우성. *Uri yet chido wa kŭ arŭmdaum* 우리 옛지도와 그 아름다움. Seoul: Hyohyŏng Ch'ulp'an, 1999.

Hargett, James M. 'Song Dynasty Local Gazetteers and their Place in the History of *Difangzhi* Writing.' *Harvard Journal of Asiatic Studies* 56, no. 2 (1996): 405-442.

Harley, J.B. 'Deconstructing the Map.' *Cartographica* 26, no. 2 (Summer 1989): 1-20.

Harley, J.B., and David Woodward. 'Preface.' In *The History of Cartography, Volume 1: Cartography in Prehistoric, Ancient, and Medieval Europe and the Mediterranean*, edited by J.B. Harley and David Woodward, xv-xxi. Chicago: University of Chicago Press, 1987.

Hegel, Robert E. *Reading Illustrated Fiction in Late Imperial China.* Stanford, CA: Stanford University Press, 1998.

Henderson, John B. 'Chinese Cosmographical Thought: The High Intellectual Tradition.' In *The History of Cartography, Volume 2, Book 2: Cartography in the Traditional East and Southeast Asian Societies*, edited by J.B. Harley and David Woodward, 203-227. Chicago and London: University of Chicago Press, 1994.

Hertz, Solange. *Rhodes of Viet Nam: The Travels and Missions of Father Alexander de Rhodes in China and Other Kingdoms of the Orient*. Westminster, MD: The Newman Press, 1966.

Hirth, Friedrich, and W.W. Rockhill. *Chau Ju-kua: His Work on the Chinese and Arab Trade in the Twelfth and Thirteenth Centuries, Entitled Chu-fan-chï*. St. Petersburg: Printing Office of the Imperial Academy of Sciences, 1911. Reprint, Taibei: Literature House, 1965.

Hŏ Yŏng-hwan 許英桓. *Chŏngdo 600-yŏn Sŏul chido* 定都600年서울地圖 / *600 Year's Maps of Seoul*. Seoul: Pŏmusa, 1994.

—. *Sŏul ŭi ko chido* 서울의古地圖. Seoul: Samsŏng Ch'ulp'ansa, 1989.

Hourani, George F. *Arab Seafaring in the Indian Ocean in Ancient and Early Medieval Times*. Revised and expanded by John Carswell. Princeton, NJ: Princeton University Press, 1995.

Hsia, R. Po-chia. *Matteo Ricci and the Catholic Mission to China: A Short History With Documents.* Indianapolis: Hackett Publishing Company, 2016.

Hsu, Mei-Ling. 'The Han Maps and Early Chinese Cartography.' *Annals of the Association of American Geographers* 68, no. 1 (March 1978): 45-60.

Hu Bangbo. *Cartography in Chinese Administrative Gazetteers of the Song Dynasty (A.D. 960-1279)*. PhD dissertation in Geography, University of Wisconsin at Milwaukee, 1994.

Huang, Ray. 'T'ang Shun-chih.' In *Dictionary of Ming Biography*, edited by L. Carrington Goodrich and Chaoying Fang, 1252-1256. New York: Columbia University Press, 1976.

Huang Shijian 黄時鑒, and Gong Yingyan 龔纓晏. *Li Madou shijie ditu yanjiu* 利瑪竇世界地圖研究. Shanghai: Shanghai guji chubanshe, 2004.

Huang, Stanley Y.C. 'Cheng Jo-tseng.' In *Dictionary of Ming Biography*, edited by L. Carrington Goodrich and Chaoying Fang, 204-208. New York: Columbia University Press, 1976.

—. 'Lo Hung-hsien.' In *Dictionary of Ming Biography*, edited by L. Carrington Goodrich and Chaoying Fang, 980-984. New York: Columbia University Press, 1976.

Hucker, Charles O. *A Dictionary of Official Titles in Imperial China*. Stanford, CA: Stanford University Press, 1985.

Hulbert, Homer B. 'An Ancient Gazetteer of Korea.' In *The Korean Repository* 4 (1897): 407-416.

Hummel, Arthur W., ed. *Eminent Chinese of the Ch'ing Period*. Washington, DC: United States Government Printing Office, 1943-1944.

Huxley, Aldous. *The Doors of Perception*. New York: Harper & Row, 1970.

Inden Emi 位田絵美. '*Wa-Kan* sansai zue ni miru taigai ninshiki: Chūgoku no *Sansai zue* kara Nihon no *Wa-Kan sansai zue* e' 和漢三才図会にみる対外認識 — 中国の『三才図絵』から日本の『和漢三才図絵』へ—. *Rekishi hyōron* 歴史評論 592 (August 1999): 58-75.

Jansen, Marius B. *China in the Tokugawa World*. Cambridge, MA and London: Harvard University Press, 1992.

Karlgren, Bernhard. *The Book of Documents*. Reprinted from the Museum of Far Eastern Antiquities, Bulletin 22. Stockholm, 1950.

Kassis, Vadim. 'Peking "Cartographers" As They Really Are.' *Socialism: Theory and Practice* (Supplement) 4 (1981): 51-54.

Kelly, Edward Thomas. *The Anti-Christian Persecution of 1616-1617 in Nanjing*. PhD Thesis, Columbia University, 1971.

Kim, Michael. *The Apparition of the Rational Public: Reading Collective Subjectivity in the Korean Public Sphere*. PhD Dissertation, Harvard University, 2004.

Kim, Sun Joo. 'A Glance at the *Chosŏn chiri soji*.' In *Studies on the Korean Materials in the Harvard-Yenching Library*, edited by Choong Nam Yoon, 173-185. Seoul: Kyŏngin Munhwasa, 2004.

Kim Tai-Jin. *A Bibliographical Guide to Traditional Korean Sources*. Seoul: Asiatic Research Center, 1976.

Kim, Yungsik. 'Problems and Possibilities in the Study of the History of Korean Science.' *Osiris*, 2nd Series, Vol. 13, 'Beyond Joseph Needham: Science, Technology, and Medicine in East and Southeast Asia' (1998): 48-79.

Kimura Tōichirō 木村 東一郎. *Kinsei muraezu kenkyū* 近世村絵図研究. Tokyo: Komiyama Shoten, 1962.

King, Gail. 'Note on a Late Ming Dynasty Chinese Description of "Ricci's Church" in Beijing.' In *Sino-Western Cultural Relations Journal* 20 (1998): 49-51.

Kish, George. 'Some Aspects of the Missionary Cartography of Japan during the Sixteenth Century.' *Imago Mundi* 6 (1949): 39-47.

Kōbe Shiritsu Hakubutsukan 神戸市立博物館. *Kochizu no sekai: Nanba Matsutarō-shi shūshū: Nihon zu, dōchūzu, sekaizu* 古地図の世界: 南波松太郎氏 收集: 日本図r道中図r世界図. Kōbe: Kōbe Kenkō Kyōiku Kōsha 神戸健康教育公社, 1983.

Kochizu ni miru sekai to Nihon: chizu wa kataru yume to roman: kaikan isshūnen kinen tokubetsuten 古地図にみる世界と日本: 地図は語る夢とロマン: 開館一周年記念特別展. Kōbe: Kōbe-shi Kenkō Kyōiku Kōsha, Shōwa 58 [1983].

Kongzi shengji tu 孔子聖蹟圖. Beijing: Zhongguo shudian, 1998.

Kornicki, Peter. *The Book in Japan: A Cultural History from the Beginnings to the Nineteenth Century.* Leiden and Boston: Brill, 1998.

Lagerwey, John. 'Wu Yüeh ch'un ch'iu.' In *Early Chinese Texts: A Bibliographical Guide*, edited by Michael Loewe, 473-476. Berkeley, CA: Society for the Study of Early China and Institute of East Asian Studies, University of California, 1993.

Lâm, Trương Bửu, 'Quelques mots en guise d'introduction.' Introduction to facsimile edition of *Hồng-Đức bản đồ.* Saigon: Publications of the Historical Research Institute, 1962.

Laures, John. *Kirishitan Bunko: A Manual of Books and Documents on the Early Christian Mission in Japan.* 3rd edition. Tokyo: Sophia University, 1957.

Ledyard, Gari. 'Cartography in Korea.' In *The History of Cartography, Volume 2, Book 2: Cartography in the Traditional East and Southeast Asian Societies*, edited by J.B. Harley and David Woodward, 235-345. Chicago and London: University of Chicago Press, 1994.

Lee, Ki-suk, Kim Shin, and Soh Jung-chul. *East Sea in World Maps.* Seoul: The Society for East Sea, 2002.

Lee, Peter H. *A Korean Storyteller's Miscellany: The P'aegwon Chapki of Ŏ Sukkwŏn.* Princeton, NJ: Princeton University Press, 1989.

Lewis, Mark Edward. *The Construction of Space in Early China.* Albany: SUNY Press, 2006.

Li Xiaocong. *Ouzhou shoucang bufen Zhongwen gu ditu xu lu = A Descriptive Catalogue of Pre-1900 Chinese Maps Seen in Europe.* Beijing: Guoji wenhua chuban gongsi, Xinhua shudian jingxiao, 1996.

Libbrecht, Ullrich. *Chinese Mathematics in the Thirteenth Century*. Cambridge, MA: MIT Press, 1973.

—. 'What Kind of Science Did the Jesuits Bring to China?' In *Western Humanistic Culture Presented to China by Jesuit Missionaries (XVII-XVIII centuries), Proceedings of the Conference held in Rome, October 25-27, 1993*, edited by Federico Masini, 221-234. Rome: Institutum Historicum S.I., 1996.

Lu Liangzhi 卢良志. *Zhongguo ditu xue shi* 中国地图学史. Beijing: Cehui chubanshe, 1984.

Luk, Bernard Hung-Kay. 'A Study of Giulio Aleni's Chih-fang wai chi.' *Bulletin of the School of Oriental and African Studies, University of London* 40, no. 1 (1977): 58-84.

Ma Biao. 'Johann Adam Schall and Chinese Traditional Philosophy.' In *Western Learning and Christianity in China: The Contribution and Impact of Johann Adam Schall von Bell, 1592-1666*, edited by Roman Malek. Sankt Augustin, Germany: China-Zentrum and Monumenta Serica Institute; Nettetal, Germany: Steyler Verlag, 1998.

Mackay, A.L. 'Kim Su-Hong and the Korean Cartographic Tradition.' *Imago Mundi* 27 (1975): 27-38.

Matsuki Satoru 松木哲. 'Kōro zu ni tsuite' 航路図について, in *Kochizu ni miru sekai to Nihon: chizu wa kataru yume to roman, kaikan isshūnen kinen tokubetsuten* 古地図にみる世界と日本: 地図は語る夢とロマン, 開館一周年記念特別展. Kōbe: Kōbe-shi Kenkō Kyōiku Kōsha, Shōwa 58 [1983].

McCune, Shannon. 'World Maps by Korean Cartographers.' *Journal of Social Sciences and Humanities* 45 (June 1977): 1-8.

McDermott, Joseph. 'The Ascendance of the Imprint in China.' In *Printing and Book Culture in Late Imperial China*, edited by Cynthia J. Brokaw and Kai-wing Chow, 55-104. Berkeley: University of California Press, 2005.

Menegon, Eugenio. 'A Different Country, The Same Heaven: A Preliminary Biography of Giulio Alenis, S.J. (1582-1649).' *Sino-Western Cultural Relations Journal* 15 (1993): 27-51.

Millinger, James F. 'Yü Ta-yu.' In *Dictionary of Ming Biography*, edited by L. Carrington Goodrich and Chaoying Fang, 1616-1618. New York: Columbia University Press, 1976.

Mish, John. 'Creating an Image of Europe for China: Aleni's Hsi-fang ta-wen.' *Monumenta Serica* 23 (1964): 1-87.

Moerman, D. Max. *The Japanese Buddhist World Map: Religious Vision and the Cartographic Imagination*. Honolulu: University of Hawaii Press, forthcoming.

Mosca, Matthew William. *From Frontier Policy to Foreign Policy: The Question of India and the Transformation of Geopolitics in Qing China*. Stanford, CA: Stanford University Press, 2013.

Mungello, David E. *Curious Land: Jesuit Accommodation and the Origins of Sinology*. Honolulu: University of Hawaii Press, 1989.

—. *The Forgotten Christians of Hangzhou*. Honolulu: University of Hawaii Press, 1994.

Muroga, Nobuo, and Unno Kazutaka, 'The Buddhist World Map in Japan and Its Contact with European Maps.' *Imago Mundi* 16 (1962): 49-69.

Murray, Julia K. 'Didactic Illustrations in Printed Books.' In *Printing and Book Culture in Late Imperial China*, edited by Cynthia J. Brokaw and Kai-wing Chow, 417-450. Berkeley: University of California Press, 2005.

Nakamura Hiroshi. 'Old Chinese World Maps Preserved by the Koreans.' *Imago Mundi* 4 (1947): 3-22.

—. 'The Japanese Portolanos of Portuguese Origin of the XVIth and XVIIth Centuries.' *Imago Mundi* 18 (1964): 24-44.

Needham, Joseph. 'Astronomy.' In *Science and Civilisation in China, Volume 3: Mathematics and the Sciences of the Heavens and the* Earth, 171-461. Cambridge: Cambridge University Press, 1959.

—. 'Geography and Cartography.' In *Science and Civilisation in China, Volume 3: Mathematics and the Sciences of the Heavens and the Earth*, 497-590. Cambridge: Cambridge University Press, 1959.

—. 'Geology (and Related Sciences).' In *Science and Civilisation in China, Volume 3: Mathematics and the Sciences of the Heavens and the Earth*, 591-623. Cambridge: Cambridge University Press, 1959.

Ōba Osamu 大庭脩, translated by Joshua A. Fogel. 'Sino-Japanese Relations in the Edo Period (Part Three: The Discovery of Banned Books).' In *Sino-Japanese Studies* 9, no. 1 (October 1996): 56-74.

—. *Kanseki yunyū no bunkashi: Shōtoku Taishi kara Yoshimune e* 漢籍輸入の文化史: 聖徳太子から吉宗へ. Tōkyō: Kenbun Shuppan, 1997.

Okamoto Yoshitomo 岡本良知. *Jūrokuseiki ni okeru Nihon chizu no hattatsu* 十六世紀における日本地図の発達. Tokyo: Yagi Shoten 八木書店, Shōwa 48 [1973].

Park, Hyunhee. *Mapping the Chinese and Islamic Worlds: Cross-Cultural Exchange in Pre-Modern Asia*. Cambridge: Cambridge University Press, 2012.

Pegg, Richard A. *Cartographic Traditions in East Asian Maps*. Honolulu: University of Hawaii Press, 2014.

Peterson, Willard. 'Why Did They Become Christians?' In *East Meets West: The Jesuits in China, 1582-1773*, edited by Charles E. Ronan and Bonnie B.C. Oh, 129-152. Chicago: Loyola University Press, 1988.

Pinto dos Santos, José Miguel. 'The "Kuroda Plot" and the Legacy of Jesuit Scientific Influence in Seventeenth Century Japan.' *Bulletin of Portuguese-Japanese Studies* 10/11 (June/December 2005): 97-191.

Paramore, Kiri. 'Hayashi Razan's Redeployment of Anti-Christian Discourse: The Fabrication of Haiyaso.' *Japan Forum* 18, no. 2 (July 2006): 185-206.

Pratt, Keith. *Korea: A Historical and Cultural Dictionary*. London: Curzon Press, 1999.

Prescott, Hilda F.M. *Friar Felix at Large; A Fifteenth-Century Pilgrimage to the Holy Land*. Westport, CT: Greenwood Press, 1967.

Ricci, Matteo. *Lettere (1580-1609)*. Macerata: Quodlibet, 2001.

Rienstra, M. Howard, ed. and trans. *Jesuit Letters from China 1583-84*. Minneapolis: University of Minnesota Press, 1986.

Robinson, Arthur H. *The Look of Maps: An Examination of Cartographic Design*. Madison: University of Wisconsin Press, 1952.

—. 'Mapmaking and Map Printing: The Evolution of a Working Relationship.' In *Five Centuries of Map Printing*, edited by David Woodward, 1-9. Chicago: University of Chicago Press, 1975.

Robinson, Kenneth. 'Centering the King of Chosŏn: Aspects of Korean Maritime Diplomacy, 1392-1592.' *The Journal of Asian Studies* 59, no. 1 (February 2000): 109-125.

Sadler, Arthur Lindsay. *The Maker of Modern Japan: The Life of Tokugawa Ieyasu*. Rutland, VT: C. E. Tuttle, 1978.

Sakanishi, Shio. 'Prohibition of Import of Certain Chinese Books and the Policy of the Edo Government.' *Journal of the American Oriental Society* 57, no. 3 (September 1937): 290-303.

Sanders, Graham. Personal communication granting permission to use his translation of an Yi Kyubo poem posted on Facebook. March 13, 2013.

Schiffer, Michael B. *Formation Processes of the Archaeological Record*. Albuquerque: University of New Mexico Press, 1987.

Schuessler, Axel, and Michael Loewe. 'Yüeh chüeh shu.' In *Early Chinese Texts: A Bibliographical Guide*, edited by Michael Loewe, 473-476. Berkeley: Society for the Study of Early China and Institute of East Asian Studies, University of California, 1993.

Shapinsky, Peter D. 'Polyvocal Portolans: Nautical Charts and Hybrid Maritime Cultures in Early Modern East Asia.' *Early Modern Japan* 14 (2006): 4-26.

Shaughnessy, Edward. 'Shang Shu,' in *Early Chinese Texts: A Bibliographical Guide*, edited by Michael Loewe, 376-389. Berkeley: Society for the Study of Early China and Institute of East Asian Studies, University of California, 1993.

Short, John Rennie. *Korea: A Cartographic History*. Chicago and London: University of Chicago Press, 2012.

Sivin, Nathan. 'Copernicus in China.' In *Science in Ancient China: Researches and Reflections*, 1-53. Aldershot: Variorum, 1995.

Skelton, Raleigh Ashlin. *Maps: A Historical Survey of Their Study and Collecting*. Chicago: Published for the Hermon Dunlap Smith Center for the History of Cartography, the Newberry Library, by the University of Chicago Press, 1975.

Smith, Richard J. 'The Jesuits and Evidential Research in Late Imperial China: Some Reflections.' *Ex/Change* (Newsletter of the City University of Hong Kong), February 2002.

—. *Mapping China and Managing the World: Culture, Cartography and Cosmology in Late Imperial Times*. New York: Routledge, 2013.

—. 'Mapping China's World: Cultural Cartography in Late Imperial Times.' In *Landscape, Culture, and Power in Chinese Society*, edited by Wen-hsin Yeh, 52-109. Berkeley, CA: Institute of East Asian Studies, 1998.

So, Kwan-wai. *Japanese Piracy in Ming China During the 16th Century*. East Lansing: Michigan State University Press, 1975.

—. 'Chao Wen-hua.' In *Dictionary of Ming Biography*, edited by L. Carrington Goodrich and Chaoying Fang, 132-136. New York: Columbia University Press, 1976.

Soothill, William Edward and Lewis Hodous. *A Dictionary of Chinese Buddhist Terms: With Sanskrit and English Equivalents and a Sanskrit-Pali Index*. London: K. Paul, Trench, Trübner & Co., 1937.

Standaert, Nicolas. 'Note on the Spread of Jesuit Writings in Late Ming and Early Qing China.' *China Mission Studies (1550-1800) Bulletin* 1 (1985): 22-32.

Steinhardt, Nancy Shatzman. *Chinese Imperial City Planning*. Honolulu: University of Hawaii Press, 1990.

Stephenson, F. Richard. 'Chinese and Korean Star Maps and Catalogs.' In *The History of Cartography, Volume 2, Book 2: Cartography in the Traditional East and Southeast Asian Societies*, edited by J.B. Harley and David Woodward, 511-578. Chicago and London: University of Chicago Press, 1994.

Sun Shangyang 孙尚扬. *Li Madou yu Xu Guangqi* 利玛窦与徐光启. Beijing: Xinhua chubanshe, 1993.

Szcześniak, Boleslaw. 'Matteo Ricci's Maps of China.' *Imago Mundi* 11 (1954): 127-136.

—. 'The Seventeenth Century Maps of China: An Inquiry into the Compilations of European Cartographers.' *Imago Mundi* 13 (1956): 116-136.

Takahashi Tadashi 高橋正. 'Tōzen seru chūsei Isurāmu sekaizu.' *Ryūkoku Daigaku ronshū* 374 (1963): 77-95.

—. 'Chūsei Isurāmu sekaizu' 中世イスラーム世界図. In *Chizu no shisō* 地図の思想, edited by Hasegawa Kōji 長谷川孝治, 53-55. Tōkyō: Asakura Shoten, 2005.

Tanaka Takeo 田中健夫. 籌海圖編の成立 (The Formation of the *Chouhai tubian*). *Nihon rekishi*日本歴史 57 (1953): 18-23.

Tang Xiaofeng. *From Dynastic Geography to Historical Geography: A Change in Perspective Towards the Geographical Past of China*. Beijing: Commercial Press International, 2000.

Teng, S.Y. 'Cheng Chiao.' In *Sung Biographies*, edited by Herbert Franke, 1: 146-156. Wiesbaden: Steiner, 1976.

Thongchai Winichakul. *Siam Mapped: A History of the Geo-Body of a Nation*. Honolulu: University of Hawaii Press, 1994.

Toby, Ronald. *State and Diplomacy in Early Modern Japan: Asia in the Development of the Tokugawa Bakufu*. Princeton, NJ: Princeton University Press, 1984.

Tooley, R.V. *Early Antarctica: A Glance at the Beginnings of Cartographic Representation of the South Polar Regions*. London: Map Collectors' Circle, 1963.

Trigault, Nicolas. *China in the Sixteenth Century: The Journals of Matthew Ricci, 1583-1610; Translated from the Latin by Louis J. Gallagher.* New York, Random House, 1953.

Üçerler, M. Antoni J. 'Gutenberg Comes to Japan: The Jesuit Enterprise and the First IT Revolution of the Sixteenth Century.' Accessed August 2014. http://www.usfca.edu/ricci/events/Ucerler.pdf.

—. 'The Jesuits in East Asia in the Early Modern Age: A New "Areopagus" and the "Re-invention" of Christianity.' In *The Jesuits and Globalization: Historical Legacies and Contemporary Challenges*, edited by Thomas Banchoff and José Casanova, 27-48. Washington, DC: Georgetown University Press, 2016.

Unno Kazutaka. 'Cartography in Japan.' In *The History of Cartography, Volume 2, Book 2: Cartography in the Traditional East and Southeast Asian Societies*, edited by J.B. Harley and David Woodward, 346-477. Chicago and London: University of Chicago Press, 1994.

—. 'Maps of Japan Used in Prayer Rites or as Charms.' *Imago Mundi* 46 (1994): 72-75.

—. *Tōzai chizu bunka kōshōshi kenkyū* 東西地図文化交渉史研究 (Monographs on the history of cartographical exchange between the East and the West). Osaka: Seibundō, 2003.

Validi, Ahmet Zeki. 'Islam and the Science of Geography.' *Islamic Culture, the Hyderabad Quarterly Review* 8 (October 1934): 511-527.

Wallis, Helen. 'The Influence of Father Ricci on Far Eastern Cartography.' *Imago Mundi* 19 (1965): 38-45.

Walraven, Boudewijn. 'Reader's Etiquette, and Other Aspects of Book Culture in Chosŏn Korea.' In *Books in Numbers: Some Papers Read at the Conference in Celebration of the Seventy-Fifth Anniversary of the Harvard-Yenching Library*. Cambridge, MA: Harvard-Yenching Library, 2003.

Wang Chung-min, 'Yang T'ing-yün.' In *Eminent Chinese of the Ch'ing Period*, edited by Arthur W. Hummel, 894-895. Washington, DC: United States Government Printing Office, 1943-1944.

Wang Qianjin 汪前进. *Xixue dong chuan di yi shi: Li Madou* 西学东传第一师: 利玛窦. Beijing: Kexue chubanshe, 2000.

Wang Yong 王庸. *Zhongguo dilixue shi* 中國地理學史. Changsha: Shangwu yinshuguan 商務印書館, [1938].

Whelan, Christal. *The Beginning of Heaven and Earth: The Sacred Book of Japan's Hidden Christians*. Honolulu: University of Hawaii Press, 1996.

Whitfield, Peter. *New Found Lands: Maps in the History of Exploration*. London: The British Library, 1998.

Whitmore, John K. 'Cartography in Vietnam.' In *The History of Cartography, Volume 2, Book 2: Cartography in the Traditional East and Southeast Asian Societies*, edited by J.B. Harley and David Woodward, 478-508. Chicago and London: University of Chicago Press, 1994.

Wiethoff, Bodo. 'Lu T'ang.' In *Dictionary of Ming Biography*, edited by L. Carrington Goodrich and Chaoying Fang, 1003-1005. New York: Columbia University Press, 1976.

Wigen, Kären. 'Orienting the Past in Early Modern Japan.' In *Time in Maps: From the Age of Discovery to Our Digital Era*, edited by Kären Wigen and Caroline Winterer, 37-62. Chicago: University of Chicago Press, 2020.

Wigen, Kären, Fumiko Sugimoto, and Cary Karacas, eds. *Cartographic Japan: A History in Maps.* Chicago and London: University of Chicago Press, 2016.

Witek, John W. 'Johann Adam Schall von Bell and the Transition from the Ming to the Ch'ing Dynasty.' In *Western Learning and Christianity in China: The Contribution and Impact of Johann Adam Schall von Bell, 1592-1666*, edited by Roman Malek, 109-124. Sankt Augustin, Germany: China-Zentrum and Monumenta Serica Institute; Nettetal, Germany: Steyler Verlag, 1998.

Wong, George H.C. 'China's Opposition to Western Science During Late Ming and Early Ch'ing.' *Isis* 54, no. 1 (March 1963): 29-49.

Wood, Denis, John Fels, and John Krygier. *Rethinking the Power of Maps.* New York: Guilford Press, 2010.

Woodward, David. 'Medieval Mappamundi.' In *The History of Cartography, Volume 1: Cartography in Prehistoric, Ancient, and Medieval Europe and the Mediterranean*, edited by J.B. Harley and David Woodward, 286-368. Chicago: University of Chicago Press, 1987.

—. 'Preface.' In *The History of Cartography, Volume 2, Book 2: Cartography in the Traditional East and Southeast Asian Societies*, edited by J.B. Harley and David Woodward, xxiii-xxvii. Chicago and London: University of Chicago Press, 1994.

Yajima Suketoshi. 'The European Influence on Physical Sciences in Japan.' *Monumenta Nipponica* 19, no. 3/4 (1964): 340-351.

Yamada Keiji 山田慶兒. 'Ryūkoku Daigaku Ōmiya Toshokan shozō shukushōgi zusetsu oyobi mokei ni tsuite' 龍谷大学大宮図書館所蔵 縮象儀-図「説および模型-について. *Bulletin of the International Research Center for Japanese Studies* 16 (September 1997): 59-71.

Yamori Kazuhiko 矢守一彦. 'Bākurē hon Nihon zu byōbu' バークレー 本日本図屏風.' *Fukui kenshi shiori* 福井県史しおり, Shiryō hen 16 ue 資料編16上 (February 28, 1990): 1-5.

Yan Ping. *China in Ancient and Modern Maps*. London: Sotheby's, 1998.

Yang Ching-yi and L. Carrington Goodrich. 'Mao Jui-cheng.' In *Dictionary of Ming Biography*, edited by L. Carrington Goodrich and Chaoying Fang, 1041-1042. New York: Columbia University Press, 1976.

Yap, Paul Teh-lu, and J.C. Yang. 'Li Chih-tsao.' In *Eminent Chinese of the Ch'ing Period*, edited by Arthur W. Hummel, 452-454. Washington, DC: United States Government Printing Office, 1943-1944.

Yee, Cordell. 'A Cartography of Introspection: Chinese Maps as Other Than European.' *Asian Art* 5, no. 4 (1992): 29-47.

—. 'Chinese Cartography Among the Arts: Objectivity, Subjectivity, Representation.' In *The History of Cartography, Volume 2, Book 2: Cartography in the Traditional East and Southeast Asian Societies*, edited by J.B. Harley and David Woodward, 128-169. Chicago and London: University of Chicago Press, 1994.

—. 'Chinese Maps in Political Culture.' In *The History of Cartography, Volume 2, Book 2: Cartography in the Traditional East and Southeast Asian Societies*, edited by J.B. Harley and David Woodward, 71-95. Chicago and London: University of Chicago Press, 1994.

—. 'Concluding Remarks.' In *The History of Cartography, Volume 2, Book 2: Cartography in the Traditional East and Southeast Asian Societies*, edited by J.B. Harley and David Woodward, 228-231. Chicago and London: University of Chicago Press, 1994.

—. 'Reinterpreting Traditional Chinese Geographical Maps.' In *The History of Cartography, Volume 2, Book 2: Cartography in the Traditional East and Southeast Asian Societies*, edited by J.B. Harley and David Woodward, 35-70. Chicago and London: University of Chicago Press, 1994.

—. 'Taking the World's Measure: Chinese Maps between Observation and Text.' In *The History of Cartography, Volume 2, Book 2: Cartography in the Traditional East and Southeast Asian Societies*, edited by J.B. Harley and David Woodward, 96-127. Chicago and London: University of Chicago Press, 1994.

—. 'Traditional Chinese Cartography and the Myth of Westernization.' In *The History of Cartography, Volume 2, Book 2: Cartography in the Traditional East and

Southeast Asian Societies, edited by J.B. Harley and David Woodward, 170-202. Chicago and London: University of Chicago Press, 1994.

—. 'Concluding Remarks.' In *The History of Cartography, Volume 2, Book 2: Cartography in the Traditional East and Southeast Asian Societies*, edited by J.B. Harley and David Woodward, 228-231. Chicago and London: University of Chicago Press, 1994.

—. *Zhongguo di tu xue shi* 中国地图学史 (History of Chinese cartography). Translated by Jiang Daozhang 姜道章. Beijing: Beijing Daxue chubanshe, 2006.

Yi Ch'an (Lee Chan) 李燦. *Han'guk ko chido* 韓國古地圖. Seoul: Han'guk Tosŏgwanhak Yŏn'guhoe 韓國圖書館學研究會, 1977.

—. *Han'guk ŭi ko chido* 韓國의古地圖 (Old maps of Korea). Seoul: Pŏmusa, 1991.

Yi, Yang-chin, and L. Carrington Goodrich. 'Mao Jui-cheng.' In *Dictionary of Ming Biography*, edited by L. Carrington Goodrich and Chaoying Fang, 1041-1042. New York: Columbia University Press, 1976.

Yonemoto, Marcia. *Mapping Early Modern Japan: Space, Place, and Culture in the Tokugawa Period, 1603-1868*. Berkeley: University of California Press, 2003.

—. 'The "Spatial Vernacular" in Tokugawa Maps.' *The Journal of Asian Studies* 53, no. 3 (August 2000): 647-666.

[Yŏngnam University Museum]. *Yŏngnamdae pangmulgwan sojang Han'guk ŭi yet chido* 영남대박물관소장韓國의옛地圖. Kyŏngsan: Yŏngnam Taehakkyo Pangmulgwan, 1998.

Yoon, Inshil Choe, trans. *Yi Chung-hwan's T'aengniji: The Korean Classic for Choosing Settlements*. Sydney: Wild Peony Press, 1998.

Zhang Cuo 張錯. *Li Madou ru Hua ji qita* 利瑪竇入華及其他. Hong Kong: Xianggang chengshi daxue chubanshe, 2002.

Zurndorfer, Harriet T. 'Science Without Modernization: China's First Encounter With Useful And Reliable Knowledge From Europe.' Paper for the Fourth Global Economic History Network Workshop. Accessed December 28, 2020. https://www.lse.ac.uk/Economic-History/Assets/Documents/Research/GEHN/GEHNConferences/conf4/Conf4-HZurndorfer.pdf

Index

Made in the USA
Monee, IL
13 September 2026

63176225R00175